Volume II: Applications

Digital Transmission Systems and Networks

ELECTRICAL ENGINEERING, COMMUNICATIONS, AND SIGNAL PROCESSING

ISSN 0888-2134

Raymond L. Pickholtz, Series Editor

1. *Computer Network Architectures**
   Anton Meijer and Paul Peeters

2. *IEEE Communication Society's Tutorials in Modern Communications**
   Victor B. Lawrence, Joseph L. Lo Cicero, and Laurence B. Milstein, Editors

3. *Spread Spectrum Communications, Volume I**
   Marvin K. Simon, Jim K. Omura, Robert A. Scholtz, and Barry K. Levitt

4. *Spread Spectrum Communications, Volume II**
   Marvin K. Simon, Jim K. Omura, Robert A. Scholtz, and Barry K. Levitt

5. *Spread Spectrum Communications, Volume III**
   Marvin K. Simon, Jim K. Omura, Robert A. Scholtz, and Barry K. Levitt

6. *Elements of Digital Satellite Communication: System Alternatives, Analyses and Optimization, Volume I**
   William W Wu

7. *Elements of Digital Satellite Communication: Channel Coding and Integrated Services Digital Satellite Networks, Volume II**
   William W Wu

8. *Current Advances in Distributed Computing and Communications**
   Yechiam Yemini

9. *Digital Transmission Systems and Networks, Volume I: Principles*
   Michael J. Miller and Syed V. Ahamed

10. *Digital Transmission Systems and Networks, Volume II: Applications*
    Michael J. Miller and Syed V. Ahamed

ANOTHER WORK OF INTEREST

Local Area and Multiple Access Networks
Raymond Pickholz, Editor

---

*These previously published books are in the *Electrical Engineering, Communications, and Signal Processing* series but they are not numbered within the volume itself. All future volumes in this series will be numbered.

Volume II: Applications

Digital Transmission Systems and Networks

MICHAEL J. MILLER
The South Australian Institute of Technology

SYED V. AHAMED
The City University of New York

COMPUTER SCIENCE PRESS

Copyright © 1988 Computer Science Press, Inc.

Printed in the United States of America.

All rights reserved. No part of this book may be reproduced in any form including photostat, microfilm, and xerography, and not in information storage and retrieval systems, without permission in writing from the publisher, except by a reviewer who may quote brief passages in a review or as provided in the Copyright Act of 1976.

Computer Science Press
1803 Research Boulevard
Rockville, Maryland 20850

1 2 3 4 5 6   Printing   Year   93 92 91 90 89 88

Library of Congress Cataloging-in-Publication Data

Miller, Michael J., 1939–
  Digital transmission systems and networks.

  Bibliography: p.
  Includes index.
  Contents: v. 1. Principles — v. 2. Applications.
  1. Digital communications. I. Ahamed, Syed V.,
1938–   . II. Title.
TK5103.7.M55   1987   621.38   86-33357
ISBN 0-88175-094-8 (v. 1)
ISBN 0-88175-176-6 (v. 2)

Dedicated to
Edith, Kristina, David, Karen
and
Ameera, Sonya, Nisha

VOLUME II
CONTENTS

Preface xvi

CHAPTER 1   DIGITAL RADIO SYSTEMS 1
1.1   Introduction 1
    1.1.1   Historical Developments 2
    1.1.2   Digital Versus Analogue Radio 3
1.2   Typical Digital Radio Systems 4
    1.2.1   Digital Radio Equipment 4
    1.2.2   Transmission Capacities and Frequency Bands 6
    1.2.3   Typical Equipment Characteristics 7
    1.2.4   Performance Objectives 10
1.3   Modulation Methods 11
    1.3.1   Spectral Efficiency 11
    1.3.2   Choice of Modulation Type 12
    1.3.3   PSK (Phase-shift-keying) Modulation Schemes 14
    1.3.4   FSK (Frequency-shift-keying) Modulation 24
    1.3.5   16QAM (quadrature-amplitude) Modulation 24
    1.3.6   Quadrature Partial Response Signalling Schemes 27
    1.3.7   Continuous-Phase FSK and Minimum Shift Keying (MSK) Systems 28
1.4   Detection 30
    1.4.1   Optimum Detector for Binary PSK, FSK, or ASK in Gaussian Noise 31
    1.4.2   Coherent Detector for QAM and M-PSK Signals 34
    1.4.3   Probability of Error 35
    1.4.4   Optimum Detectors for Channels with ISI and Noise ... 39
    1.4.5   Combining Modulation and Coding 40
    1.4.6   Non-Coherent Detectors 41
1.5   Radio Link System Design 41
    1.5.1   Introduction 41
    1.5.2   Free Space Calculations for Single Hops 44
    1.5.3   Flat Fade Margins 48
    1.5.4   Percentage Outage Prediction—Vigant's Formula 49
    1.5.5   Frequency Selective Fading Model 51

		1.5.6	Intersymbol Interference Resulting from Frequency Selective Fading	54
		1.5.7	Space Diversity	56
		1.5.8	Adaptive Equalization	57
		1.5.9	System Signature	58
	1.6	Hybrid Radio Systems		59
		1.6.1	Data in Voice (DIV) Systems	60
		1.6.2	Data Above (DAV) and Data Above Video (DAVID) Systems	60
		1.6.3	Data Over Voice (DOV) Systems	61
		1.6.4	Data Under Voice (DUV) Systems	61
	1.7	Point-to-Multipoint Subscriber Radio Systems		61
	1.8	Problems		63
	1.9	References		65
CHAPTER 2		DIGITAL TELEPHONE NETWORKS		68
	2.1	Digital Switching		68
		2.1.1	Local Networks	69
		2.1.2	Concentrators	71
		2.1.3	Digital Group Selectors	75
		2.1.4	Advantages of Concentrators with Centralized Exchanges	76
		2.1.5	Digital Switching Principles	77
		2.1.6	Exchange Congestion	81
	2.2	Network Synchronization		83
		2.2.1	Synchronization Requirements—Slips	83
		2.2.2	Causes of Slips	85
		2.2.3	Approaches to Network Synchronization	88
		2.2.4	Plesiochronous Networks	92
		2.2.5	Master-Slave Synchronization	94
		2.2.6	Mutual Synchronization	96
		2.2.7	Comparison of Synchronization Methods	97
	2.3	Frame Synchronization		98
		2.3.1	Introduction	98
		2.3.2	Frame Alignment Systems	100
		2.3.3	State Diagrams and Designs Principles	103
		2.3.4	Choice of Frame Alignment Signal	107
	2.4	Problems		108
	2.5	References		109
CHAPTER 3		COMPUTER NETWORKS		111
	3.1	Introduction		111

3.2	Classification of Computer Networks	112
	3.2.1 Terminology	112
	3.2.2 Network Classification	112
3.3	Computer Network Structures	115
	3.3.1 Data Networks	115
	3.3.2 Computer Networks	116
	3.3.3 Circuit, Message, and Packet Switching	116
3.4	The ISO Architectural Model for Open Systems Interconnection (OSI)	118
3.5	The Physical Layer	120
3.6	The Data Link Layer	122
	3.6.1 Character-Oriented Protocols	122
	3.6.2 The Binary Synchronous Communications Protocol	125
	3.6.3 Byte Count Protocols	130
	3.6.4 Bit-Oriented Protocols	130
3.7	Network Layer	133
	3.7.1 Virtual Circuits and Datagrams	134
	3.7.2 Routing	136
	3.7.3 Congestion Control in Networks	138
3.8	The X.25 Interface	140
	3.8.1 The X.25 Packet Characteristics	140
	3.8.2 Services Provided by X.25	144
	3.8.3 The Transaction-Oriented Features of X.25	146
3.9	The X.75 Inter-Network Protocol	148
3.10	Higher Levels of the ISO Reference Model	149
3.11	Some Aspects of Data Network Design	150
3.12	Problems	159
3.13	References	161
CHAPTER 4	**ERROR CONTROL IN DIGITAL NETWORKS**	**163**
4.1	Introduction	163
4.2	Errors and Erasures	165
4.3	Error Detection using Block Codes	168
	4.3.1 Single Bit Parity Detection	168
	4.3.2 Weight Distribution of a Code	170
	4.3.3 Error Detection Reliability of the Single-Parity Code	171
	4.3.4 Linear Block Codes for Error Detection	173
	4.3.5 Minimum Distance of a Code	174
4.4	Cyclic Codes for Error Detection	175
	4.4.1 Polynomial Representation	175
	4.4.2 Generator Polynomial	176
	4.4.3 Generation of Parity (Encoding)	177

		4.4.4	Encoder for Rec. X.25 Frame Check Sequence	179

		4.4.4	Encoder for Rec. X.25 Frame Check Sequence	179
		4.4.5	Decoding for Error Detection	180
		4.4.6	Error Detection for the CCITT Rec. X.25 Code	182
		4.4.7	Variable Block Lengths—Shortened Cyclic Codes	183
		4.4.8	Probability of Undetected Error	184
	4.5	Forward Error Correction		185
		4.5.1	Types of Codes	185
		4.5.2	Soft Decision Decoding	187
		4.5.3	Decoding Techniques	189
	4.6	Automatic-Repeat-Request (ARQ) Systems		190
		4.6.1	ARQ Procedures	192
		4.6.2	Throughput of Go-back-N ARQ	196
		4.6.3	Other ARQ Procedures	198
	4.7	Hybrid ARQ Schemes		205
		4.7.1	Parity Retransmission ARQ Strategy	206
		4.7.2	Retransmission Protocols	208
		4.7.3	Choice of Error Correction Code	208
		4.7.4	Throughput Analysis	209
	4.8	Problems		209
	4.9	References		212

CHAPTER 5 INTEGRATED SERVICES DIGITAL NETWORKS (ISDN) ... 214

	5.1	Introduction		214
		5.1.1	Fundamental Concepts of ISDN	214
		5.1.2	ISDN and Telephone Networks Features	216
		5.1.3	An Overview and Impact of ISDN	218
		5.1.4	The Data Rates for ISDN	220
		5.1.5	The Modes for ISDN Data Transmission	221
	5.2	The Global Status of ISDN		222
		5.2.1	ISDN in United States and Canada	224
		5.2.2	ISDN in Other Countries	228
	5.3	ISDN and the Subscriber Loop Environments		231
		5.3.1	Physical Characteristics of Loop Plants	232
		5.3.2	Electrical Characteristics of Subscriber Networks	240
	5.4	The Major Limitations for Loop Data Transmission		252
		5.4.1	Limitations Resulting from the Physical Design Rules	254
		5.4.2	Limitations from Environmental Conditions	256
		5.4.3	Limitations Resulting from Electrical Interference	257
		5.4.4	Impulse Noise	269
	5.5	Evolving Trends in ISDN		271
	5.6	References		273

Volume II Contents xi

CHAPTER 6	DIGITAL SUBSCRIBER SYSTEMS (DSS)	275
6.1	Recent Growth of Digital Subscriber Systems	275
	6.1.1 ISDN and Digital Subscriber Systems	275
	6.1.2 Developments in DSS Design	277
6.2	Data Transmission Systems and their Components	278
	6.2.1 The TCM System and Its Components	278
	6.2.2 The Adaptive Echo Canceller System and Its Components	281
6.3	Design and Implementation of Digital Subscriber Systems	285
	6.3.1 The Choice of Line Codes	285
	6.3.2 The Role of Simulation	287
	6.3.3 Design Rules for Loop Selection	299
6.4	The Component Optimization Procedures	301
	6.4.1 Analysis and Design Optimization of Equalizers	301
	6.4.2 The Filters for Digital Subscriber Systems	318
	6.4.3 Timing Recovery Circuits	321
6.5	Global Overview of the Status of the DSS	324
	6.5.1 TCM Systems	325
	6.5.2 The Hybrid Echo Canceller (AEC) Systems	329
	6.5.3 Facts in Favor of the TCM System	332
	6.5.4 Facts in Favor of the Hybrid Echo Duplex (AEC) System	334
6.6	Summary	336
6.7	References	337

Index to Volume II		339
Index to Volume I		355

VOLUME I
CONTENTS

CHAPTER 1   THE DEVELOPMENT OF DIGITAL NETWORKS 1
1.1   Introduction to Digital Telecommunications 1
      1.1.1   The Digital Revolution 1
      1.1.2   Digital Network Development 4
      1.1.3   Pulse Code Modulation Fundamentals 6
      1.1.4   Growth in Data Transmission Demands 9
1.2   The Components of a Communication System 10
      1.2.1   System Functions 10
      1.2.2   Comparison of Analogue and Digital Telephone
             Systems 11
1.3   Digital Data Transmission 14
      1.3.1   Voice-band Data Transmission 14
      1.3.2   Public Data Networks 17
      1.3.3   Nonswitched Data Networks 19
      1.3.4   Packet Switched Data Networks 20
      1.3.5   Local Area Networks 25
1.4   Integrated Services and Hierarchies 26
      1.4.1   PCM Hierarchies 27
      1.4.2   Bit Rates for Other Services 28
1.5   Problems 29
1.6   References 31

CHAPTER 2   BASEBAND DIGITAL TRANSMISSION SIGNALS ... 33
2.1   Introduction 33
      2.1.1   A Digital Transmission System 33
      2.1.2   Baseband Signals 34
2.2   Baseband Line Transmission Systems 35
2.3   Algebraic Representation of Line Signals 37
2.4   Encoding and Pulse Shaping 39
      2.4.1   System Elements 39
      2.4.2   Alternate-Mark-Invasion (Bipolar) Code 40

xii

2.5	Line Waveforms		40
2.6	Line Code Selection		46
	2.6.1	Desirable Code Characteristics	46
	2.6.2	AMI Code Properties	47
	2.6.3	Manchester Code (Twinned Binary, Split Phase)	50
	2.6.4	Differential Diphase Code	51
2.7	Methods for Calculating Frequency Spectra		51
	2.7.1	Spectra of Periodic Signals	52
	2.7.2	Spectra of Aperiodic Signals	54
	2.7.3	Spectra of Random Waveforms	55
2.8	Power Spectral Density of Line Codes		62
	2.8.1	Spectral Density of the Line Signal	62
	2.8.2	Autocorrelation Function of Coded Sequences	64
	2.8.3	Spectrum of AMI Coded Signal	69
	2.8.4	Cyclostationary Signals	71
2.9	Other Ternary Line Codes		73
	2.9.1	High Density Bipolar (HDBn) Codes	74
	2.9.2	MBNT Codes	77
	2.9.3	4B3T, 5B4T, 7B5T, 8B6T, and 10B7T Codes	79
2.10	Problems		79
Appendix 2.1 Table of Fourier Transform Pairs			84

CHAPTER 3 INTERSYMBOL INTERFERENCE AND PULSE SHAPING 88

3.1	Introduction		88
3.2	Nyquist Pulse Shaping		95
	3.2.1	Maximum Rate Pulses	95
	3.2.2	Symbol Packing Rate	97
	3.2.3	Nyquist Vestigial Symmetry Criterion for Zero ISI	97
	3.2.4	Raised Cosine Spectrum for Zero ISI	99
	3.2.5	Pulse Shaping Circuits	101
3.3	Multilevel Signalling		104
3.4	Correlative (Partial-Response) Signalling		106
	3.4.1	Elementary Duobinary Scheme	107
	3.4.2	Nonbinary Inputs	112
	3.4.3	Duobinary Decoding and Error Propagation	113
	3.4.4	Precoding	114
	3.4.5	Regeneration and Decoding for the Duobinary Scheme with Precoding	115
	3.4.6	Generalized Correlative Encoding	117
	3.4.7	Modified Duobinary (Class-4 Partial Response) Scheme	118

3.5	Problems		120
3.6	References		122

CHAPTER 4 SIGNAL REGENERATION 123
- 4.1 Introduction 123
- 4.2 Regenerative Repeaters 124
  - 4.2.1 Functions 124
  - 4.2.2 Clock Recovery 126
  - 4.2.3 Sampling and Decision Circuits 128
- 4.3 Equalization 129
  - 4.3.1 Equalizer Functions 129
  - 4.3.2 Transmit Pulse Shapes 131
  - 4.3.3 Typical Transmission Line Characteristics 132
  - 4.3.4 Frequency-domain Characteristics of Equalizers 133
  - 4.3.5 Transversal Equalizers 135
  - 4.3.6 Automatic Equalizers 140
  - 4.3.7 Computer Simulation 143
- 4.4 Bit-Error Rate Calculations 155
  - 4.4.1 Mathematical Models 155
  - 4.4.2 Probability of Error—Wideband Gaussian Noise Case 162
  - 4.4.3 Allocation of Transmit and Receive Filtering 166
- 4.5 Problems 169
- 4.6 References 172

CHAPTER 5 MEASUREMENT TECHNIQUES 173
- 5.1 Introduction 173
- 5.2 Eye-Diagrams 174
  - 5.2.1 Measurement Procedures 174
  - 5.2.2 Important Features of Eye-Patterns 176
  - 5.2.3 Effects of Intersymbol Interference 179
  - 5.2.4 Effects of Noise and Crosstalk 181
- 5.3 Near End Crosstalk Noise Figure 183
  - 5.3.1 Regenerator Performance Measurement 183
  - 5.3.2 Input Signal to NEXT Noise Ratio 185
- 5.4 Pseudorandom Binary Test Signals 185
  - 5.4.1 Introduction 185
  - 5.4.2 Searching for a Random Sequence 186
  - 5.4.3 Feedback Shift Register Generators 187
  - 5.4.4 Properties of Pseudorandom Binary Signals 192
  - 5.4.5 Applications of Pseudorandom Sequences 196

Volume I Contents xv

5.5	Error Rate Measurements	197
	5.5.1 Bit-Error Rates	197
	5.5.2 Error-Free Seconds	202
5.6	Regenerator Fault Location Tests	205
	5.6.1 Triples Test Signal	205
	5.6.2 Fault Location Procedure	207
5.7	Error Performance Objectives	208
	5.7.1 Hypothetical Reference Connection (HRX)	208
	5.7.2 Error Performance Specifications	209
5.8	Questions and Problems	213
5.9	References	214

CHAPTER 6 DIGITAL CODING OF SPEECH SIGNALS 216

6.1	Introduction	216
	6.1.1 Coding Methods	216
	6.1.2 Distortion Criteria	218
6.2	Impairments Resulting from Sampling	221
	6.2.1 Flat-top Samples	222
	6.2.2 Anti-aliasing Filter	225
	6.2.3 Reconstruction Filter	225
	6.2.4 Switched Capacitor Filter Methods	227
6.3	PCM System Performance	229
	6.3.1 Quantization Noise	231
	6.3.2 Companding Techniques	237
	6.3.3 CCITT Standards for Quantization Noise	241
6.4	Coding Techniques for Reduced Bit Rates	243
	6.4.1 Principles of Data Compression	243
	6.4.2 Adaptive Pulse Code Modulation (APCM)	246
	6.4.3 Differential Pulse Code Modulation (DPCM)	246
	6.4.4 Delta Modulation (DM)	248
	6.4.5 Adaptive Differential Pulse Code Modulation (ADPCM)	254
	6.4.6 Adaptive Predictive Coding (APC)	257
	6.4.7 Other Waveform Coding Techniques	258
6.5	Problems	258
6.6	References	261

PREFACE

With the rapid growth of digital techniques in telecommunications networks, there is a need for a revised approach to the study of the principles and the applications of digital transmission systems and networks. The purpose of this two-volume text is to provide an introductory treatment of principles involved in digital communications (Volume I) and to integrate these principles into the applications environment (Volume II). The engineering aspects of integrated communications systems for telephony, television, data, and other network services are discussed in the second volume.

These texts should be useful for senior undergraduate students in the communication sciences, graduate students, and practicing engineers. They should also prove valuable to computer science students and software designers who wish to understand networks and their control. Network control and the design of special purpose software, associated with the management of networks under normal and abnormal conditions needs a firm grasp of the principles discussed in Volume I and their applications discussed in Volume II.

The two volumes are intended for a sequence of two semester courses in digital transmission systems and networks. Some familiarity with the principles of analogue communication systems will give the reader an appreciation of the digital techniques discussed in the two volumes. Prior exposure to basic probability theory related to random processes would be helpful.

Volume I, in particular, is concerned with the evolution of the digital networks, the different types of signals encountered in baseband digital transmission, intersymbol interference and pulse shaping, signal regeneration, measurement techniques, and digital encoding of speech. Considerable effort has been made to make the material useful to senior undergraduate students and to practicing engineers in the telecommunications industry. Emphasis is therefore given to a careful presentation of the essential concepts and typical engineering solutions with computer programs and without becoming unnecessarily concerned with too many theoretical proofs. The rapidly changing VLSI environment preempts our detailed presentation of the hardware realizations of the basic building blocks of these baseband systems.

Volume II is concerned with implementation of the principles (discussed in Volume I) in the physical networks for the transmission of data. In particular, digital radio, telephone, and computer networks are presented together with the implementation of error control in such networks. Integrated Services Digital

Networks (ISDN) and Digital Subscriber Systems (DSS) are also presented in considerable detail in the light of their capabilities, design, optimization, trade-offs, and the potential impact on the telephone networks around the world. We have delineated the steps and the optimizations undertaken in the early designs and progress of the 144 kbit/s facility and the 56 kbit/s circuit switched digital capability introduced by AT&T in the early eighties as a precursor to the ISDN-like network services now being offered by certain telephone operating companies in the United States. This volume is directed towards graduate students in telecommunications and practicing network and design engineers. It also provides an insight into the devices necessary to realize the network terminal functions. System programmers and software designers will find these discussions attractive in implementing the network control strategies.

The emphasis in these texts is on a pedagogic approach to understanding essential engineering concepts. Often this is achieved by examining a number of specific cases which illustrate an idea rather than by attempting to develop proofs for the general case. For many practicing engineers several years out of graduation, some basic tools of mathematical analysis have lain dormant and need to be revitalized. For a professional engineer working in a rapidly developing field such as telecommunications, an analytical appreciation of new ideas is essential. The purely descriptive approach is too limiting. The use of many problems with worked solutions in the text is intended to foster the reader's understanding and development of these analytical skills.

We anticipate that these texts should fill a gap in the current literature on the subject particularly with its emphasis on evolving digital networks. It can provide the basis of courses for senior students and for professional development courses for practicing engineers.

We are particularly grateful to Telecom Australia for supporting the initial preparation of this material. The result reflects the benefits of practical experience gained by each of us as employees of Telecom Australia and Bell Telephone Laboratories Inc., respectively. It is almost certain that this work would never have been commenced were it not for John Grivell, a senior engineer in Telecom Australia, who attended one of our lecture courses and recorded and annotated lecture notes which were much more clear and comprehensive than the original source material. We also wish to sincerely thank Dr. Teresa Buczkowska of the New South Wales Institute of Technology for contributing the chapter on computer networks and Professor Shu Lin of Texas A&M University for his influence on the section on error control. Thanks also go to Elaine Milsom, Judy Duval, and Isobel Keegan for their assistance with typing.

We also thank President Edmond L. Volpe, and Dr. Barry Bressler, Vice President for faculty and instruction at the College of Staten Island, City University of New York, for their support and encouragement. The authors are also indebted to the management of Morris Research and Engineering Center of the

Bell Communications Research at Morristown, New Jersey, where the original research work from AT&T Bell Laboratories was considerably extended. The opportunity to carry on the basic ISDN and CSDC work was initiated under the direction of Eric E. Sumner of the AT&T Bell Laboratories and by Frederick T. Andrews, now of Bell Communications Research. Dr. Ralph W. Wyndrum, Dr. Barry Bosik, Dr. Harold Seidel, and Dr. Peter Bohn of AT&T Bell Laboratories have also influenced the direction and findings we present in the last two chapters of the second volume. Dr. N. S. Jayant, William L. Shafer, and Albert J. Schepis of the AT&T Bell Telephone Laboratories, Joseph F. Urich, Rein R. Laane, Dr. Richard A. McDonald, Wilhelm H. von Aulock of Bell Communications Research have provided constructive comments based upon the manuscript we had presented to them.

Michael J. Miller,
South Australian Institute of Technology

Syed V. Ahamed
The City University of New York

July 1986

Chapter 1

DIGITAL RADIO SYSTEMS

1.1 INTRODUCTION

Digital radio systems operating in the microwave frequency bands are an important medium for digital transmission. In many countries they presently appear to be one of the most economic ways of providing a "back-bone" of high-capacity digital bearers interconnecting large population centers over long routes.

In most countries, analogue frequency-modulated microwave radio links have already been well established for high capacity transmission of signals such as frequency-division-multiplexed (FDM) telephony and video signals. The rapidly developing demand for long-haul transmission of digital data and voice signals has led to the evolution of new radio-relay systems designed specifically to meet the needs of digital transmission. Long-haul high capacity digital radio systems extending over many thousands of kilometers are currently being developed in many countries such as Japan, USA, Canada, Italy, and Australia.

Digital radio systems are distinguished by their use of digital modulation methods and digital regenerator techniques. These digital radio systems will be the main subject of this chapter. We will examine typical modulation methods, transmitter configurations, and receiver configurations. Methods for evaluating and comparing the performance of systems using different forms of modulation will be discussed. Particular attention will be given to the problems associated with multi-path fading and its resultant frequency-selective effects. This is likely to be a dominant factor in the design of system configurations and path lengths.

Extensive high capacity digital radio systems, such as those envisaged for transmission at 140 Mbit/s rates, are still in the development stages in most countries. In the interim, some countries have found it necessary to use the existing analogue microwave radio networks to provide a means of transmitting medium and high-speed digital signals until replacement digital radio networks can be developed. Hybrid systems for this purpose include the use of a 12-channel telephone FDM group band frequency spectral allocation or a hybrid system such as "Data under FDM voice" (DUV) or "Data above FDM voice" (DAV). During this transitional period, existing analogue and new digital radio systems will be operated side-by-side on many transmission routes. System designs must be such as to ensure this compatibility.

In addition to high-capacity long distance applications, digital radio systems are likely to become increasingly important in short-haul and local networks. This includes their use in point-to-multipoint digital radio concentrator systems for telephone subscribers in remote areas. Distributed digital radio networks, including point-to-multipoint systems and packet-radio networks, have been developed in conjunction with high-speed local-area data networks to link groups of computer users in various locations of city areas. These networks are often referred to in the context of planning for the "office of the future." Here the transmission requirement differs from the long-haul application, where high bit rates have to be transmitted over long distances. In short-haul and local radio networks it is a matter of setting up multi-directional nodes for widely differing sizes of message groups, often with the smallest possible angular spacing between radio paths.

1.1.1 Historical Developments

A *digital radio system* accepts signals directly in digital format (for example, HDB3 coded signals) and after removal of the line coding, uses these signals to modulate a radio carrier. The term *digital radio* encompasses any modulation scheme which alters the characteristics of a carrier in a discrete fashion. The resultant modulated signal can assume only a finite number of changes in state. These states may relate to frequency, phase, amplitude, or some combination.

Digital microwave radio systems are rapidly being developed in many countries to become a significant component of voice and data transmission networks. In the 1960s and early 1970s, analogue microwave radio relay systems dominated in the provision for systems for long-distance transmission of FDM speech, voice-band data, and TV relay. These analogue radio-relay systems have made extensive use of the frequency bands which have been allocated for fixed services in the vicinity of 2, 4, 6.1 and 6.7 GHz. In these systems, the analogue frequency division multiplexed (FDM) multichannel telephony signals frequency modulate the microwave carrier.

In some countries, notably in North America, digital transmission of PCM signals in metropolitan areas via pair cable has been developing over the past two decades. However, efforts to extend these digital systems to the long-distance intercity networks have been limited. This has been due primarily to the cost of cable and repeaters.

In the early 1970s, a small number of digital microwave radio systems were in operation, most of them on an experimental or development basis. Since the

Introduction

mid-1970s, the picture has changed rapidly. Now it appears likely that, over the next decade in many countries, digital radio link networks will be rapidly developed to provide for a significant percentage of the high capacity long-haul digital transmission requirements.

1.1.2 Digital versus Analogue Radio

It is interesting to note that, at the present time, analogue radio link costs seems likely to remain lower than digital radio costs per voice channel mile. The reason for the change from analogue to digital radio is not related to the cost of the radio links themselves. Rather, it is related to the significant performance improvement and cost reduction that can be potentially obtained in the total telephone system when digital techniques are used. In the overall transmission costs, the less expensive digital channel bank and multiplexing carrier equipment offset the somewhat higher digital radio system costs. Dinn (1) suggests that relative transmission costs per circuit remain lower for digital radio transmission out to distances of the order of 500 km as illustrated in Figure 1.1. See also, Bailey (3).

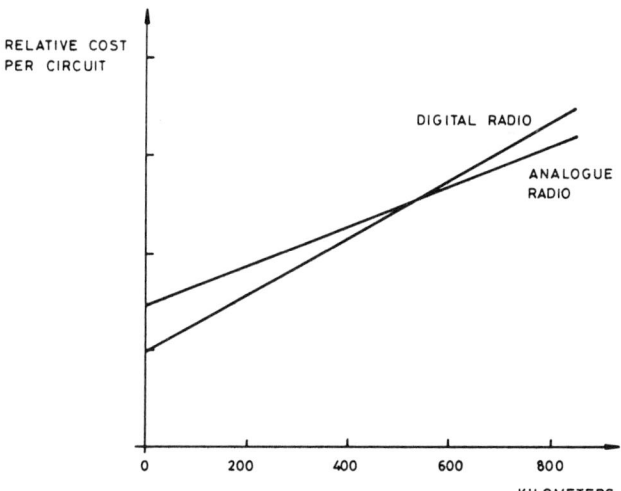

Figure 1.1 Relative transmission costs.

As digital networks and digital switching become more common, digital radio systems will be used extensively for high capacity intercity interconnection of digital traffic. That is, the development of digital radio networks is proving necessary to enable the realization of the full economic advantages of digital switching using TDM techniques.

Other advantages of digital radio systems compared to analogue include:

(1) the ability to regenerate the digital signal at each radio repeater with the result that the overall circuit long-term bit error rate performance becomes essentially independent of length
(2) a higher immunity to noise and interference so that operation is possible at higher frequency bands where attenuation because of rain and snow is significant.
(3) the benefit in respect of spectrum congestion resulting from the ability to use frequency allocations higher than the bands commonly used for analogue systems.
(4) the tendency to smaller equipment size making such equipment easier to install and transport.

On the other hand, it is currently apparent that digital radio systems appear likely to suffer certain disadvantages. These include the following:

(1) the sensitivity of high capacity systems to frequency selective fading which results in the need to use shorter paths and/or incorporate adaptive equalizers and space diversity systems. (Space diversity refers to the provision, for each radio channel, of a second receiving antenna with associated waveguide, a second receiver, and monitor/combiner equipment.) Virtually all high capacity digital radio systems seem likely to require space diversity to combat the effects of fading.
(2) the absence of a "sub-baseband" traffic capability which makes it more costly to drop and insert small numbers of telephone circuits typically used for wayside traffic, telemetry, or control. (In analogue microwave radio systems, the FDM telephone input baseband signal typically extended from a lower limit of 60 kHz. Input frequencies below this limit are referred to as sub-baseband).
(3) the higher supply power requirement for digital radio repeaters when compared to currently available low drain analogue radio equipment. Steele (2) suggests that it may be uneconomic to provide power for remote repeaters by solar cell arrays as is currently done for many analogue radio systems.

1.2 TYPICAL DIGITAL RADIO SYSTEMS

1.2.1 Digital Radio Equipment

Figure 1.2 illustrates a typical digital radio system terminal in a long-haul network. The input digital signal modulates an intermediate-frequency (IF) carrier,

Typical Digital Radio Systems

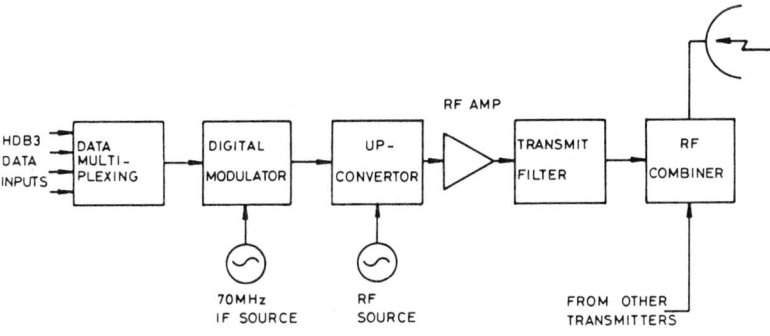

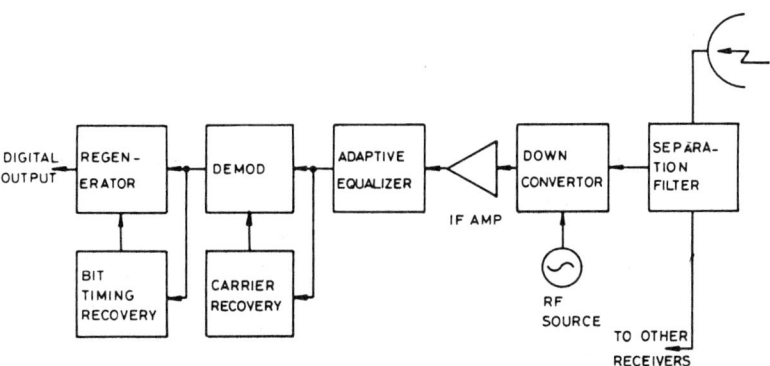

Figure 1.2 Digital radio terminal.

typically at 70 MHz, using a digital modulation technique, such as 8-PSK (8-point phase-shift-keying) or 16-QAM (16-point quadrature-amplitude-modulation). Digital modulation methods will be discussed in the next section.

The resultant IF signal is upconverted to the allocated transmit radio-frequency (RF) signal value. Special care is taken with filtering the RF signal before transmission to limit the transmitted spectrum. This is necessary in order to minimize the likelihood of co-channel interference to other systems operating in adjacent frequency channels.

At the receiver, the complementary processing takes place. The signal is downconverted, amplified, and then equalized to compensate for amplitude and phase distortion resulting from the RF equipment or propagation effects. Then demodulation and regeneration takes place, resulting in the received baseband digital signal. Space diversity equipment is omitted from the figure for clarity.

By way of comparison, in an analogue radio link transmitter, the FDM input signal is connected to a pre-emphasis unit and frequency-modulator which generates a 70 MHz IF signal. The remainder of the transmitter is essentially the same as for the digital system except that less stringent requirements may be placed in the transmit filter because of the nature of the FDM-FM spectrum.

In digital radio equipment for short-haul networks, the digital signal is often directly modulated on the RF carrier. This leads to reduced space and power supply requirements. The radio equipment is usually designed for accommodation in weather-proof housings for outdoor installation, if required.

It is true to say that far more "intelligence" is incorporated in digital than in analogue radio systems. In digital radio relay systems, the digital signal undergoes a number of processing steps such as pre-coding, scrambling, adaptive equalization, and regeneration. These processes represent more signal processing complexity than in analogue systems in which the delivered analogue signal remains essentially unchanged during transmission.

1.2.2 Transmission Capacities and Frequency Bands

Digital radio systems in public communication networks must be designed to accommodate transmission rates compatible with digital network standards. In Chapter 1, Volume 1, we examined the hierarchical development of transmission rates. As shown in Table 1.1, for the 30-channel PCM hierarchy as used in Europe and Asia, a single voice channel is encoded into 64 kbit/s. The primary multiplex 30-channel rate is 2.048 Mbit/s, which for brevity is often referred to

Table 1.1
Digital hierarchical bit rates

EUROPE AND ASIA					
Voice Channels	30	120	480	1920	7680
Bit rate (Mbit/s)	2.048	8.448	34.348	139.264	*
Nominal rate (Mbit/s)	2	8	34	140	565
CCITT Rec.	G.732	G.742	G.751	G.751	*
*Undetermined					
USA AND CANADA					
Voice Channels	24	96	672	4032	
Bit rate (Mbit/s)	1.544	6.312	44.736*	274	
U.S. Category	DS-1	DS-2	DS-3	DS-4	
CCITT Rec	G733	G743	G753A		
JAPAN					
Voice Channels	24	96	480	1440	5760
Bit rate (Mbit/s)	1.544	6.312	32.064	97	397
CCITT Rec	G.733	G.743	G.753(3)		

*Note that 90 Mbit/s systems that accommodate two DS-3 signals (1344 voice channels) are common in North America.

as a nominal 2 Mbit/s rate. Higher order multiplex rates are multiples of this rate, namely nominal rates of 8, 34, and 140 Mbit/s. We also noted that different hierarchies, based on 24-channel primary multiplexing, are used in North America and Japan. For convenience the bit-rate hierarchies are summarized in Table 1.1. The relevant CCITT recommendation numbers are also shown.

The transmission capacities of digital radio systems are normally limited to integer multiples of the PCM hierarchical bit rates shown in Table 1.1. Digital radio systems are classified as low, medium, or high capacity systems depending on the overall bit rate per RF channel as summarized in Table 1.2.

Table 1.2
Capacity Classification of Digital Radio Systems.

Digital Radio Capacity Classification	Bit rates	Examples of Typical digital radio system capacity
Low capacity	0–10 Mbit/s	2,4,8 Mbit/s
Medium capacity	10–100 Mbit/s	17,34,70,78,90 Mbit/s
High capacity	100– Mbit/s	105,140 Mbit/s

1.2.3 Typical Equipment Characteristics

Digital radio systems for the long-haul network operate chiefly with frequency allocations below about 12 GHz. In these bands, radio hop lengths of approximately 50 km can be used. On the other hand, systems for short-haul and local networks operate in practically all available frequency ranges from 0.5 to 30 GHz. Since the limitation of range plays a subordinate role with these systems, it is possible to use the frequency ranges between 12 and 30 GHz.

Table 1.3 lists a selection of some of the frequency allocations recommended by the International Radio Consultative Committee (CCIR) for analogue FDM and digital radio systems. See CCIR (6) for more information.

Some popular radio systems in North America do not conform with the frequency use recommendations of Table 1.3. For example, in the 5.925–6.425 GHz band, provision is made for frequency allocations with 30 MHz channel spacing

Table 1.3
Typical CCIR Microwave Frequency Use Recommendations.

CCIR Rec.	Frequency Band (GHz)	Frequency Range (MHz)	Channel Spacing (MHz)	FDM Channel Cap.	Band Cap.	Digital Channel Cap.	Band Cap.
382-3	2	1900–2300	29	120–960	6	120–960	6
384–2	6.7	6430–7110	40	2700	8	1920	8
387–1	11	10700–11700	40	1800	12	480–960	11
497	13	12700–13250	28	960	8	960	8

for use by 90 Mbit/s systems carrying the equivalent of 1344 voice channels. Also, Canada uses frequency allocations in the 8 GHz band for 91 Mbit/s coast-to-coast digital radio systems.

Note that in Table 1.3, the band capacity refers to the number of go and return RF channels that are envisaged. Also, for the 13 GHz frequency band, the recommended digital radio channel capacity allows the use of two 480-channel radio bearers on the same frequency allocation but using polarization separation. That is, one system uses vertical polarization and the other, horizontal polarization to provide the necessary channel separation. At this point in time, systems using this type of cross polarization separation are still in the development phase.

As an example of equipment availability, Figure 1.3 illustrates a possible radio terminal hierarchical configuration suggested by one company (Nippon Electric Co. Ltd.) for various 30-channel multiplex levels.

As shown in Figure 1.3, digital radio equipment suitable for low bit rates is available in the 2 GHz hand, for medium rates in the 13 GHz band, and for high rates in the 6.7 GHz band.

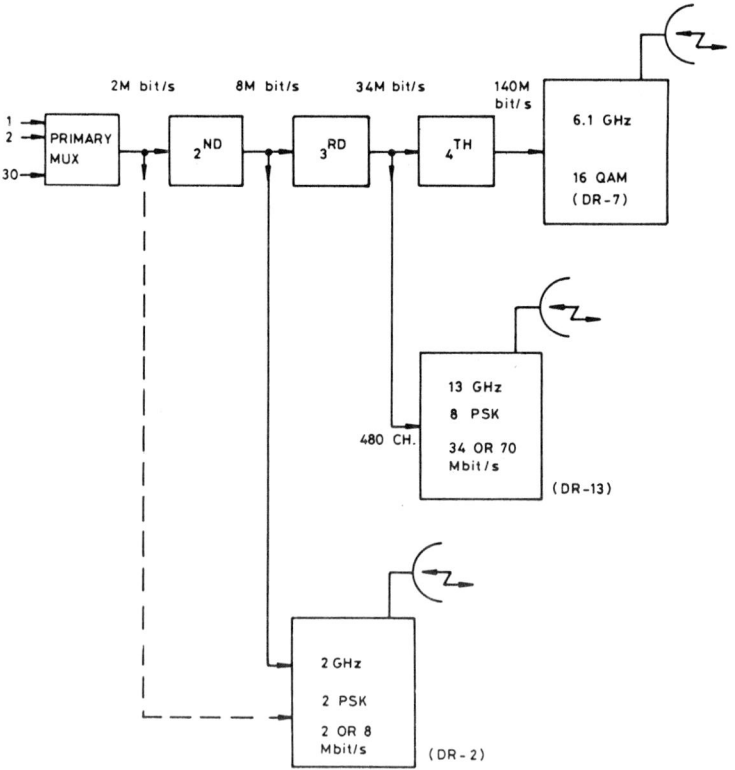

Figure 1.3 Typical digital radio bit rates and frequency bands.

Typical Digital Radio Systems

Many public communication network operators are currently introducing digital radio systems into their networks. In an illustrative scenario, the following developments are typical of those required to meet the needs of a national network.

(1) A long-haul high capacity (140 Mbit/s) digital radio relay network, operating in the 6.7 GHz band, may be installed to link major cities. These systems will, in the main, be installed on existing radio relay routes, thereby using the existing towers and buildings. The resultant path lengths dictate the use of IF-combining space diversity equipment together with adaptive IF equalizers on almost all paths to meet the desired performance objectives. This is discussed in a later section.

(2) Medium capacity 34 Mbit/s digital radio systems in the 13 GHz band may be used within cities for the provision of inter-exchange junction circuits and wideband customer services.

(3) Low capacity digital radio subscriber systems may be used in rural and remote areas to connect customer telephones into the telephone network.

High capacity 140 Mbit/s digital radio equipment is currently undergoing development and field trials in the U.S., U.K., Australia, France, West Germany, Italy, Switzerland, Sweden, and Norway. Frequency bands in use include the 4, 6.7, and 11 GHz bands.

Figure 1.4 shows the CCIR frequency plan for the upper 6 GHz (or 6.7 GHz) band which is likely to be used in many countries for long-haul high-capacity digital radio systems. This frequency plan enables a radio link system to be expanded to accommodate eight two-way RF channels for 140 Mbit/s operation.

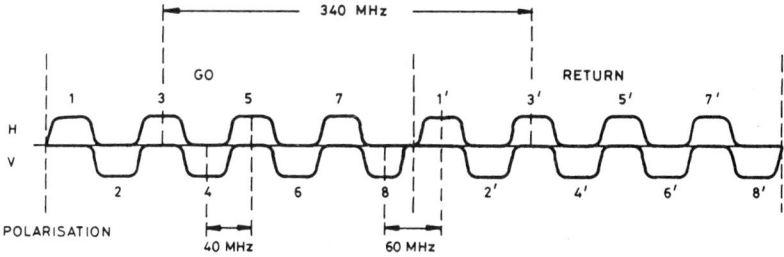

Figure 1.4 CCIR Frequency plan for 6.7 GHz band.

One of these channels would normally be used as the standby channel. Adjacent channels may be used provided they use orthogonal polarization to provide extra isolation to supplement channel filtering. That is, one channel is transmitted horizontally polarized and the other vertically polarized.

In the U.S. networks, digital radio allocations are available primarily in bands in the vicinity of 2, 6, and 11 GHz. In the 2 GHz band, available bandwidths are small compared to other bands so lower bit rates are achievable, typical rates falling in the 6–12 Mbit/s range. Maximum authorized radio channel bandwidths are 30 MHz in the 6 MHz band and 40 MHz in the 11 GHz band. Systems operating in these bands in the U.S. typically provide 90 Mbit/s per radio channel, representing 1344 voice channels.

As can be seen by reference to Table 1.1, a transmission rate of 90 Mbit/s accommodates two DS-3 signals. Each DS-3 signal corresponds to a rate of 44.736 Mbit/s representing 672 digitized voice channels. More details are available in Dinn (1), Feher (5), and Coutts (7). In the 6 and 11 GHz bands, 140 Mbit/s systems are also being installed in North American networks.

1.2.4 Performance Objectives

The performance and availability objectives for digital radio-relay systems are specified in CCIR Recommendations. These objectives refer to a 2,500 km 64 kbit/s Hypothetical Reference Digital Path (HRDP). The HRDP has nine digital radio sections of equal length with demultiplexing down to primary level PCM at two intermediate points.

The digital radio system is considered to be unavailable if one or both the following conditions occur in each of 10 or more consecutive seconds:

(1) the bit error rate (BER) is greater than 10^{-3}
(2) the digital signal is interrupted

The *availability* objective describes the percentage of time the performance of the HRDP is required to exceed these criteria. As specified in CCIR Recommendation 557, the HRDP should be designed to be available for at least is 99.7 percent of the time measured over a period of at least one year. That is the circuit should be unavailable for less than 0.3 percent of a year.

The *performance* objectives, refer to the quality of the HRDP during the periods when the system is considered to be available. The long-term and short-term CCIR objectives are that the BER should not exceed:

(1) 1×10^{-7} for more than 1 percent of any month, and
(2) 1×10^{-3} for more than 0.05 percent of any month.

These values take into account fading, interference, and all other sources of degradation, but exclude contributions from multiplex equipment. The specification of the higher error rate of 1×10^{-3} limits the amount of short interruptions because of fading, and the lower error rate of 1×10^{-7} controls the system performance for the majority of the time when the main sources of errors are noise and interference.

Modulation Methods

There remains the problem of deriving the performance objectives for individual radio paths within the 2500 km HRDP. Let the path length in km be d. Then the bit error rate should not exceed

(1) $(d/2500) \times 10^{-7}$ for more than $d/2500$ percent of any month
(2) 1×10^{-3} for more than $(d/2500) \times 0.05$ percent of any month.

See Godier (25) for further discussion of the application of error rate specifications to real digital radio paths.

1.3 MODULATION METHODS

1.3.1 Spectral Efficiency

Traditional analogue high-capacity FDM-FM radio relay systems for long-haul applications have usually been designed to carry either 960, 1800, or 2700 analogue voice channels per RF carrier. With the introduction of frequency allocations for digital radio systems, radio channels are at a premium. As a result, regulating authorities such as the CCIR and the FCC (U.S.) have required digital radio systems to carry roughly the same number of voice channels per unit bandwidth as the analogue systems.

This requirement is expressed in terms of the radio system *spectral efficiency* which is defined

$$\text{Spectral efficiency} = \frac{\text{Input digital rate}}{\text{Allocated RF channel bandwidth}} \quad \text{(bit/s/Hz)} \quad (1.1)$$

High spectral efficiency can be achieved by the choice of an appropriate digital modulation scheme. In evaluating the theoretical spectral efficiency required for a digital radio system, it is usual to assume that the available RF bandwidth is equal to the channel spacing. In a later section we will examine practical limits on spectral shape.

Exercise 1.1

Use the values in Table 1.3 to determine the spectral efficiency required of a digital radio system working in the 6.7 GHz band to be comparable with that of an analogue FDM-FM radio system.

Solution. The parameters for both systems to be equivalent are

$$\text{Channel bandwidth} = 40 \text{ MHz}$$
$$\text{Number of voice channels} = 2700$$

For the digital radio system, assuming a bit rate of 2048 kbit/s per 30 voice channels, we obtain

$$\text{Spectral efficiency} = \frac{2700 \times (2048 \times 10^3/30)}{40 \times 10^6} = 4.4 \text{ bit/s/Hz}.$$

It is a major objective for digital radio systems to achieve a spectral efficiency of 4.4 bit/s/Hz. The use of 4.4 bit/s/Hz radio systems leads to the following advantages when new digital radio systems replace existing analogue networks:

(1) the same capacity per digital radio bearer is maintained as the existing 1800 channel systems
(2) the replacement of existing FM analogue radio systems by digital systems is possible within the same frequency allocation, and
(3) digital and analogue radio systems are able to use the same frequency plan.

We will next examine some modulation schemes to determine whether this spectral efficiency goal is achievable.

1.3.2 Choice of Modulation Type

The choice of modulation scheme for a digital radio system must be made with the following considerations in mind:

(1) *spectral efficiency* should be a maximum. As we have seen, for high capacity systems, a spectral efficiency of 4.4 bit/s/Hz is a goal. In order to achieve this it is necessary to use:

   (a) phase modulation or amplitude modulation or some combination. These methods have better spectral efficiency than frequency shift keying (FSK).
   (b) multilevel modulation schemes where each carrier symbol interval represents more than one binary bit. For example, in 16-QAM modulation (16-valued quadrature amplitude modulation), each carrier symbol represents four bits ($2^4 = 16$).

(2) *susceptibility to noise and interference* should be minimized. In general, increasing the number of modulation levels results in a worse bit error rate because of noise and interference.

(3) *Susceptibility to fading*—During fading as a result of multipath propagation (see Section 1.5.5 for details) error rate performance and sensitivity to adjacent channel interference many differ for different modulation types. For example, it has been found that in fading conditions, adjacent channel interference effects are worse for 8-PSK systems than for 16-QAM modulation.

Modulation Methods

(4) *Susceptibility to nonlinearities*—In general, modulation schemes which result in a constant modulation envelope are less sensitive to transmitter and receiver nonlinearities.

To see why nonlinearities can be so important, consider a nonlinear system with narrowband input and bandpass filtering of the output signal. The input signal for any modulation type, can be written

$$a(t) = R(t) \cos [\omega_c t + \phi(t)] \qquad (1.2)$$

which can be represented in complex notation form as

$$A(t) = R(t) e^{j[\omega_c t + \phi(t)]}. \qquad (1.3)$$

Let the nonlinear system output be written in complex notation form as

$$B(t) = F\{R e^{j\beta}\}$$

where for brevity

$$R = R(t)$$
$$\beta = \omega_c t + \phi(t)$$

and $F\{\cdot\}$ is a complex memoryless nonlinear function. Now $F\{Re^{j\beta}\}$ is periodic in β and can, therefore, be represented as a Fourier Series form expansion in β, namely

$$F\{Re^{j\beta}\} = \sum_{n=-\infty}^{\infty} C_n(R) e^{jn\beta}. \qquad (1.4)$$

There will be a bandpass filter at the nonlinear system output which filters out all components except those near ω_c so that we obtain for the output

$$Z(t) = BPF\{B(t)\}$$
$$= C_1(R(t)) e^{j[\omega_c t + \phi(t)]} \qquad (1.5)$$

with

$$C_1(R(t)) = f(R(t)) e^{jg(R(t))}. \qquad (1.6)$$

The function

$f(R(t))$ is called the AM/AM conversion function

and

$g(R(t))$ is called the AM/PM conversion function.

Now $Z(t)$ is the output signal written in complex notation so we can write the real output signal as

$$z(t) = f(R(t)) \cos[\omega_c t + g(R(t)) + \phi(t)]. \qquad (1.7)$$

The following special cases should be noted

(1) Linear system: $f(R(t)) = K R(t)$, K a constant
$g(R(t)) = 0$.

(2) Hard limiter: $f(R(t)) = K$
$g(R(t)) = 0$.

Hence, for the general nonlinear system case, if the input is a *constant envelope* signal

$$R(t) = R_o$$

where R_o is a constant. Then the system output is

$$z(t) = f(R_o) \cos[\omega_c t + g(R_o) + \phi(t)] \qquad (1.8)$$

where $f(R_o)$ and $g(R_o)$ are constants. That is, the output also has a constant envelope. The constant phase term $g(R_o)$ can be easily removed so the resultant signal remains unaffected by the nonlinearity. This will not be so for modulation types which result in varying envelopes since $R(t)$ is not constant. In such cases, it is often necessary to operate transmitter power amplifiers at power levels below their maximum value (say 3–10 dB "back off") to avoid the intersymbol interference that would result from nonlinear distortion.

We will now examine several modulation schemes and compare their performance in relation to the above factors (1)–(4).

1.3.3 PSK (Phase-shift-keying) Modulation Schemes

PSK modulation schemes represent one of the most important classes of modulation schemes for digital radio systems. The PSK methods of interest are

(1) *binary PSK* sometimes referred to as 2-PSK, BPSK or 2ϕPSK) in which a binary data stream modulates a constant amplitude, constant frequency carrier in such a way that two phase values differing by 180° represent the binary symbols 0 and 1, respectively. The waveform

$$s_0(t) = -A \cos \omega_c t$$

represents a 0, and

$$s_1(t) = A \cos \omega_c t$$

represents a 1. The binary PSK waveform can be represented by

$$v_p(t) = d(t)[A \cos \omega_c t] \qquad (1.9)$$

where $d(t)$ is a low-pass pulse waveform often consisting of binary rectangular pulses taking values 1 or -1, respectively. Sometimes Nyquist pulse shaping is carried out prior to modulation in which case $d(t)$ will be a positive or negative Nyquist pulse of the form discussed in Chapter 3, Volume 1.

(2) *four-phase PSK* (also known as QPSK, 4-PSK or quadrature PSK) in which the phase of the carrier can take on one of four values 0°, 90°, 180°, or 270°. Each transmitted symbol represents two input bits as follows:

Input bits	Transmitted symbols
00	$A \cos \omega_c t$
01	$A \cos (\omega_c t + 90°)$
11	$A \cos (\omega_c t + 180°)$
10	$A \cos (\omega_c t + 270°)$

(3) *eight-phase PSK* (also called 8-PSK) in which a transmitted carrier symbol takes on one of eight possible values, each representing three input bits.

In general, it is possible to subdivide the input binary signal into M-bit blocks and represent each M-bits by one of 2^M possible transmitter symbols (phases). This is known as M-ary PSK. Note that for a given input bit rate, as the value of M increases, so the transmitter symbol rate decreases. As a result, we would expect an equivalent reduction in the width of the resulting spectrum or, in other words, an increase in spectral efficiency. This can only be done at the expense of bit-error rate performance.

Figure 1.5 shows signal space vector diagrams for 2-PSK, 4-PSK, and 8-PSK signals. In PSK modulation schemes, the conversion from binary symbols to phase angles is usually done using Gray coding. The essential idea is to permit only one binary number change in the assignment of binary symbols to adjacent phase angles. This minimizes the number of bit errors that result from a demodulation error that incorrectly selects a symbol adjacent to the correct one.

For the kth M-tuple say, in M-ary PSK, the phase value is modulated on the carrier for the symbol interval T_s seconds, resulting in the M-PSK signal

$$s(t) = A \cos (\omega_c t + \phi_k), \qquad 0 \le t \le T_s \qquad (1.10)$$

where

$$\phi_k \in \{0, 2\pi/M, 4\pi/M, \ldots, 2(M-1)\pi/M\}.$$

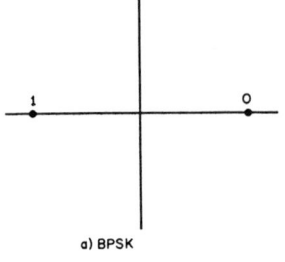

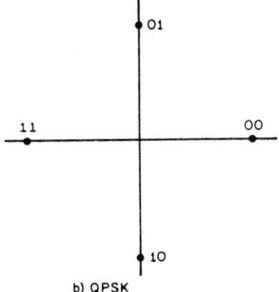

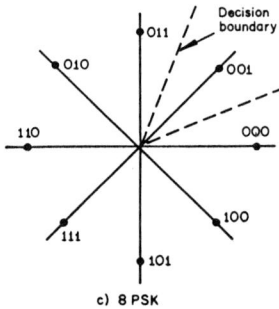

Figure 1.5 BPSK and QPSK signal spaces.

Figure 1.6(a) shows a block diagram of a typical 4-PSK modulator. The serial-to-parallel converter converts the binary input stream into two parallel, half rate signals referred to as the I (in-phase) and Q (quadrature) signals. These balance modulate two IF carriers at 0° and 90° relative phase shift, respectively. The waveforms and the table shown in Figure 1.6(b) illustrate the process.

Modulation Methods

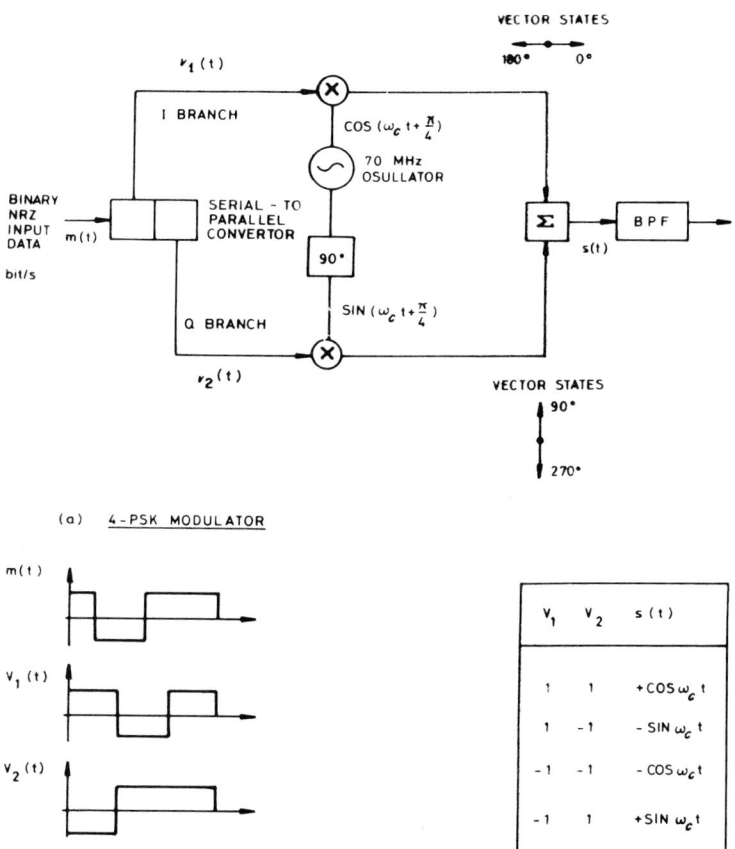

Figure 1.6 4-PSK modulation technique.

Exercise 1.2

Show that for the 4-PSK modulator in Figure 1.6, if in the first time interval $v_1 = 1$ and $v_2 = 1$, then the output $s(t)$ is given by

$$s(t) = \sqrt{2} \cos w_c t$$

Solution. The in-phase branch multiplier output is

$$s_i(t) = \cos(\omega_c t + \frac{\pi}{4})$$

and the quadrature branch multiplier output is

$$s_q(t) = \sin(\omega_c t + \frac{\pi}{4}).$$

Then the modulator output is

$$s(t) = s_i(t) + s_q(t)$$

$$= \cos(\omega_c t + \frac{\pi}{4}) + \sin(\omega_c t + \frac{\pi}{4}).$$

Using the trigonometric identities

$$\cos(\omega_c t + \frac{\pi}{4}) = \cos \omega_c t \cos \frac{\pi}{4} - \sin \omega_c t \sin \frac{\pi}{4}$$

and

$$\sin(\omega_c t + \frac{\pi}{4}) = \sin \omega_c t \cos \frac{\pi}{4} + \cos \omega_c t \sin \frac{\pi}{4}$$

we obtain

$$s(t) = \cos \omega_c t (\cos \frac{\pi}{4} + \sin \frac{\pi}{4}) + \sin \omega_c t (\cos \frac{\pi}{4} - \sin \frac{\pi}{4})$$

$$= \sqrt{2} \cos \omega_c t.$$

A typical 8-PSK modulator is shown in Figure 1.7. This modulation procedure has been used in medium capacity digital radio systems for the 6 GHz and 11 GHz bands. See, for example, Yokoyama (8).

A serial to parallel converter generates three parallel data streams. Two are used to modulate a 4-PSK modulator of the form illustrated in Figure 1.6. The third is connected to a 2-PSK modulator. The output of the 70 MHz 4-PSK

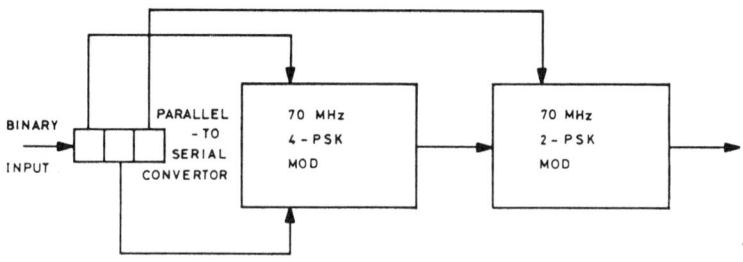

Figure 1.7 8-PSK modulator schematic.

Modulation Methods

modulator is fed to the 2-PSK modulator which shifts the input phase by $\pi/8$ radians if a 1 is present on the third input bit stream.

2-PSK, 4-PSK, 8-PSK Spectra

An M-ary PSK signal can be represented by Equation (1.9) if $d(t)$ is an M-level pulse waveform. For rectangular pulse shapes, it can be shown that the power spectral density for a PSK signal with any value of M is given by

$$S(f) = A^2 T_s \left\{ \frac{\sin[\pi T_s(f-f_c)]}{[\pi T_s(f-f_c)]} \right\}^2 \quad (1.11)$$

where T_s is the transmitter symbol duration. This assumes the input data sequence is random. If this is not so, such as during the transmission of simple idling patterns of all 1's or alternating 1's and 0's, then the spectrum can consist of large discrete spectral lines. In a later section, we will see how scramblers are used to prevent this.

If T_b is the input binary bit interval, then it is easy to see that

$$T_s = T_b \log_2 M. \quad (1.12)$$

In Equation (1.11), f_c is the unmodulated carrier frequency and A is the amplitude. The power spectral density is shown plotted on a linear scale in Figure 1.8(a) and on a dB scale in Figure 1.8(b).

Some important remarks should be made about PSK spectra.

(1) *Spectral efficiency*—Note that for a given transmitter symbol rate ($1/T_s$), the power spectrum remains the same, independent of the number M of symbol levels being used. That is 2-PSK, 4-PSK, and 8-PSK have the same spectral shape if T_s is the same in each case. In an M-ary PSK scheme, each transmitted symbol represents $\log_2 M$ bits. Therefore, the spectral efficiency of 8-PSK, for example, will be three times as great as for 2-PSK, but this is at the expense of increased error probability.

(2) *Nyquist smoothing*—The PSK spectrum shape is the same as the spectrum for the input baseband data pulse waveform. That is, the original baseband spectrum is simply translated along the frequency axis to be centered on the carrier frequency f_c. It, therefore, follows that if the input premodulation baseband signals are smoothed by low-pass filters, the spectral shape of the transmitted signal can be significantly modified. Recall from Chapter 3, Volume 1, that ideal Nyquist filtering ($\alpha = 0$) can restrict the baseband symbol bandwidth to one half the symbol rate. Practical cosine roll-off filtering ($\alpha = 0.3$ say) results in a bandwidth equal to 30 percent greater than the ideal Nyquist value.

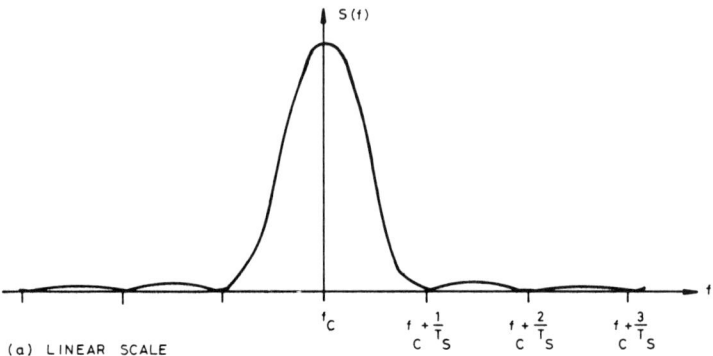

(a) LINEAR SCALE

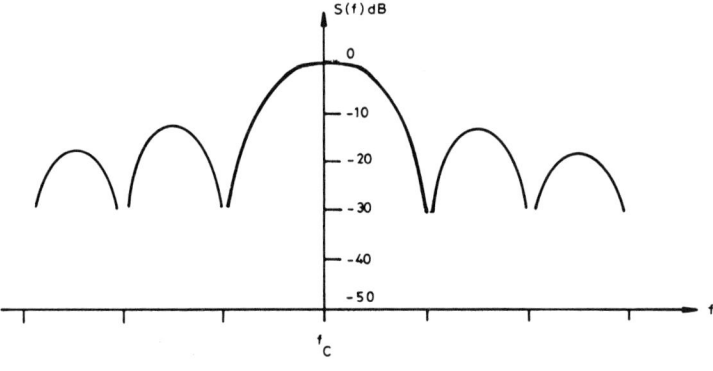

(b) dB SCALE

Note: These diagrams apply to 2-PSK, 4-PSK, 8-PSK providing T_s is taken as the symbol interval.

Figure 1.8 Power spectral density of PSK waveforms.

Hence, in a PSK radio system, ideal Nyquist prefiltering results in a spectrum that is centered on f_c, constant over the band

$$B = 1/T_s \qquad (1.13)$$

and zero outside that band. For this case, the transmitted bandwidth for an M-PSK signal can also be written

$$B = \frac{1}{T_b \log_2 M}. \qquad (1.14)$$

It follows that for M-PSK, with $\alpha = 0$ Nyquist filtering, the spectral efficiency is

$$r_b/B = \log_2 M \quad \text{(bit/s/Hz)}. \qquad (1.15)$$

Spectral shaping can be achieved by means of premodulation low-pass filtering, postmodulation intermediate-frequency (IF), or radio-frequency (RF) bandpass filtering or a combination of these.

It is important to bear in mind that when a constant envelope PSK signal is bandlimited by filtering, the envelope becomes nonconstant. To confirm this, consider the 2-PSK signal represented by Equation (1.9). If no filtering is used, $d(t)$ is a series of rectangular pulses taking values 1 or -1. Then the 2-PSK waveform has a constant envelope of value A, since in any symbol interval

$$v_p(t) = \pm A \cos \omega_c t.$$

However, if Nyquist filtering is used, $d(t)$ has the form of a sequence of positive or negative Nyquist shaped pulses, as discussed in Chapter 3, Volume 1. In this case, it is clear that the envelope $|A\, d(t)|$ will vary considerably with time.

Premodulation filtering may be easier to implement than IF or RF filtering. However, it has the disadvantage that, if a nonlinear power amplifier follows the modulator, then the spectrum is broadened and at least partially restored to its original shape.

(3) *Filtering requirements*—In practice, the emission spectra of digital radio signals must fall within authorized limits set by regulatory bodies to minimize inter-channel interference effects. For example, the spectrum must fall within the envelope (or mask) shown in Figure 1.9 for a 40 MHz authorized band-

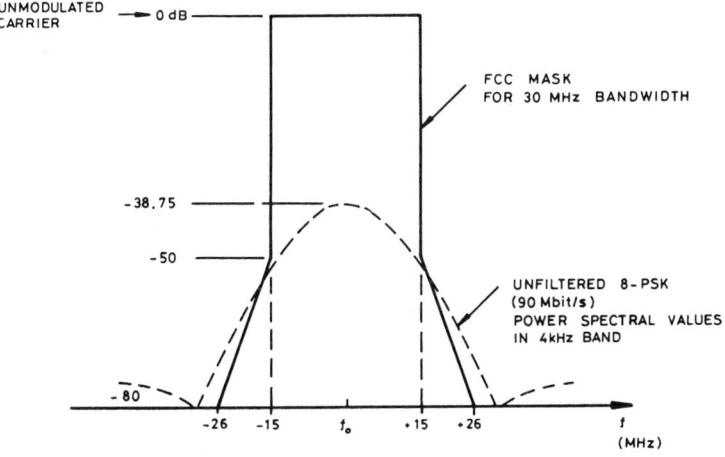

Figure 1.9 FCC Spectrum limit and PSK spectra.

width. For details refer to Feher, (5). Also shown in Figure 1.9 is the unfiltered spectrum for an 8-PSK signal operating at 34 Mbit/s.

After filtering, the spectral shape of an 8-PSK signal will be almost constant across the passband and will fall off sharply near the band limits, as shown in Figure 1.10. Note that, in general, the power spectral density of a digital radio system is more intense at the band edges than for analogue FDM-FM radio relay systems. This imposes severe restrictions on systems where digital and analogue radio channels share a common frequency band because of potential interference from digital radio bearers into analogue bearers on adjacent frequencies. When digital and analogue radio systems share a given band within the existing CCIR radio plan, it is often necessary to separate digital and analogue channels with a third unused channel. Refer to Ellershaw (9) for details.

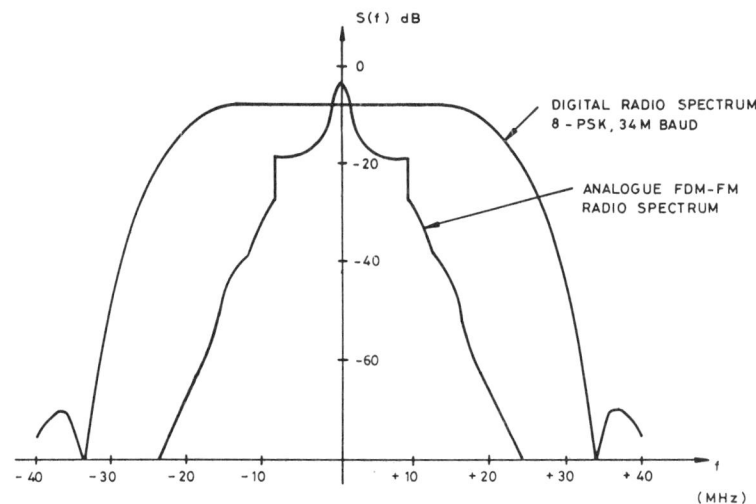

Figure 1.10 Comparison of digital and analogue radio spectra.

Exercise 1.3

Given that an input binary bit rate of 34 Mbit/s is to be transmitted at 13 GHz in a channel bandwidth of 28 MHz, determine which PSK scheme is appropriate.

Solution. From Equation 1.13, for ideal Nyquist filtering, the transmitted signal bandwidth is

$$B = \frac{1}{T_s} = \frac{1}{T_b \log_2 M}.$$

Modulation Methods

For 2-PSK, $T_s = T_b$ so

$$B = \frac{1}{T_b} = 34 \text{ MHz}.$$

For 4-PSK, $T_s = 2T_b$ so

$$B = \frac{1}{2T_b} = 17 \text{ MHz}.$$

and for 8-PSK

$$B = \frac{1}{3T_b} = 11.3 \text{ MHz}.$$

Hence, if we choose 8-PSK, we allow a margin for attenuation of the sideband components at frequencies beyond $f_c \pm B/2$. Note that the "ideal" maximum spectral efficiency for 8-PSK is

$$r_b/B = 3 \text{ bit/s/Hz}.$$

DPSK (Differential Phase Shift Keying)

Differential phase shift keying makes use of an ingenious technique to avoid the need for a coherent carrier reference signal at the receiver. The DPSK system uses changes in carrier phase to indicate data values. For example, in binary DPSK, if a 0 is to be transmitted, the transmitter changes the carrier phase from its present value ϕ_0 say, to a value $\phi_0 + \pi$. If a 1 is to be transmitted, the phase is not changed from ϕ_0 for the 1 symbol interval. Table 1.4 illustrates a typical DPSK sequence.

Table 1.4
DPSK input sequence and transmitted phase values.

Input sequence $\{d_k\}$		0	1	1	0	1	0	0	0
Transmitted phase	0	π	π	π	0	0	π	0	π

At the receiver, the carrier phase reference for the demodulator may be derived from the received signal during the preceding symbol interval. The demodulator then decides which symbol has been transmitted by comparing the received signal phase in the current symbol interval to the phase in the previous interval.

Differential encoding is used in most practical digital radio systems. That is, the digital information is represented as phase differences of the modulated carrier. This is because differential encoding can be used to resolve phase ambiguities in detection. Detection systems and the problem of phase ambiguity will be examined in a later section.

1.3.4 FSK (Frequency-shift-keying) Modulation

FSK modulation is sometimes used in low capacity digital radio systems. One application is in subscriber radio concentrator systems for rural areas operating at a rate of approximately 700 kbit/s. Another application is in local distribution point-to-multipoint systems used in cellular networks in metropolitan areas where bit rates up to 2 Mbit/s are typical.

The use of FSK modulation stems mainly from hardware advantages. Binary FSK can be generated simply using a voltage-controlled oscillator (VCO) circuit. It can be detected using a simple discriminator, involving no requirement for carrier synchronization at the receive end.

It is of interest to compare FSK with binary PSK signals with premodulation smoothing. Binary FSK uses the waveforms

$$s_0(t) = A \cos(\omega_c - \omega_d)t$$

and

$$s_1(t) = A \cos(\omega_c + \omega_d)t$$

to convey the data values 0 and 1, respectively. A composite binary FSK waveform is a continuous phase, constant amplitude FM waveform where two instantaneous frequencies $(f_c - f_d)$ and $(f_c + f_d)$ are used. The binary FSK waveform can be represented mathematically in the form

$$v_F(t) = A \cos \left\{ \omega_c t + \omega_d \int_o^t d(t') dt' \right\} \tag{1.16}$$

where $d(t)$ is a binary pulse waveform taking the values $+1$ or -1. As can be seen from the last term in Equation (1.16), binary FSK is equivalent to binary PSK modulation of a carrier by an integrated version of the binary pulse waveform. In one sense, this is a form of PSK with a particular form of premodulation smoothing.

FSK schemes have inferior spectral efficiencies and bit-error rate performance to binary PSK schemes. For details, refer to Shanmugan (12).

1.3.5 QAM (Quadrature-Amplitude) Modulation

As the name implies, QAM is a combined amplitude and phase modulation scheme. For example, 16-QAM is a modulation scheme with 16 transmitter states (illustrated in Figure 1.11). The 16-QAM signal can be represented by

$$v_Q(t) = x(t) \cos \omega_c t + y(t) \sin \omega_c t \tag{1.17}$$

where

$$x(t), y(t) \in \{-3a, -a, a, 3a\}, \quad 0 \leq t \leq T_s.$$

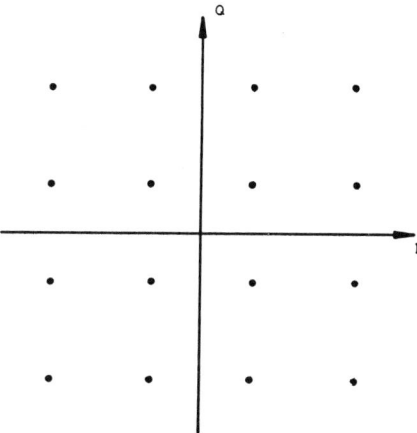

Figure 1.11 Signal states for 16-QAM modulation.

T_s is the symbol interval (four times the bit interval). Note that $x(t)$ and $y(t)$ are four-level PAM (pulse-amplitude modulation) signals.

Note that the symbol amplitude and phase states in Figure 1.11 differ from those used for 16-QAM modulation in voice-band modems, as discussed in Chapter 1, Volume 1. This is because signalling over the voice band is more susceptible to phase irregularities. The states are chosen to maximize phase separation between each symbol.

For the 16-QAM signal used in digital radio systems, the modulated carrier can be viewed as having two vector components, the in-phase I, and the quadrature Q, each of which may take one of four possible values. From Figure 1.11, it is clear that the resultant signal is not constant in amplitude. As mentioned in the previous section, this imposes the need on equipment designers to take special care to minimize transmitter and receiver nonlinearities.

It can be shown that, for a given symbol rate, the 16-QAM spectrum is the same as that given by Equation (1.11) for M-ary PSK.

The benefits of 16-QAM are that

(1) better spectral efficiency can be obtained than for 8-PSK since in 16-QAM each symbol represents four bits of information.
(2) 16-QAM is less sensitive to noise and interference than 16-PSK since, as can be seen from Figure 1.11, the spacings between symbols are larger than they would be if all 16 points were on a single circle (for the same maximum symbol amplitude in each case).

Figure 1.12 shows a schematic for a 140 Mbit/s 16-QAM radio terminal system including modulator, demodulator and associated signal processing units. The

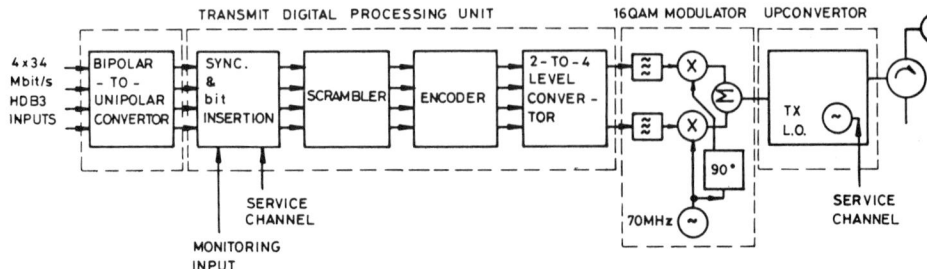

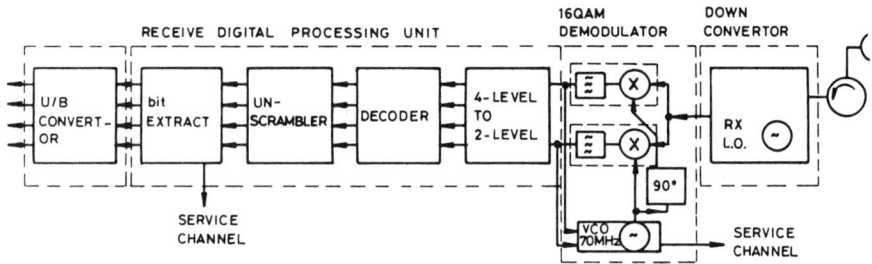

Figure 1.12 Typical modulator and demodulator sections of a 16-QAM system.

input consists of four 34 Mbit/s HDB3 coded baseband data streams. See, for example, NEC (10). The operation of the modulators and demodulators shown in Figure 1.12 is as follows:

Terminal Transmit Modulator—Four synchronous 34.368 Mbit/s HDB3 bit streams are accepted at the bipolar to unipolar converter to be converted from a ternary into a binary stream. They are then fed to the transmit digital processing unit where the functions of synchronization insertion, scrambling, encoding, and two-level to four-level conversion are undertaken. The synchronizer inserts framing, parity, and service channel bits into the streams and ensures timing of the composite stream. The scrambler is required to break up periodic pulse streams which produce strong spectral lines, and provides sufficient timing information for the receiver clock recovery circuits. The encoder converts the four streams from a binary representation to a Gray code representation so that noise-induced changes of state cause single bit errors. It then differentially encodes them so that the signal demodulation can be undertaken independently of the phase of the recovered reference carrier.

The two- to four-level converter multiplexes the four binary streams into two four-level Pulse Amplitude Modulated streams. These become the input to the 16-QAM modulator.

Modulation Methods

Within the modulator the two streams are filtered with separate cosine roll-off filters with $\alpha = 0.5$. They are then used to amplitude modulate two orthogonal 70 MHz carriers which are summed as in Equation (1.17) resulting in the IF signal to the up-converter and transmit amplifier.

Terminal Receiver Demodulator—The 16-QAM demodulator takes the amplified IF signal from the receiver down converter and multiples it with two orthogonal 70 MHz carriers in two synchronous detectors. The carriers are derived from the incoming IF signal in a phase locked loop. The loop has the additional function of extracting a frequency modulated service channel signal.

The two four-level streams coming from the synchronous detectors are then reconverted to four binary streams and the transitions are used to extract a local symbol clock. The received digital processing unit then undertakes the reverse processes to that of the transmit digital processing unit by decoding, descrambling, and extracting framing parity and service channel bits. The four lines are then converted back to HDB3 and fed to the line equipment or demultiplex equipment as appropriate.

M-QAM modulation—Other QAM schemes that use $M = 64$ and $M = 256$ symbols are also under development. The use of higher values of M allows the transmission of higher data rates in a given bandwidth. This is at the expense of poorer bit error rate performance. Systems using 64-QAM to transmit 140 Mbit/s are currently available for the 6 and 11 GHz frequency bands.

1.3.6 Quadrature Partial Response Signalling Schemes

In Chapter 3, Volume 1, we discussed the duobinary and other forms of partial response signalling. These correlative techniques deliberately introduce a limited amount of intersymbol interference over a span of one, two, or more digits. The benefit is that for a given bandwidth, correlative techniques can lead to better spectral efficiency than Nyquist-type zero memory systems.

A duobinary technique known as QPRS (quadrature partial response signalling) has been implemented in a number of digital radio systems. See, for example, Feher (5) and Anderson (11).

The QPRS signal state vector diagram is shown in Figure 1.13. Recall from Chapter 3, Volume 1, that a duobinary signal is a ternary (three-level) signal. For the QPRS signal, there are nine possible amplitude-phase vector states. The QPRS signal can be regarded as two AM-PSK three-level duobinary signals in quadrature. The modulator schematic is very similar to the implementation of 4-PSK except that duobinary filtering must be incorporated.

The correlative three-level QPRS signalling scheme spectral efficiency in practice has been found to be 2–2.25 bit/s/Hz. This is better than for 4-PSK systems

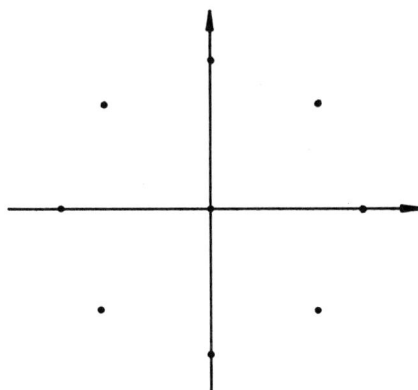

Figure 1.13 QPRS modulation using duobinary encoding.

in which the spectral efficiency is a maximum of two bit/s/Hz (for ideal Nyquist filtering) and 1.3 bit/s/Hz in practice (for $\alpha = 0.5$ Nyquist rolloff as is generally used in practice). However, the three-level QPRS scheme requires three dB more signal power than the 4-PSK for the same bit-error rate performance. See Feher (5) for details.

1.3.7 Continuous-Phase FSK and Minimum Shift Keying (MSK) Systems

There has been a continuing search for modulation schemes that are bandwidth efficient and have constant envelope. As we have seen, constant envelope schemes offer significant advantages in the presence of nonlinearities and permit the operation of transmitter power amplifiers much closer to their saturation region. M-PSK systems have constant envelope in their unfiltered form. However, M-PSK signals are too wideband as the amplitude of the spectral sidelobes falls off rather slowly so that Nyquist filtering is usually required to limit the transmission bandwidth. Filtering results in the signal having a nonconstant envelope.

Continuous-phase modulation techniques have been developed in which the information carrying phase of the transmitted signal is maintained as a smooth continuous function of time. (In M-PSK modulation, the phase variation with time is a discontinuous function.) Continuous-phase modulation methods can lead to a constant envelope signal with rapid falloff of the spectral sidelobes.

Continuous-phase frequency shift keying (CPFSK) modulation schemes are a class of continuous phase modulation techniques which result in power spectra that have much smaller sidelobe values than inherent in M-PSK systems. One class of these modulation techniques is similar to 4-PSK but the phase transitions at each symbol interval are made smoother. This is achieved by limiting the

Modulation Methods

transitions to 90° instead of the 90° and 180° phase shifts used for 4-PSK. Also the phase changes occur as smooth transitions over the symbol intervals. The result is a constant envelope signal with less spectral spread than for 4-PSK.

In general, for continuous phase modulation, the waveform for an input sequence a_k is represented by

$$x(t) = A \cos [\omega_c t + 2\pi h \sum_k a_k \int_0^t g(\tau - kT_b) d\tau] \qquad (1.18)$$

where ω_c = the unmodulated carrier frequency

h = modulation index

$a_k \in (-1, 1)$.

A schematic modulator is shown in Figure 1.14(a). The binary $\{a_k\}$ data stream at the input can be thought of as a sequence of positive and negative unit impulses. Then $g(t)$ represents the impulse response of a smoothing filter and is nonzero over a finite interval $0 \leq t \leq jT_b$ where j is an integer. The frequency modulator is a voltage controlled oscillator for which the output is given by Equation (1.18). This follows from the fact that the term in squared brackets in Equation (1.18)

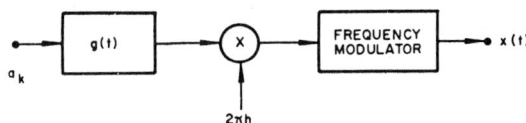

(a) CONTINUOUS PHASE MODULATION

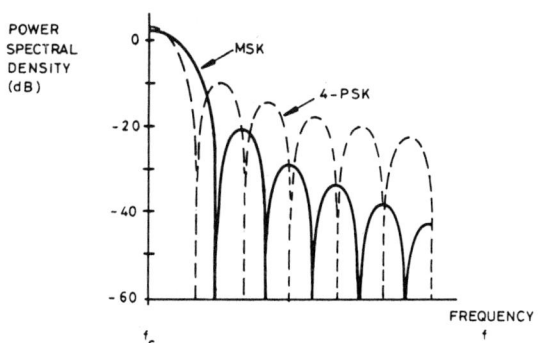

(b) SPECTRUM OF MSK (AND 4-PSK FOR COMPARISON)

Figure 1.14 Minimum shift keying modulator and spectrum.

represents the total instantaneous phase of the modulated signal and this signal is obtained as the integral of the instantaneous frequency of the voltage controlled oscillator.

An appropriate choice of the pulse shaping function $g(t)$ leads to one of several related modulation schemes such as CPFSK, tamed frequency modulation, or j-interval raised cosine (jRC). See Sundberg (36) and Korn (37) for details.

For CPFSK modulation, $g(t)$ is a shaping function of the form

$$g(t) = \begin{cases} \dfrac{1}{2jT_b}, & 0 \leq t \leq jT_b \\ 0, & \text{otherwise} \end{cases} \quad (1.19)$$

where j is an integer constant.

When $j = 1$ and $h = 0.5$, this modulation is known as *minimum shift keying* (MSK). For this case, the modulated waveform is

$$x(t) = A \cos\left[\omega_c t + \frac{\pi t}{2T_b} a_k + \phi_k\right] \quad (1.20)$$

where $\phi_k = \phi((k-1)T_b)$ is the phase value at the beginning of the kth bit interval.

The MSK signal spectrum is shown in Figure 1.14(b), in comparison with the 4-PSK spectrum (constant envelope case). Note that MSK has a wider main lobe than for QPSK, but the spectrum for MSK has lower sidelobes. The modulation and demodulation circuits for MSK are somewhat more complex than for QPSK. Both classes of modulation scheme can provide the same probability of error performance.

Continuous phase modulation schemes are still in the development phase. Their use in digital radio equipment may become more widespread in the future.

1.4 DETECTION

In a digital radio system, the term *detection* is used to refer to the receiver filtering, demodulation, and regeneration operations. If a binary modulation scheme is used, it is the detector's function to distinguish between two transmitted signals $s_0(t)$ and $s_1(t)$ say, in the presence of noise. For an M-ary modulation scheme, the receiver must distinguish between M transmitted signals in a given symbol interval. Various classes of detectors have been developed to carry out this process for different modulation schemes.

In this section, optimum detectors will be summarized together with their probability of error performance in the presence of noise. In order to make the evaluation of P_e mathematically tractable, it is usually necessary to work with

Detection

simplified models of the channel and noise sources. The results permit comparative evaluation of different modulation and detection schemes. Because of the simplifying assumptions involved, care must be taken in interpretation of the results obtained.

Detection errors occur at a receiver because of noise and the effects of distortion such as intersymbol interference. Statistical analysis is an essential element in the search for optimum detectors and the evaluation of their P_e performance. This type of engineering study is based on "Classical Detection Theory" (also called "Hypothesis Testing"). Detailed information sources on this subject include Whalen (13) and Proakis (20). We will examine some of the results here without considering the detailed derivations involved.

1.4.1 Optimum Detector for Binary PSK, FSK or ASK in Gaussian Noise

A detector is said to be optimum if it yields the minimum probability of error for a given set of conditions. The conditions to be specified in advance include the nature of the modulation scheme, channel noise, and transmission distortion. As an example, consider the problem of finding an optimum detector for the case of binary PSK, FSK, or ASK (amplitude-shift keying) transmission in the presence of noise. In particular, we consider the following:

(1) *Modulation Source*—Each random binary input data bit is assumed to be represented in any bit interval by one of two transmission waveforms, $s_o(t)$ representing a 0 and $s_1(t)$ representing a 1. The form of the waveforms depends on the modulation being used, as follows:

$$\text{PSK:} \quad s_o(t) = -A \cos\omega_c t$$
$$s_1(t) = A \cos\omega_c t$$
$$\text{FSK:} \quad s_o(t) = A \cos(\omega_c - \omega_d)t$$
$$s_1(t) = A \cos(\omega_c + \omega_d)t$$
$$\text{ASK:} \quad s_o(t) = 0$$
$$s_1(t) = A \cos\omega_c t.$$

We will first examine the optimum receiver for the general binary signalling case, that is, in terms of $s_o(t)$ and $s_1(t)$. This can then be implemented in turn for each of the specific modulation schemes by reference to the above list of waveforms.

(2) *Channel*—The modulator output is considered to be passed through a bandpass channel to the detector. For purposes of analysis, the channel is assumed

to have sufficient bandwidth so that the modulated signals pass through without suffering any distortion other than propagation delay.

(3) *Noise and Interference*—Noise and interference is assumed present at the detector input. This noise $n(t)$ is assumed to be a zero mean, stationary, white Gaussian noise signal, with known power spectral density $S_n(f)$.

Then it can be shown that the detector in Figure 1.15 is optimum for the above conditions in that it will minimize the probability of making an error in detection.

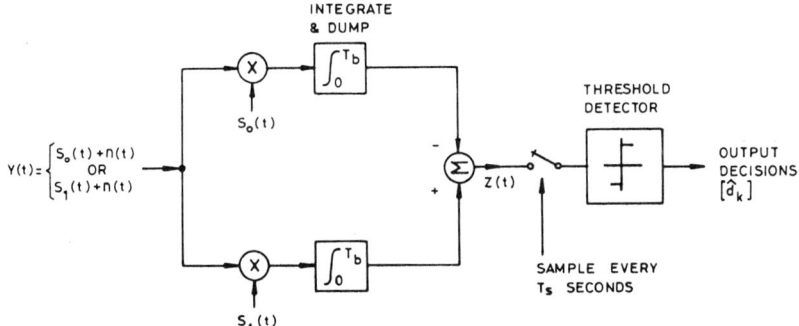

Figure 1.15 Optimum Binary Detector (PSK, FSK, ASK).

The optimum detector multiplies the noisy input signal with a copy of each of the known transmitter waveforms $s_o(t)$ and $s_1(t)$. The analogue products are passed through separate integrators. Each integrator must be reset to zero at the beginning of each bit interval in order to avoid intersymbol interference. The difference between the integrator output waveforms is sampled at the end of each T_b sec bit interval. The detector then applies this sample to a threshold detector to make a decision as to whether a 0 or a 1 is represented by the received waveform in that T_b sec interval. This detector is called a *coherent detector*. This is because it requires local carrier reference signals $s_o(t)$ and $s_1(t)$ having the same phase and frequency as the transmitter waveforms.

There is more than one form of coherent detector. The one shown in Figure 1.15 is known as a *correlation detector* because of the multiplication and integration operations it performs. Another equivalent form of receiver is the *matched-filter detector*, which will not be dealt with here. Refer for example, to Shanmugan (12) for details. In practice, the correlation detector is the one most often used. Figure 1.16 shows how the correlation detector in Figure 1.15 can be implemented for the case where 2-PSK modulation is used. Note that the circuit simplification follows from the fact that in Figure 1.15, the input to the sampler switch is given by

Detection

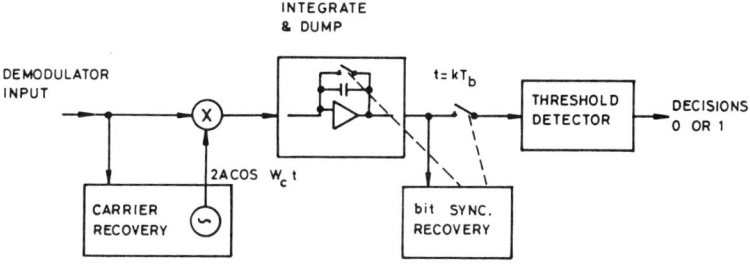

Figure 1.16 Coherent detector for binary PSK.

$$z(t) = \int_0^{T_b} y(t) \, [s_1(t) - s_o(t)] \, dt \qquad (1.21)$$

with

$$s_1(t) = -s_o(t) = A \cos \omega_c t.$$

The most complex part of a coherent detector relates to the provision of a coherent reference signal in phase with the modulator oscillator. A number of carrier recovery systems have been proposed for this purpose, notably those based on the phase-locked loop. In addition, a symbol timing recovery circuit is required to ensure the sampling and resetting instants are correctly chosen. Design techniques for carrier recovery and synchronization subsystems are described in Spilker (14).

Carrier recovery systems used in coherent detection usually suffer from a problem known as phase ambiguity. This phenomenon is associated with the carrier recovery circuit. The task of that circuit is to generate a local oscillator signal and to lock it in phase with the carrier associated with the received signal. In such systems, the phase locking mechanism usually cannot distinguish between two conditions, the one when the local oscillator is in phase with the received carrier and the other when they are exactly 180° out of phase.

This problem of phase ambiguity is overcome by the use of differential encoding (described in Section 1.3.3). To illustrate this, consider the sequences shown in Table 1.4. This shows a sequence of transmitted phase values that would represent the given data sequence in a differentially encoded 2-PSK system. Now if the detector local oscillator at the receiver is locked in phase with the incoming carrier, then the detector performs as follows. It produces an output of 1 if the phase of the current received symbol is the same as that of the preceding symbol. However, if there has been a phase reversal, then the output is 0.

Now consider the case where the local oscillator is locked 180° out of phase with the received carrier. That is, the detector behaves as though all of the phase values in Table 1.4 were reversed. A little consideration will show that the detected output sequence will be unaffected by this phase ambiguity.

On the other hand, if 2-PSK modulation had been used without differential precoding, the 180° phase error would have resulted in inversion of all the received information bits. That is, transmitted 1's would have been detected as 0's, and vice-versa.

Exercise 1.4

Sketch a block diagram of a coherent receiver suitable for binary FSK transmission. This exercise is left for the reader.

1.4.2 Coherent Detector for QAM and M-PSK Signals

The block diagram of a coherent detector for QAM signals is shown in Figure 1.17. Recall that QAM is equivalent to the inphase and quadrature modulation of a carrier by two separate 4-PAM (four-level pulse amplitude modulation) signals. Therefore, the coherent detector first separates out the two 4-PAM signals by synchronous demodulation. The three threshold comparators estimate which of the four PAM levels is represented by each sample. Then simple logic hardware is required for four-level to binary conversion, differential decoding, and multiplexing of the two resultant bit streams.

Figure 1.17 also incorporates the necessary elements of a coherent M-ary PSK detector. That is, with only minor changes in the signal processing circuitry after

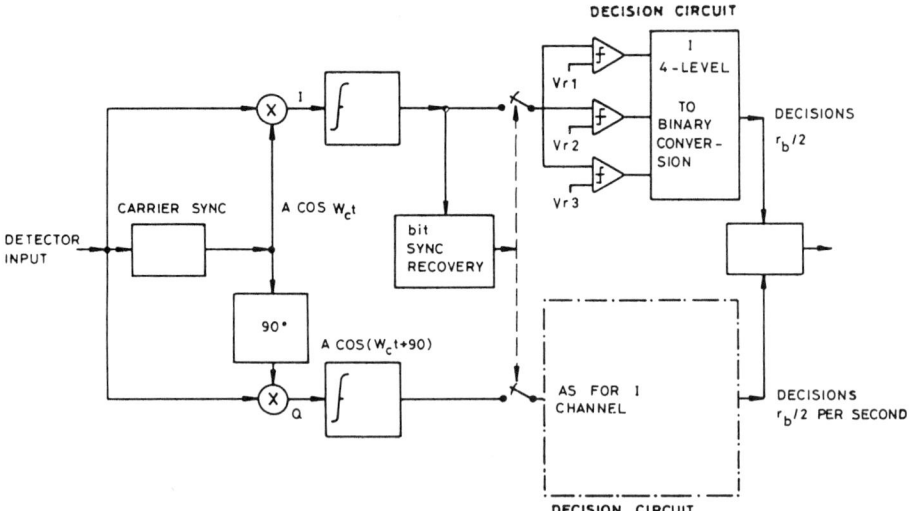

Figure 1.17 Coherent detector for 16-QAM.

the decision circuits, this demodulation structure can be used for the coherent detection of 8-PSK signals.

1.4.3 Probability of Error

The bit error probability at the output of a coherent detector can be evaluated for the case where a wideband additive Gaussian noise channel model is assumed. In that case, the detector input is assumed to consist of an undistorted version of the transmitted signal together with noise. Algebraic expressions can be obtained for the probability of error P_e for a given modulation type and given detector characteristics.

In general, the derivation of P_e is based on the concept of a signal vector space, as illustrated for example in Figure 1.5. At the detector, the regenerator must estimate which symbol is represented by the received waveform in each symbol interval. The decisions must be made in terms of decision regions, thresholds for which are illustrated by dotted lines in Figure 1.5. An error occurs when a given symbol is transmitted and the noise or intersymbol interference causes the received signal value to lie outside the decision region associated with the transmitted symbol.

The probability that the detector chooses an incorrect symbol is referred to as the *symbol error probability* P_s. In an M-ary modulation scheme with $M = 2^n$ bits, each symbol represents n bits. The most probable symbol errors are those that choose an incorrect symbol adjacent to the correct one. If Gray coding is used, then in the most likely cases, only one bit error results from a symbol error. If only these types of errors occur, the bit error probability would be obtained using

$$P_e = P_s/n.$$

In a complete analysis, other types of errors must also be taken into account. In practice, if $P_s \ll 1$ then the other types of errors can often be neglected.

For PSK systems, it can be shown that when the signal-to-noise ratio at the receiver is greater than one, the bit error probability is given to a good approximation by

$$P_e = Q\{\sqrt{(2C/N)}\} \quad \text{for 2-PSK} \tag{1.22}$$

and

$$P_e = (2/\log_2 M)\, Q\{\sqrt{(2C/N)\sin^2(\pi/M)}\} \tag{1.23}$$

for M-PSK, $M > 2$. The function $Q(x)$ is defined in Equation (4.33) of Chapter 4, Volume 1 and is plotted in Figure 4.24. For convenience, Equation (4.33) is reproduced here.

$$Q(x) = \frac{1}{\sqrt{2\pi}} \int_x^\infty e^{-t^2/2}\, dt.$$

An upper bound on $Q(x)$ which is useful for high C/N values is

$$Q(x) \leq \frac{1}{2} e^{-x^2/2}, \qquad x \geq 0.$$

It follows that for M-PSK systems another useful approximation which is accurate to within 1 dB for C/N ratios in dB greater than 15 dB is

$$P_e = \frac{1}{\log_2 M} \exp\{-(C/N)\sin^2(\pi/M)\}. \tag{1.24}$$

The C/N term in each of the above equations represents the mean carrier power to noise power ratio where the noise is specified in the double-sided Nyquist bandwidth that equals the symbol rate. That is, if the symbol rate is $r_s = 1/T_s$ then the double-sided Nyquist bandwidth will be

$$B = r_s.$$

For DPSK systems with differentially coherent phase detection, the bit error probability is approximated by

$$P_e = \exp\{-(2C/N)\}. \tag{1.25}$$

For a 16-QAM scheme, the bit error probability is given to a good approximation by

$$P_e = 0.75 \, Q\{\sqrt{(C/5N)}\}. \tag{1.26}$$

For a 2-FSK system with coherent detection

$$P_e = Q\{\sqrt{C/N}\}. \tag{1.27}$$

For an MSK system with deviation ratio $h = 0.5$ and a phase decoding process based on two successive bit intervals of the received signal, the bit error probability is approximated by

$$P_e = 2 \, Q\{\sqrt{(2C/N)}\}. \tag{1.28}$$

Figure 1.18 illustrates some of the above results. Note that a 2-FSK system performs three dB worse than a 2-PSK system with coherent detection. For M-MSK systems, as the number of symbol vectors M increases, so the regenerator decision regions become smaller and the probability of error increases. For M-QAM, the P_e performance for coherent detectors can be shown to be better than for M-PSK by an amount equal to an increase in the C/N ratio in dB of

$$D(M) = 10 \log \{3M^2/2(M-1)\pi^2\}. \quad \text{(dB)} \tag{1.29}$$

For $M = 16$, we obtain $D(M) = 4.1$ dB. That is, the 16-PSK system requires an average carrier power 4.1 dB greater than the 16-QAM system for the same

Detection

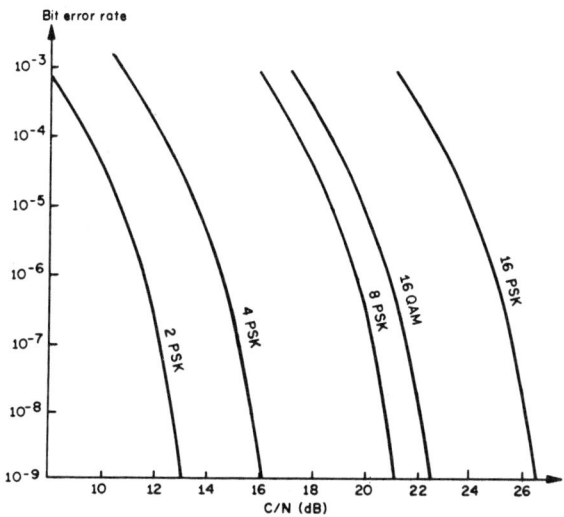

Figure 1.18 Bit error rate performance for coherent detection.

performance. This is because, for the same power, the signal vectors for 16-PSK are clustered closer together than for 16-QAM.

For details of the methods of derivation see Feher (5), Shanmugan (12), Whalen (13) or Proakis (20).

Exercise 1.5

Compare the performance of 8-PSK, 16-PSK, and 16-QAM for transmission at 2 Mbit/s with a bit error rate less than or equal to 10^{-6}. In each case, determine the minimum transmission bandwidth and carrier-to-noise power required.

Solution. The symbol rate for 8-PSK is $f_s = 2 \times 10^6/3 = 0.67 \times 10^6$ symbol/s. For 16-PSK for 16-QAM, it is $2 \times 10^6/4 = 0.5 \times 10^6$ symbol/s. Hence, the minimum bandwidths for Nyquist filtering ($\alpha = 0$) are

$$0.67 \text{ MHz for 8-PSK, and}$$

$$0.5 \text{ MHz for 16-PSK and 16-QAM.}$$

The required carrier-to-noise power in each case is obtained as follows:

8-PSK

$$10^{-6} < (2/3) \, Q\{\sqrt{(0.293C/N)}\}.$$

This yields
$$C/N = 75 \quad (19\text{dB}).$$

16-PSK
$$10^{-6} \leq 0.5 \, Q\{\sqrt{(0.076 \, C/N)}\}.$$

This yields
$$C/N = 280 \quad (24\text{dB}).$$

16-QAM
$$10^{-6} \leq 0.75 \, Q\{\sqrt{(C/5N)}\}$$

This yields
$$C/N = 120 \quad (21\text{dB}).$$

It should be noted that for terrestrial radio systems, the probability of error P_e is most commonly specified in terms of the C/N ratio as in Equations 1.22 to 1.28. On the other hand, satellite systems performance is usually specified in terms of E_b/N_o. Here E_b is the energy per bit and N_o is the single-sided noise spectral density with units V^2/Hz. The energy per bit is

$$E_b = C \, T_s/(\log_2 M). \tag{1.30}$$

For white noise in a bandwidth equal to the double-sided Nyquist bandwidth $B = 1/T_s$ it follows that $N = N_o B$ and, hence,

$$E_b/N_o = (C/N)(1/\log_2 M). \tag{1.31}$$

In many practical systems, the receiver noise bandwidth B_R is greater than the Nyquist bandwidth. In order to compare the P_e performance of the wider bandwidth systems with that of the minimum bandwidth theoretical system, the following relation is useful. Let C/N_R be the carrier-to-noise ratio in the receiving system of noise bandwidth B_R, and let C/N be the carrier-to-noise ratio in the Nyquist bandwidth system. Then,

$$C/N_R = (C/N)\{(B_R \log_2 M/r_b\} \tag{1.32}$$

where r_b is the bit rate.

Exercise 1.6

For a 2 Mbit/s 16-QAM system, it was shown in Exercise 1.5 that the two-sided Nyquist bandwidth was equal to $B = 0.5$ MHz. For that case the minimum

C/N ratio for $P_e \le 10^{-6}$ is 21 dB. What is the required carrier-to-noise ratio if the receiver bandwidth is increased to $B_R = 1$ MHz?

Solution. From Equation (1.32) the required carrier-to-noise ratio is

$$C/N_R = (C/N)(B_R \log_2 M / r_b)$$

which becomes in decibels

$$10 \log(C/N_R) = 21 + 10 \log 2 = 24 \text{ dB}.$$

1.4.4 Optimum Detectors for Channels with ISI and Noise

The above coherent detectors are only optimum if their underlying assumptions are valid. One assumption was that the channel filtering (transmitter, receiver, or multi-path propagation effects) does not cause pulse distortion. In most practical systems, it is not possible to guarantee that whenever the digital signal passes through a filter, its transfer function will be constant over the bandwidth of the signal. With a nonideal filter, intersymbol interference (ISI) will occur just as it did for the baseband transmission case discussed in Chapter 3. The narrower the channel bandwidth, the more severe the ISI.

When ISI occurs, a detector input signal in any one bit interval is affected by signals from other bit intervals. It is, therefore, not surprising that it is not possible to design an optimum detector system that makes independent decisions about each data bit based on just the received signal in that bit interval. Rather, it is necessary to either use adaptive equalization to attempt to minimize the ISI or otherwise, to examine and process long sequences of "observables," that is, the outputs from a correlator and sampler (or matched filter).

A convenient procedure for performing the necessary operations to choose the optimum sequence is based on the Viterbi algorithm. This is a signal processing procedure originally developed for decoding convolutional codes. The algorithm is discussed in Chapter 4. It can be shown that the Viterbi algorithm is an optimum detection algorithm for demodulating data sequences that pass through any channel which can be modelled as an arbitrary (possibly nonlinear) noiseless part with ISI followed by a noisy part with no ISI. Omura (15) provides an excellent discussion of the demodulation problem. Integrated circuits are currently being developed which can perform the Viterbi algorithm at increasing speeds. They may well be an important component of future digital systems. The algorithm requires an estimate to be available of channel transmission properties at each symbol interval. This could be achieved by making use of the adaptive equalizers already provided in many digital radio systems. Adaptive equalizers will be discussed later in Section 1.5.8.

Viterbi detection schemes can only be used with systems with memory. That is, there must be interdependence between symbols. This property may come about because of the choice of modulation scheme, because of ISI resulting from channel characteristics, or because memory is introduced via trellis codes. The latter will be discussed next.

1.4.5 Combining Modulation and Coding

Most conventional digital radio modulation schemes involve the transmission of independent M-ary symbols. As we have seen, the BER performance of these schemes is a function of received carrier-to-noise ratio. It can also be affected by limitations in channel bandwidth and by nonlinearities.

It is possible to use forward error correction (FEC) codes to mitigate against errors when they occur. The traditional coding techniques involve the addition of redundant parity bits to the input message sequence before transmission. Then, at the receiver end, the redundant parity bits are used to correct the errors that have occurred during transmission. FEC coding techniques will be discussed in detail in Chapter 4, Volume 2.

The insertion of redundant bits associated with FEC coding can result in a reduction of the effective spectral efficiency of the system. It is possible, however, to combine coding and modulation techniques to improve the error performance of digital radio systems without sacrificing effective data rates or requiring more bandwidth. The most effective schemes for doing this have been trellis coding schemes with signal set partitioning proposed by Ungerboeck (26) and studied by Wilson (27), Forney (28), Wei (29), Miyaki (30), Pizzi (35), and Rhodes (31).

To illustrate the principles involved in these combined modulation and coding techniques, consider a scheme described in Miyaki (30) for 120 Mbit/s time-division multiple-access (TDMA) satellite communications. The scheme involves using 8-PSK modulation at a symbol rate $r_s = 60$ Msymb/s. Of the three bits represented by each 8-PSK symbol, two are information bits and one is a parity bit generated by an FEC code. At the receiver end, the redundant parity bits in the transmitted sequence are used to carry out error correction using a decoding scheme based on the Viterbi algorithm. The demodulator and decoder functions at the receiver use a technique called "soft-decision decoding." That is, a coherent demodulator is used except that unlike conventional demodulators, it does not attempt to make "hard" decisions in each symbol interval about which 8-PSK symbol was transmitted. Instead, the demodulator output consists of the inphase and quadrature signal outputs from the correlators as described in Section 1.4.2. Each output is quantized into three bits. These signals are connected to the FEC decoder, which uses the Viterbi algorithm to form the best estimate of the symbol sequence that was transmitted. In doing so, it is able to provide significantly better BER performance than a conventional demodulator. The overall system provides a BER performance equivalent to approximately 3 dB increase in C/N ratio. That is, we say it provides a coding gain of 3 dB. Let us compare this with the situation if FEC coding had not been used.

The system could have used the same symbol transmission rate ($r_s = 60\ M$ symb/s) for the same input data rate, by using 4-PSK modulation rather than 8-PSK. As discussed in Section 1.3.3, this would not affect the transmitted signal spectrum at all. However, it would result in a BER rate improvement of approximately 5 dB (refer to Section 1.4.2).

Consider now the 8-PSK coded system. The use of an 8-PSK signal constellation instead of a 4-PSK scheme incurs 5 dB performance penalty. However, the use of the FEC code with soft decision decoding provides a gain of approximately 8 dB (for channel error rates below approximately 1×10^{-2}). Hence, an overall coding gain of 3 dB is achieved by the coded 8-PSK system over the uncoded 4-PSK system. The cost in terms of input data rate and transmitter spectrum width is nil! Of course, we can never gain something for nothing in the real world. The cost is in demodulator/decoder complexity. In this microelectronic/microcomputer age, complexity is generally not an intolerable price to pay.

1.4.6 Non-Coherent Detectors

In optimum coherent detection described in previous sections, it is assumed that a subsystem is provided at the receiver to generate a phase-coherent local carrier. This may be complex and expensive. Noncoherent detector schemes do not require a phase-coherent local reference. They involve some form of filtering, rectification, and low-pass filtering. A typical example is the use of a discriminator circuit for noncoherent detection of 2-FSK. The bit-error rate performance of noncoherent detectors is, of course, inferior to the coherent detector schemes. The noncoherent 2-FSK detector performs approximately 2 dB worse than the coherent detector. Noncoherent detector systems are only used when receiver simplicity is of prime concern.

1.5 RADIO LINK SYSTEM DESIGN

1.5.1 Introduction

The design of a digital radio link system involves selection of repeater spacing, antenna size and height above ground, modulation method, equalizer design, space diversity, and standby protection procedures. These decisions have to be made in the light of CCIR performance objectives, which were discussed in Section 1.2.4.

Also, typical performance parameters for available digital radio equipment can be obtained from manufacturers. This will include information about available modulation types, transmitter power, and receiver sensitivity. The following information on typical digital radio-relay equipment available in the various fixed-services frequency bands is taken from Steele (2). This relates particularly to the 30-channel PCM hierarchy.

The bands below 10 GHz have been extensively used in the past by analogue radio-relay systems for the transmission of FDM telephony and TV relay. Within these bands, RF channel arrangements have been formulated in various CCIR recommendations to suit FDM/FM systems ranging in capacity from 60 to 2700 channels. Digital radio systems operating in these bands have to be compatible with the same RF channel plans whenever these are still being used by analogue radio-relay systems. Table 1.5 summarizes a range of typical equipment that is available or is being developed in the various frequency bands below 10 GHz.

Table 1.5
Typical digital radio systems operating below 10 GHz.

Frequency Band CCIR Rec.	RF Channel Separation MHz	No. of RF Bothway Channels	FDM Channel Capacity	Digital Capacity Mbit/s	Modulation Method
UHF					
(790–960 MHz)	2.5			4	4PSK
(1.42–1.53 GHz)	2.5			4	4PSK
2 GHz					
(1.7–1.9)	14	6I	60–300	8	2PSK
Rec. 283-4				17	4PSK
(1.9–2.3)	29	6I	960	8/17	2PSK
Rec. 382-3				34	4PSK
4 GHz					
(3.8–4.2)	29	6I	960–1800	34	4PSK
Rec. 382-3				70	8PSK
6.1 GHz					
(5.925–6.425)	29.65	8I	1800	34	4PSK
Rec. 283-2				70	8PSK
6.7 GHz					
(6.425–7.110)	40	8I	1800–2700	105	PSK
Rec. 384-3		8I		140	16QAM, 8PSK
7 GHz					
(7.125–7.425)	14	10I	300	17	4PSK
Rec. 385					
8 GHz					
(7.725–8.275)	29.65	8I	960	34	4PSK
Rec. 386-2				70	8PSK
Annex 1					

*Under Development   I = Interleaved   C = Co-channel

Radio Link System Design

With RF channel spacings of 29 or 29.65 MHz, 4PSK-34 Mbit/s, and 8-PSK-70 Mbit/s digital radio systems can operate compatibly on the same route with 1,800 channel FDM-FM systems. In general, with the exception of the 6.7 GHz band, the existing RF channel arrangements are suitable for small and medium capacity digital radio systems. Moves have also been made by some administrations to formulate new RF channel plans in the 4, 8, and 6.1 GHz bands to make them suitable for high capacity 140 Mbit/s.

Table 1.6 shows those frequency bands above 10 GHz for which equipment is available or is being developed. An important feature of the higher frequency bands is the availability of large bandwidths. This permits the use of simpler modulation methods, and hence, lower equipment costs.

Table 1.6
Typical radio systems operating above 10 GHz.

Frequency Band CCIR Rec.	RF Channel Separation MHz	No. of RF Channels	Digital Capacity Mbit/s	Modulation Method
11 GHz	40	12I	70	8PSK
Rec. 387-3	40	11I	140	16QAM, 8PSK
	,	,	105	8PSK
Rep. 782-1	60	8I	140	8PSK
Rep. 782-1	67	6I	140	4PSK
13 GHz	14	16C	8	2PSK
(12.75–13.25)	28	8C	8/17/34	4PSK
Rec. 497-2			17	2PSK
Do. Annex I	35	6I	70	8PSK
15 GHz				
(14.5–15.35)	14	8C	8	2PSK, 2FSK
120 MHz HI and LO	,	,	8/17	4PSK
240 MHz HI and LO	28	8C	34	4PSK
Full band	40	8C	105	8PSK
18 GHz				
(17.7–19.7)	220	7I	280	4PSK*
Draft Rec. AB/9	110	15I	140	4PSK*
	27.5	35C	34	4PSK*
30 GHz				
(27.5–29.5)			2	2FSK
			70	2PSK*
40 GHz				
(38.6–40)			2	2FSK
Infra Red				
(850 nm)			100	2ASK

*Under Development   I = Interleaved   C = Co-channel

1.5.2 Free Space Calculations for Single Hops

The design of individual digital radio links must be such as to ensure bit error rate and availability performance objectives are met. These objectives are stated in statistical terms taking into account the variable nature of the propagation medium and also the possible degradation or failure of equipment. As an example of a single link design, consider a path of length $d = 50$ km. From Section 1.2.4, the link availability objective becomes

The BER should not be greater than 10^{-3} for more than 0.001% of any month.

The design of the radio link to achieve this objective may have to commence with certain assumptions about the equipment likely to be available from manufacturers. For example, if a 140 Mbit/s digital radio link is under consideration, typical selected parameters for available equipment might be (quoted from NEC 6.7 GHz, Digital Radio Description Equipment):

(1) *modulation type*: 16-QAM
(2) *transmitter power*: 1 Watt (+30 dBm) effective value after back-off at test points. (Because of the nonconstant envelope of 16-QAM modulation, transmitter nonlinearity must be avoided, as it may lead to BER deterioration and spectral spreading.)
(3) *transmit frequency*: in the 6.7 GHz band.
(4) *receiver noise figure*: 3.5 dB or less.
(5) *receiver IF bandwidth (3 dB points)*: ± 24 MHz ± 5 MHz.
(6) *demodulation method*: coherent detection.

Next, the path losses and antenna gains must be considered to determine the expected BER performance of the link when no fading occurs. That is, we compute the receiver C/N ratio and, hence, BER assuming free-space loss. This BER performance must be significantly better than 10^{-3} to provide an appropriate margin against the effects of fading.

Consider the single hop radio system represented in Figure 1.19. Assuming that antenna heights are chosen to provide free-space propagation under normal atmospheric conditions, it is straightforward to compute the ratio of transmit antenna input power P_i to receive antenna output power P_o (6). This is referred to as the free-space loss L_{dB} which is defined

$$L_{dB} = 10 \log_{10} (P_i/P_o). \qquad (1.33)$$

For a single hop this can be calculated by using (6):

$$L_{dB} = 92.4 + 20 \log_{10} d + 20 \log_{10} f_c - G_{T\ dB} - G_{R\ dB} \qquad (1.34)$$

Radio Link System Design

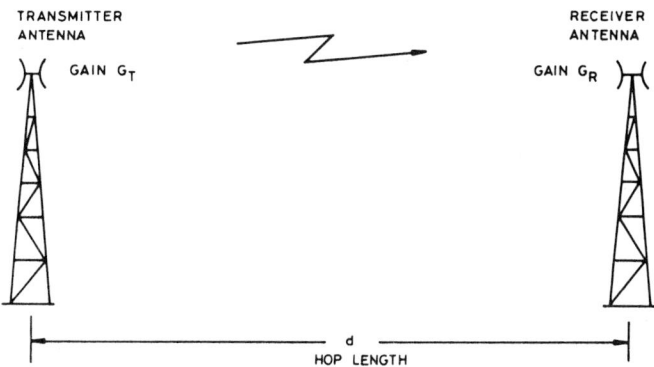

Figure 1.19 A single hop radio link.

where d = hop length (km)

f_c = carrier frequency (GHz)

$G_{T\,\text{dB}}$ = transmitter antenna gain (dB)

$G_{R\,\text{dB}}$ = receiver antenna gain (dB)

The antenna gains are functions of antenna size and frequency. A useful expression for the gain G_{dB} of a transmitting or receiving antenna at a frequency f_c (in GHz) is

$$G_{dB} = 10 \log_{10} \frac{4\pi}{\lambda^2} A_r = 10 \log_{10} \frac{4\pi f_c^2 A_r}{0.09} \tag{1.35}$$

where λ = wavelength (m) and A_r is the effective receiving area. For a parabolic reflector type of microwave antenna, the effective receiving area is given approximately by

$$A_r = 0.6\,A \tag{1.36}$$

where A is the actual reflector area.

It is instructive to examine the derivation of Equation (1.34). For P_i (watts) power input to the transmitter antenna, then at a receiver d_m (m) away, the power per unit area (power density) of the received space wave can easily be computed. If the transmitter antenna were an isotropic source, it would radiate power equally in all directions over the surface of a sphere in space. The surface area of a sphere of radius d_m is $4\pi d_m^2$. The transmit antenna directive gain G_T is defined as the ratio of the power density D_p radiated by the antenna at the receiving antenna to the power density that would be radiated by an isotropic antenna. It follows that the power density at the receiver is

$$D_p = P_i G_T/(4\pi d_m^2) \quad (W/m^2). \tag{1.37}$$

The output power P_o from the receive antenna is equal to the product of the antenna's effective area A_r and the power density. We, therefore, obtain

$$P_o = P_i G_T A_r/(4\pi d_m^2) \tag{1.38}$$

and using Equation (1.35) this can be written

$$P_o/P_i = \frac{G_T G_R \lambda^2}{16\pi^2 d_m^2}. \tag{1.39}$$

Now by substituting $\lambda = 0.3/f_c$, Equation (1.34) follows using (1.33). Now the receiver input carrier power for free-space propagation is given (in dBm) by

$$(P_{Rx})_{dBm} = (P_{Tx})_{dBm} - L_{dB} - A_{dB} \tag{1.40}$$

where A_{dB} = system losses caused by feeders, branching filters, and other transmission line components, and

$$(P_{Tx})_{dB} = \text{transmitter power level}$$
$$= 10 \log_{10} (P_{Tx}/10^{-3}).$$

P_{Tx} is the transmitter power in Watts.

Exercise 1.7

Compute the receive power level $(P_{Rx})_{dB}$ for a 50 km hop at 6.7 GHz with antennas of three meter diameter at each end, for a transmitter power output of 1 Watt and assuming $A_{dB} = 9$ dB.

Solution. From Equations (1.35) and (1.36), the antenna gains are

$$G_{T\ dB} = G_{R\ dB} = 44.2 \quad (dB).$$

Then from Equation (1.34), the free space loss is

$$L_{dB} = 92.4 + 34.0 + 16.5 - 44.2 - 44.2$$
$$= 54.5 \quad (dB)$$

and using Equation (1.40) we obtain the received power level as

$$(P_{RX})_{dBm} = 30 - 54.5 - 9$$
$$= -33.5 \text{ (dBm)} \quad (= 0.45 \ \mu W).$$

We are now in a position to determine the carrier-to-noise ratio C/N at the receiver detector input. We consider the receiver noise in a bandwidth

$$B = 1/T_s$$

Radio Link System Design

where T_s is the symbol (baud) rate. If we assume that receiver noise is the dominant source of noise, then the carrier to noise ratio at the detector input is given by

$$\frac{C}{N} = \frac{P_{RX}}{FkTB} \tag{1.41}$$

where P_{RX} = received power (W)

F = receiver noise figure

k = Boltzmann's constant = 1.37×10^{-23}

T = absolute temperature (°K)

B = receiver bandwidth equal to the baud rate.

Once the value of C/N is known, we can determine the average BER from Figure 1.18.

Exercise 1.8

Determine the minimum receiver input power $(P_{RX})_{dBm}$ required to ensure that BER $\leq 10^{-3}$ for a digital radio link with the following parameters:

140 Mbit/s bit rate, 16-QAM modulation

6.7 GHz carrier frequency

50 km hop length

3 m antennas at transmitter and receiver

feeder and branching losses $L = 9$ dB

transmitter power = $+30$ dBm

receiver noise figure = 4 dB.

Solution. From Figure 1.18 we obtain the minimum C/N value (in dB) for BER = 10^{-3} for a 16-QAM system as

$$(C/N)\text{dB} = 17 \text{ dB}. \tag{1.42}$$

From Equation (1.41) we have

$$(C/N)\text{dB} = (P_{RX})_{dBm} - (F)_{dB} - 10 \log_{10}(kTB_r) - 30.$$

Note that the -30 factor converts from dBW to dBm. For 16-QAM modulation the baud rate is one quarter the bit rate. Assuming an ambient temperature of 20°C ($T = 290$°K), we obtain

$$(C/N)\text{dB} = 17 = (P_{RX})_{\text{dB}m} - 4$$
$$- 10 \log_{10} (1.37 \times 10^{-23} \times 290 \times \frac{140}{4} \times 10^6) - 30$$

and, hence,

$$(P_{RX})_{\text{dB}m} = 17 + 4 - 128.6 + 30 = -77.6 \text{ (dB}m\text{)}.$$

1.5.3 Flat Fade Margins

Many radio link systems are subject to fading, particularly in the early mornings and evenings during summer months when hot still weather conditions or inversions result in irregularities in the refractive index of the atmosphere. Single hop system design must, therefore, provide for a margin of performance so that BER objectives can still be met even when fading occurs. Two types of fading must be considered when operating margins of digital radio systems are being determined:

(1) *Flat fading*—a reduction in the received signal level (and hence, a decrease in *C/N*) which affects equally all the spectral components in the signal. That is, a flat fade is equivalent to the insertion of attenuation at the receiver input. Heavy rain is a cause of flat fading for systems above 10 GHz.

(2) *Frequency selective fading*—the reduction in received signal levels such that the attenuation values vary with frequency across the channel band. This is illustrated in Figure 1.20. Frequency selective fading occurs as a result of multipath propagation, that is, when the received signal consists of two or more components, which have propagated over differing path lengths between the transmitting and receiving antennas. Frequency selective fading tends to be the limiting factor in the fade margins of digital radio systems operating below 10 GHz. This will be considered in more detail in the next section.

The flat fade margin $(M)_{\text{dB}}$ of a digital radio system is defined as the difference between the nominal receive power $(P_{RX})_{\text{dB}m}$ and the threshold value $(P_t)_{\text{dB}m}$ where the threshold value corresponds to a given BER threshold. It is assumed here to be 10^{-3}. Hence,

$$(M)_{\text{dB}} = (P_{RX})_{\text{dB}m} - (P_t)_{\text{dB}m}. \tag{1.43}$$

For a typical system, the flat fade margin is designed to be about 40 dB. From the results of Exercises 1.7 and 1.8 we have for the 140 Mbit/s, 16-QAM link with the parameters given, that

$$(M)_{\text{dB}} = -33.4 - (-77.6)$$
$$= 44.2 \text{ (dB)}.$$

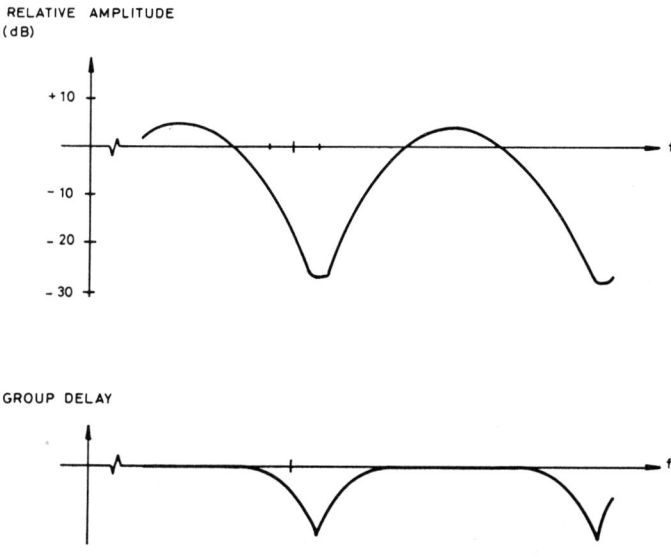

Figure 1.20 Channel transfer function during frequency selective fading.

1.5.4 Percentage Outage Prediction—Vigant's Formula

The main reason for such a large fade margin for radio systems is the drastic effect of fading. Typically, these fading effects are most severe in the summer months. Traditionally, for analogue FDM-FM radio system design, empirical fading distributions have been used to estimate the proportion of time that a single hop will operate below a given C/N ratio. The relative occurrence of fading is a complex function of the path length, terrain type, and localized meteorological conditions.

From propagation experiments carried out in different countries, empirical formulae have been developed that can be used to estimate the percentage of time a fade margin M is exceeded. One such formula given by CCIR (6) is known as *Vigant's formula*. This formula estimates the probability of exceeding a margin M (dB) in the worst month as

$$P = k_1 C f_c d^3 \, 10^{-M/10} \qquad (1.44)$$

where

$k_1 = 6 \times 10^{-7}$

C = a terrain/weather factor

$$C = \begin{cases} 1/4 \text{ for mountainous terrains with dry climates} \\ 1 \text{ for average terrain with average inland climate} \\ 4 \text{ for paths over water in hot humid areas} \end{cases}$$

f_c = carrier frequency (GHz)

and d = hop length (km).

Exercise 1.9

For the 140 Mbit/s, 6.7 GHz system operating over a 50 km hop as discussed in Exercises 1.7 and 1.8, estimate the percentage of time a fade margin of 44.2 dB will be exceeded, assuming mountainous terrain with dry weather conditions.

Solution. Vigant's formula gives

$$P = 6 \times 10^{-7} \times 1/4 \times 6.7 \times (50)^3 \times 10^{-4.42}$$
$$= .000005.$$

That is, the system is estimated to operate with a BER $\geq 10^{-3}$ for 0.0005% of the worst month (approximately 13 seconds.)

Note that the outage probability given by Equation (1.44) is proportional to the cube of the path length. Clearly, the use of shorter hops can have a significant effect on the outage times. Also, the multipath fading shows a 10 dB/decade relationship between the probability that the received signal amplitude fades below a given level and the nonfading level. This is also predicted if the signal level variations are modelled using the Rayleigh probability distribution function used in statistics. The Rayleigh probability density function is represented by

$$P_R(r) = \frac{r}{N_o} e^{-r/2N_o}, \quad N_o = \text{constant}, r \geq 0. \quad (1.45)$$

The Rayleigh model has often been used to simulate fading processes or to estimate their effects.

The dependence of outage probability on fade margin M indicates that the reliability of a link may be considerably improved by increasing M. This can be achieved by use of higher transmitter power, larger antennas, or by use of shorter paths. Unfortunately, experience has shown that, for digital radio systems, the use of a flat fade margin and Vigant's formula does not adequately estimate the percentage outage. This is because of the frequency selective effects of multipath fading on digital radio signals. We will examine these effects next.

1.5.5 Frequency Selective Fading Model

If flat fading was the only propagation problem, then the system performance of the 50 km hop system considered in Exercise 1.7–1.9 would be adequate. Flat fading implies that as the received signal amplitude drops, errors occur as a result of thermal noise effects (that is, as a result of a decrease in carrier-to-noise ratio).

However, fading of digital radio systems has been found to give rise to error rates that far exceed those predicted using flat-fade calculations. Outages (BER>10^{-3}) vastly exceeding accepted objectives have been measured on radio hops unprotected by space diversity systems. See for example, Giger (15) for details. This was found to be especially true for some digital radio systems overlaid on existing analogue FDM/FM systems with relatively long paths 50–60 km in length.

For radio systems operating at the relatively high frequency of 11 GHz, rain and multipath fading has been found to cause higher transmission outages than expected from flat-fading calculations. In part, this was found to be related to a loss of cross-polarization discrimination between channels. The systems transmitted separate signals on each of two polarizations (V and H) on the same frequency. Loss of cross-polarization discrimination was found to be predominantly associated with multipath fading, particularly on long hops. As a result of these studies, it is no longer as popular to design multichannel digital radio systems using only cross-polarization techniques to maximize efficiency of spectrum use. Yet experiments have shown that the avoidance of cross-polarization techniques will not solve the fading problems. Still another manifestation of fading becomes evident. As a result of multipath propagation effects, amplitude and phase distortion occurs within the channel bandwidth, which in turn gives rise to intersymbol interference (ISI). Consequently, this form of fading is referred to as *frequency-selective fading*. Its effect on ISI is usually more significant than the increase of thermal noise.

As a result of frequency-selective fading, bit error rates may be considerably worse for a given C/N ratio than they would be if flat fading occurred. That is, the minimum receive power P_t required for BER = 10^{-3} will be considerably higher under frequency-selective fading conditions than as predicted for flat fading. Hence, the fade margin given by Equation (1.23) will be reduced. This is illustrated by the BER values plotted versus C/N ratio in Figure 1.21. These results are taken from measurements for a 4 GHz, 45 Mbit/s, 8-PSK radio system operating over a 46 km hop in Georgia, USA. See Giger (15). For that system, the BER values measured with flat fades (various amounts of attenuation of the receiver input signal) showed that for a BER = 10^{-3}, the normal free-space signal had to be reduced by 41 dB. That is, the flat fade margin of the links was 41 dB. The scatter plot values represent the measurements taken under actual

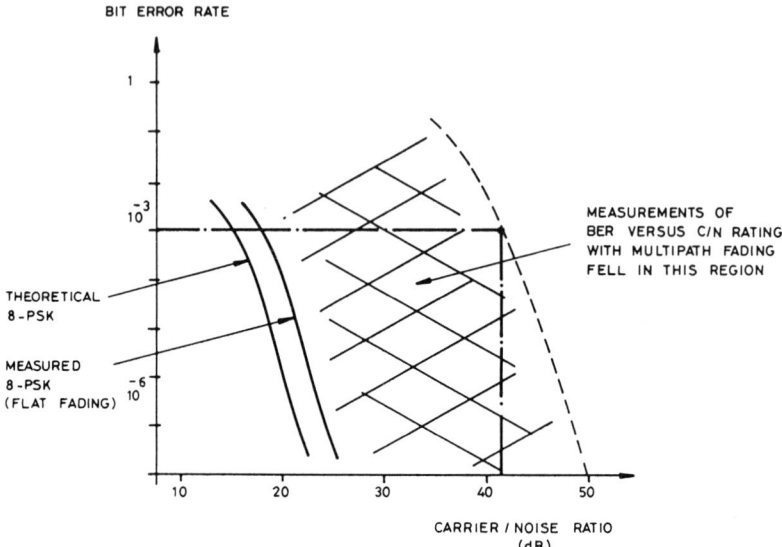

Figure 1.21 Measured errors resulting from frequency-selective multipath fading.

fading conditions. During fading conditions, the received signal levels (and hence, fading depths) were measured and compared with monitored BER values.

The results show that the flat-fading mechanism is certainly not confirmed as the dominant cause of errors. In most cases, the measurements show that a BER of 10^{-3} was reached long before the signal had faded by 41 dB. In some cases, only 19 dB of fading was sufficient. For this reason it is only of limited value to refer to the traditional flat-fade margin in digital radio systems.

Furthermore, it becomes apparent that something other than transmitter power has to be employed to make digital radio systems meet their availability and performance requirements. Techniques such as adaptive equalization and space diversity are required. However, before these are discussed, let us examine the frequency selective fading phenomenon in more detail.

Several statistical models have been advanced to characterize the frequency selective phenomenon associated with multipath propagation. One of these provides a simple and useful starting point. It is a three path, single echo model as illustrated in Figure 1.22(a).

The received signal is assumed to be composed of two components, one delayed τ (sec) with respect to the other. It is also assumed that each of the rays has also been attenuated to some extent by a third pseudo-ray such that this third ray produces a reduction of the amplitude levels of the other two rays. This is referred to as a "median depression." See for example, Rummler (16).

Radio Link System Design

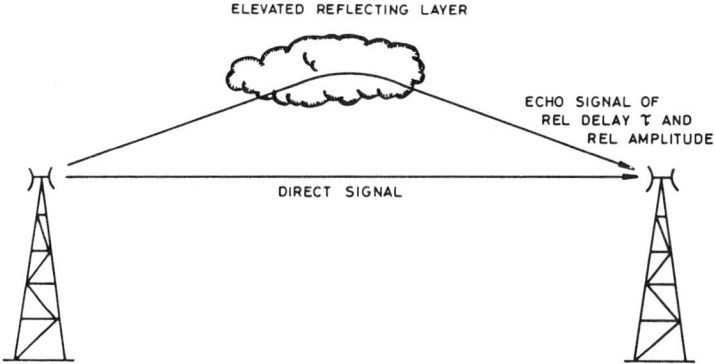

Figure 1.22 Three-path fading.

As illustrated in Figure 1.22(b), the transfer function of the three-path model channel is given by

$$C(f) = a(1 + be^{-j2\pi f\tau}) \tag{1.46}$$

where

a = the flat relative gain of the received signal with respect to the nonfaded signal (that is, nonfrequency selective).
b = the relative amplitude of the reflected signal, and
τ = the relative delay of the reflected signal.

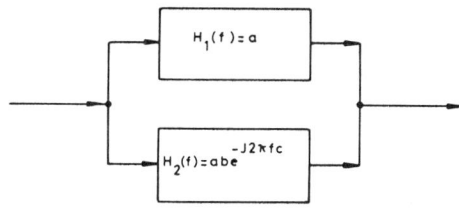

Figure 1.23 Transfer function model for frequency-selective fading.

The resultant composite channel's amplitude-frequency response can be obtained by noting that Equation (1.46) can be rewritten

$$C(f) = ab\, e^{-j\pi f\tau}(e^{j\pi f\tau} + e^{-j\pi f\tau}) + a(1 - b)$$
$$= ab\, e^{-j\pi f\tau}\, 2\cos\pi f\tau + a(1 - b)$$

so that the amplitude frequency response is

$$|C(f)| = 2\,ab|\cos \pi f\tau| + a(1 - b). \tag{1.47}$$

This is shown plotted in Figure 1.24.

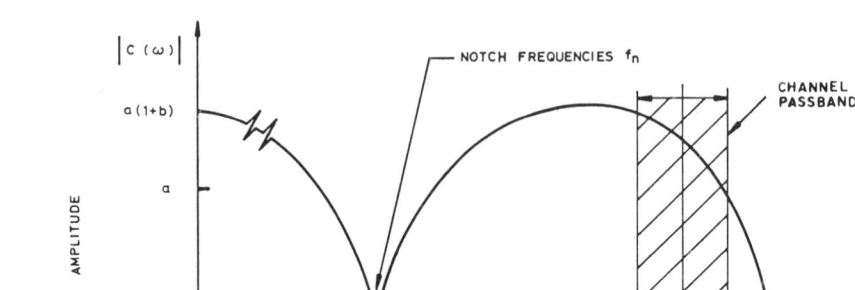

Figure 1.24 Amplitude response of frequency-selective fade model.

It becomes obvious, from Figure 1.24, why multipath propagation results in frequency dependent amplitude variations across the channel. The response shows "notches" or minima spaced $1/\tau$ (Hz) apart. In typical high capacity digital radio systems that have an operating bandwidth of less than 60 MHz, only one notch can affect the performance of a receiver at a given time.

Although multipath fading may be far more complex than this would suggest, such a single-echo model superimposed on a median depression has been found to be sufficiently accurate for system performance studies. It also provides a guide to the design of appropriate amplitude and phase equalizers to compensate for the frequency-selective effect.

1.5.6 Intersymbol Interference Resulting from Frequency Selective Fading

Consider the parameters a, b, and τ in the two path model to vary at a much slower rate than the digital signal information. Then using a quasi-stationary approach, we can consider the channel as a linear time-invariant system and examine its effect on the digital signal waveforms.

Consider a 16-QAM signal. For a nonfaded channel ($a = 1$, $b = 0$) the received 16-QAM digital radio signal $z(t)$ can be expressed as

$$z(t) = x(t) \cos 2\pi f_c t + y(t) \sin 2\pi f_c t \qquad (1.48)$$

Radio Link System Design

where

$x(t)$ = the I-channel four level data signal,

$y(t)$ = the Q-channel four-level data signal, and

f_c = the carrier frequency.

We assume the transmitter and receiver filtering have been designed for zero ISI in nonfading conditions. For example, $\alpha = 0.5$ raised-cosine filtering could be used with filtering split equally between transmitter and receiver. Then, if we neglect the effect of noise, for the two-ray fading channel, the impulse response is obtained from Equation (1.46) as

$$c(t) = a\,[\delta(t) + b\,\delta(t-\tau)]. \qquad (1.49)$$

Then with fading, but neglecting noise, the received signal is given by

$$r(t) = z(t) * c(t) \qquad (1.50)$$

where * represents convolution. Convolution with delta functions was discussed in Chapter 2, Volume 1. It follows from Equations (1.49) and (1.50) that the received signal is

$$r(t) = a\,[z(t) + b\,z(t-\tau)]. \qquad (1.51)$$

Consider coherent demodulation with a carrier

$$e_c(t) = \cos(2\pi f_c t + \phi)$$

with phase offset error ϕ.

Then the demodulated I-channel signal is given by

$$x(t) = \text{LPF}[2r(t)\cos(2\pi f_c t + \phi)] \qquad (1.52)$$

where LPF [·] represents low pass filtering. From Equations (1.48), (1.51), and (1.52), we obtain after some trigonometric manipulation, the I-channel output as

$$x(t) = a\,[x(t)\cos\phi - bx(t-\tau)\cdot\cos(2\pi f_0 t + \phi)$$
$$- y(t)\sin\phi + by(t-\tau)\cdot\sin(2\pi f_0 t + \phi)]. \qquad (1.53)$$

The frequency

$$f_0 = f_c - f_n \qquad (1.54)$$

is the notch offset defined as the frequency difference between the carrier frequency f_c and the nearest notch frequency f_n where

$$f_n = \frac{2n+1}{2\tau}, \quad n = 1,2,3,\dots\,.$$

For zero carrier phase offset, Equation (1.53) becomes

$$x(t) = a[x(t) - bx(t-\tau) \cdot \cos 2\pi f_0 t + by(t-\tau) \cdot \sin 2\pi f_0 t]. \quad (1.55)$$

In Equation (1.55), it is possible to identify the first term as the wanted signal and the remaining two terms as ISI. The first ISI term represents ISI from the I-channel and the second term represents co-symbol interference from the quadrature channel signal. Note that the quadrature channel ISI goes to zero when the fade notch is in the center of the transmission band ($f_0 = 0$). This means that the worst error probability occurs when the fade notch is on either side of the center frequency. A phase offset can be introduced at the receiver to reduce this effect.

The ISI terms are the main cause of system outage for high capacity digital radio systems rather than thermal noise effects. Unless the ISI can be reduced, its effect is to close the received eye diagram to such an extent that high error rates occur irrespective of the presence of noise. As a result of these effects, the *fade margin* of the system may typically be 20 dB less than that calculated for flat fading and the percentage unavailability, therefore, increases significantly. This must be taken into account in outage prediction.

An outage prediction technique developed by Campbell and Coutts (24) is based on two parameters, namely

(1) the probability of frequency selective fading in the worst month
(2) the mean echo delay τ during fading

Both these parameters can be obtained from recordings of AGC and pilot levels on existing analogue radio systems in the same location. The scheme uses the concept of a "system signature," which enables comparisons of performance for different modulation schemes and equalizer implementations. System signatures will be discussed in Section 1.5.9. Outage prediction techniques for route design are discussed further in Campbell (33).

1.5.7 Space Diversity

To overcome the effects of frequency-selective fading on system availability performance, space diversity and/or adaptive equalizer techniques are used. Space diversity receiver systems involve the use of two antennas vertically spaced approximately 10 m apart on the receiving tower. This is intended to ensure that the fading effects on the two signals are uncorrelated. The two signals can then be combined using either of the following methods:

(1) *co-phase diversity combiner* at IF or RF frequencies in which the relative phase of the two received signals is automatically adjusted for maximum resultant signal level. Fading on any one receiver signal will then have a

much reduced effect on the combined signal, resulting in greatly improved system availability. In addition, cophasing combiners tend to "condition" the overall channel response so that adaptive equalization may be more effective. This is because, for the diversity system, there is a low probability of a fading notch occurring in the channel band.

(2) *hitless switching combiners* which are simpler, less expensive baseband switch systems. A "hitless switch" is a baseband switch in which the two demodulated bit streams are switched after they are adjusted to be in synchronization.

Typically, space diversity combining without equalization can reduce the outage time by about a factor of 10 over the uncombined signal. The improvement factor depends on the baud rate and multipath delay. For more details, see Coutts (17).

1.5.8 Adaptive Equalization

An adaptive equalizer is a variable network which reduces the channel distortion that results from frequency selective fading. Two types of equalizer have been proposed.

(1) *IF amplitude compensation equalizers* are commonly used in high capacity digital radio systems. Variable gain-frequency response networks are used with center frequency at 70 MHz. Figure 1.25 shows a block schematic of an IF equalizer used by Nippon Electric Co. (NEC). This incorporates two sections with variable gain-frequency response. The combined effect of these two sections permits the notch frequency to be varied as well as the equalizer shape.

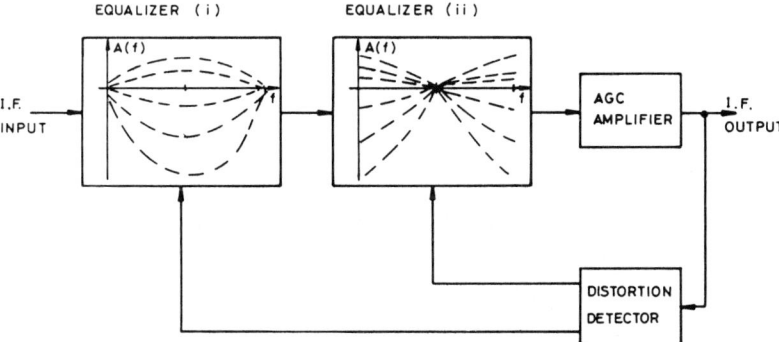

Figure 1.25 IF equalization to compensate for frequency-dependent distortion as a result of multipath effects.

(2) *Time domain equalizers* such as a transversal equalizer or decision feedback equalizers. These types of equalizers are more complex and are still in the early stages of development because of the need for very high speed digital components. See for example, Dudeck (18).

The effect of IF equalizers and space diversity is indicated in Figure 1.26. Without diversity, an IF equalizer can improve the outage time by a factor of about 2–3. However, with diversity, one obtains a "synergistic" effect wherein a combined improvement of typically 10–30 is obtained.

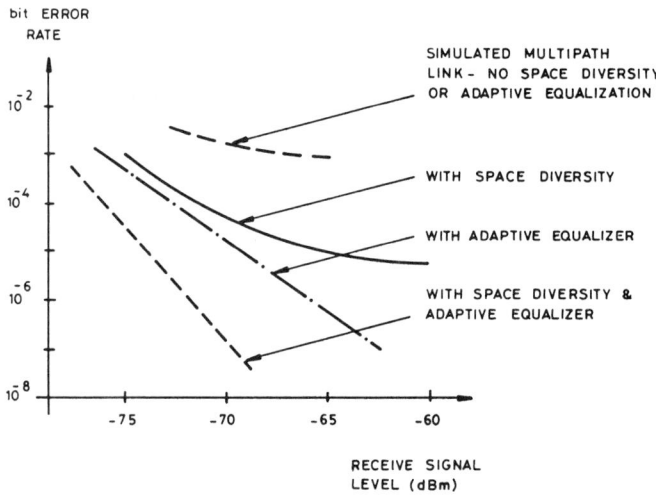

Figure 1.26 The effect of an equalizer and space diversity on system performance.

1.5.9 System Signature

One method used to characterize the sensitivity of particular digital equipment to frequency selective fading is to measure the system signature. This is a method of characterizing radio link equipment in the laboratory by emulating a single echo, as represented by Equation (1.46).

The transmitter output signal is fed to the receiver via two parallel paths. One path consists of a direct connection which emulates a direct signal. This effectively sets the value of the parameter a in Equation (1.46) to $a = 1$. The other path consists of an attenuator and a delay network to emulate a single echo. The attenuator and delay network are used to set the values of the parameters b and τ in Equation (1.46). The value of τ determines the relative offset frequency f_0

Hybrid Radio Systems

between the carrier frequency and that of the notch that results from the echo signal.

The notch is first located at the carrier frequency. That is, the value of f_0 is set to zero. Then the other echo parameter b is set to the value at which the measured BER = 10^{-3}. The above procedure is repeated for other values of notch offset f_0 so that a locus of fade-notch depths and positions corresponding to a 10^{-3} bit error rate is identified. This plot is known as the system signature.

Two typical system signatures are shown in Figure 1.27. On the abscissa is the notch offset frequency and on the ordinate is the notch depth required to produce BER = 10^{-3} (system outage). Note from Figure 1.24 that the critical notch depth in dB is obtained using $B = 20\log(1-b)$. Figure 1.27 shows system signatures that are typical of systems with and without adaptive equalization, respectively. It can be seen that the use of adaptive equalization decreases the sensitivity of the system to frequency-selective fading.

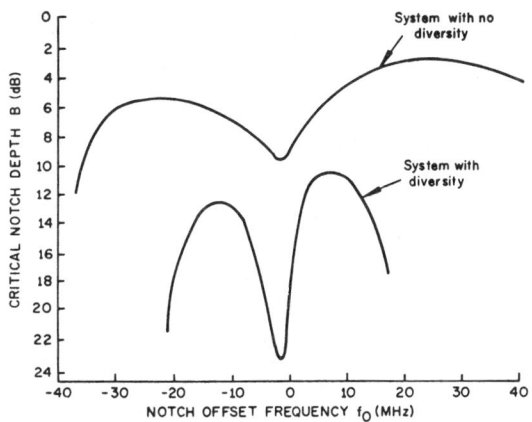

Figure 1.27 System signature.

1.6 HYBRID RADIO SYSTEMS

It is sometimes desirable to use analogue radio link systems for simultaneous transmission of FDM voice signals and TDM digital signals. These signals share the baseband of a microwave radio system which was originally designed to carry analogue FDM-FM signals. Such situations commonly occur during the phasing-in and overlaying of digital networks on existing analogue networks. For example, in some countries, new Digital Data Network (DDN) facilities have necessitated 2048 kbit/s digital links between capital cities before dedicated digital radio systems can be provided. Hybrid radio systems are used to meet this need.

Hybrid systems can be of several types. The different arrangements are given the abbreviations DIV, DAV, DOV, DUV, and DAVID. We will examine these briefly in the next sections.

1.6.1 Data In Voice (DIV) Systems

DIV systems utilize the bandwidth of two or three supergroups in the telephony baseband for 2048 kbit/s digital transmission. (A supergroup represents 60 telephone channels and occupies 240 kHz.) That is, the data occupies 480 or 720 kHz, depending on whether transmission at 4.3 or 2.8 bit/s/Hz is used. This means that efficient multilevel modulation schemes must be used. As a result, DIV modems are often more expensive and sensitive to transmission imperfections than other methods that use less spectrally efficient modulation techniques.

1.6.2 Data Above Voice (DAV) and Data Above Video (DAVID) Systems

DAV Systems use the unoccupied frequency spectrum space above the FDM telephony baseband for digital transmission. This is perhaps the most popular form of hybrid transmission. Figures 1.28(a) and (b) show the resultant frequency spectra for typical DAV systems on 960 channel and 1800 channel telephony bearers, respectively. Likewise, a DAVID system is one where a television bearer is used. These systems use 4-PSK or QAM modulation of a subcarrier at 5.9, 7.5, or 10.3 MHz. More details are given in Michaelides (19).

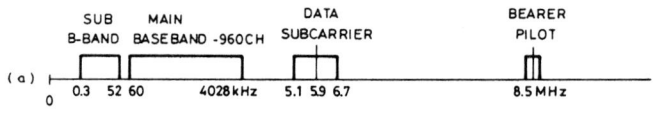

DAV ON 960 CHANNEL TELEPHONY BEARER

DAV ON 1800 CHANNEL TELEPHONY BEARER

Figure 1.28 Baseband spectra for data above voice (DAV) systems.

1.6.3 Data Over Voice (DOV) Systems

The terms DOV and DAV represent the same modulation scheme except that they are applied to coaxial cable and radio systems, respectively.

1.6.4 Data Under Voice (DUV) Systems

DUV systems can, in principle, be used to transmit high-speed data in the sub-baseband of high capacity FDM links. Often, however, the sub-baseband (below 300 kHz for an 1800 channel FDM-FM radio system) is occupied by service channels and subscriber wayside channels. In that case, a DUV system cannot be used.

A 1.544 Mbit/s DUV system developed by Bell Telephone Laboratories in the U.S. uses seven-level partial response (correlative) encoding techniques occupying the baseband spectra up to 470 kHz. Feher (5) gives details.

1.7 POINT-TO-MULTIPOINT SUBSCRIBER RADIO SYSTEMS

Transmission systems involving digital radio and message concentrators have been developed to connect rural and remote subscribers to public telephone networks. They are currently being developed in several countries. For example, one point-to-multipoint digital radio concentrator system design (21) is characterized by the following:

(1) the encoding of speech into 32 kbit/s using Adaptive Differential Pulse Code Modulation (ADPCM).

(2) the use of Time Division Multiple Access (TDMA) to provide 15 trunk speech channels and one signalling channel for connection of subscribers in remote areas to a telephone exchange using a cellular radio configuration. This is sufficient to provide up to 128 telephones services together with up to 28 telex services.

(3) the use of regenerative repeaters up to a maximum of 15 in tandem. This will allow the service area of a system to be extended up to 600 km from an exchange without noticeably degrading the transmission performance of circuits.

(4) digital radio transmission in bursts at a rate of 704 kbit/s using 2-FSK modulation with discriminator detection.

Figure 1.29 illustrates a possible configuration of a digital radio concentrator system. It shows the basic system elements consisting of an exchange unit, a repeater unit, a subscriber unit, and a subscriber drop out unit.

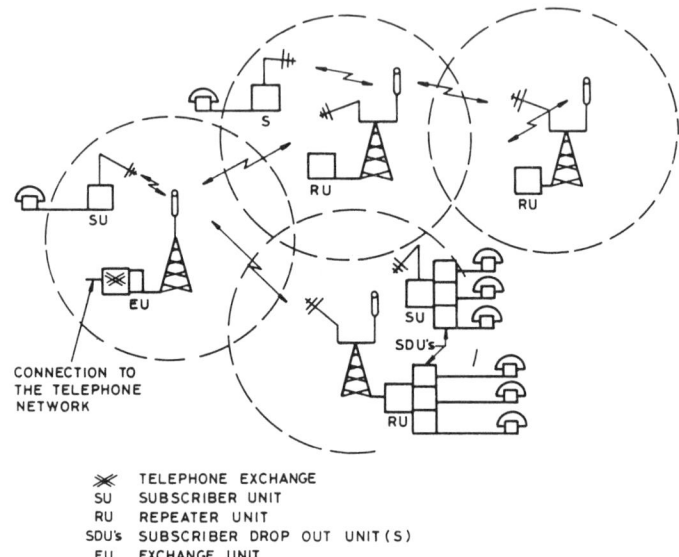

Figure 1.29 A digital radio concentrator system configuration.

The subscriber radio systems are arranged in cells, each with its own radio base station. For some cells, the base may be located at an existing telephone exchange. For other cells remote from an exchange, the base acts as a regenerative repeater to connect the subscriber unit signals to an exchange unit in another cell.

Each cell operates as a point-to-multipoint system using only two radio frequencies. The system uses digital encoding of speech, time division multiplexing with multiple access on demand and digital radio transmission in bursts. The radio baseband signal is structured in frames, each typically 4 msec long. Each frame is divided into 16 time slots, 15 of which are available for speech transmission and 1 for control. An active subscriber unit when assigned one of these 15 time slots for speech transmission to the base, turns its transmitter on and sends its previously stored 256-bit digital speech signal in that time slot in each frame. The other time slots are then available for use by other subscribers. A similar frame structure is used for the TDM transmissions from the exchange or repeater units to the subscribers. This signal is received by all subscriber units in the cell. A single go and return frequency pair can, therefore, serve all users in a cell.

Point-to-multipoint digital subscriber radio concentrator systems are currently under development in several countries for providing telephony and data services for remote subscribers in rural areas. For more details, see for example Sargeant (21), Couesnongle (22), Telecom Australia (23), or Wu (34).

1.8 PROBLEMS

1.1 A binary PSK radio system has a carrier frequency of 2 GHz. The input binary sequence consists of an alternating sequence of 0's and 1's at a bit rate of 8 Mbit/s. No Nyquist spectral filtering is used.
Find algebraic expressions for and sketch the following:
(1) the transmitted waveform
(2) the transmitted signal spectrum

1.2 Consider the PSK signal referred to in Problem 1.1. If Nyquist filtering is used with roll-off factor $\alpha = 0.5$, sketch the following:
(1) the transmitted waveform
(2) the transmitted signal spectrum

1.3 The input to a digital radio system with carrier frequency of 6 GHz is a 140 Mbit/s random binary input sequence. Sketch the transmitted signal spectrum for the following types of modulation. (For each case, plot spectra for no Nyquist filtering and for $\alpha = 0.5$ roll off, respectively.)
(1) 4-PSK
(2) 8-PSK
(3) 16-PSK
(4) 16-QAM

1.4 Compute the spectral efficiency for each of the modulation types in Question 1.3. Assume Nyquist filtering with $\alpha = 0.5$.

1.5 An input binary bit stream at a rate of 90 Mbit/s is to be transmitted in a channel bandwidth of 30 MHz using Nyquist filtering with $\alpha = 0.3$. Determine which modulation schemes could be used.

1.6 A signal state diagram for 16-QAM modulation is shown in Figure 1.11.
(1) Use a Gray encoding scheme to assign four-bit binary symbols to each signal vector.
(2) For an input sequence 1101 use your assignment scheme to write an algebraic expression for the resultant 16-QAM signal vector.

1.7 A 16QAM modulator is shown schematically in Figure 1.12. Let the output signals from the two filters in any symbol interval $0 - T_s$ be d_I and d_Q, respectively, where d_I and d_Q can take any one of the values $\{-3, -1, +1, +3\}$.

Write expressions for each of the 16 possible modulated output waveforms in any symbol interval.

1.8 Explain with appropriate analysis the operation of the 16-QAM demodulator illustrated in Figure 1.12.

1.9 Draw a signal states vector diagram for 64-QAM.

1.10 Propose a schematic diagram similar to Figure 1.12 to show how a 64-QAM modulator could be constructed.

1.11 Sketch a schematic of a modulator suitable for the QPRS signalling scheme using duobinary encoding. The QPRS signal state vector diagram is given in Figure 1.13.

1.12 Consider a minimum shift keying (MSK) system with input sequences

$$a_k = 1 \; -1 \; -1 \; 1$$

(1) Determine the phase values ϕ_k after each input bit, assuming that initially $\phi_o = 0$.

(2) Sketch the output MSK waveform.

1.13 Design a coherent detector system (block schematic) suitable for detection of a 4-PSK signal.

1.14 For the 2-PSK detector of Figure 1.15, consider the situation where a 1 is being transmitted and the input signal over an interval $0 - T_b$ is

$$y(t) = A \cos \omega_c t + n(t).$$

Assume $n(t)$ is Gaussian noise with mean zero and spectral density $N_0/2$. Let $z(T_b)$ be the sampler output value at time $t = T_b$.

(1) Find the value of $z(T_b)$ for the noiseless situation, that is $n(t) = 0$.

(2) Assuming noise is present, find the variance of the Gaussian random variable $z(T_b)$.

(3) Given that the detector threshold is set at 0 volts, show that an incorrect decision will be made with probability

$$P(\hat{d}_k = 0 \mid d_k = 1) = Q(\sqrt{A^2 T_b/N_0})$$

where $N_0/2$ is the two-sided power spectral density of the noise.

1.15 Assuming free-space propagation, compute the receiver carrier power obtained in a radio system operating over a 50 km path at 11 GHz with parabolic transmitting and receiving antennas of 1 meter diameter. Assume the transmitter power output is 5 Watts and system losses resulting from feeder and filters total 6 dB.

1.16 How could Figure 1.17 be modified to enable coherent detection of an 8-PSK signal?

1.17 Determine the flat-fade margin for a 70 Mbit/s 8-PSK digital radio system operating at 11GHz over a 40 km path. At the transmitter and receiver, parabolic antennas are used with diameter of 1 m and with effective receiving area equal to 60 percent of the actual aperture area. The transmitter power is 2 Watts and the receiver noise figure is 6 dB. Assume the receiver regenerator characteristic is as shown in Figure 1.18.

1.18 Use Equations (1.23) and (1.26) to show that a 16-QAM system requires approximately 10 dB more power than 2-PSK for the same bit error rate.

1.19 The 16-QAM signalling vectors used in digital radio systems are illustrated in Figure 1.11. By contrast, Figure 1.9 of Chapter 1, Volume 1 shows an alternative 16-QAM signal set used in 9600 bit/s data modems.

The important parameters of the signal sets are:

(1) the minimum distance (d_n) between the phasors which is a measure of immunity against additive noise
(2) the minimum phase difference (ϕ) which is a measure of immunity against phase distortion or phase jitter
(3) the ratio of peak to average power (r) which is a measure of immunity to non-linear distortion.

Assuming that the average power is to be the same for both signal sets, compare the d_n, ϕ and r parameters for each.
Comment.

1.9 REFERENCES

1. N. F. Dinn, "Digital Radio: Its Time Has Come," *IEEE Communications Magazine*, Vol. 18, pp. 6–12, Nov. 1980.
2. J. Steele, "Digital Radio Systems Overview," Telecom Australia Seminar on Digital Radio Systems, Melbourne, Sept. 1981.
3. C. C. Bailey, "Digital Radio Economics," *IEEE Communications Magazine*, Vol. 19, pp. 6–8, July 1981.
4. L. M. Nirenbergand, C. T. Wolverton, "Microwave Transmission Systems for Local Data Networks," *Telecommunications*, June 1980.
5. K. Feher, *Digital Communications: Microwave Applications*, Prentice-Hall, 1981.
6. CCIR (International Radio Consultative Committee), Recommendations of the XII'th Plenary Assembly, Geneva, 1974.
7. R. P. Coutts and A. L. Martin, "High Capacity, Long Haul Digital Radio," *Journal of Electrical and Electronics Engineering*, Australia, Vol. 1, No. 2, June 1981.

8. S. Yokoyoma, K. Kinoshita, Y. Tan, T. Ryu, S. Omo, H. Hashimoto, and R. Mitchell, "An Eight-level PSK Microwave Radio for Long-Haul Data Communications," *Proc. of IEEE ICC-75 Conference*, June 1975.
9. J. Ellershaw, "Interference Aspects of Digital Radio," *Proc. of IREE Conference*, pp. 264–266, Sydney, 1981.
10. Nippon Electric Co., "Handbook for NEC 6.7 GHz 140 Mbit/s 16QAM Digital Radio Equipment," 1981.
11. C. W. Anderson and S. G. Barber, "Modulation Considerations for a 91 Mbit/s Digital Radio," *IEEE Trans. Comm.*, pp. 523–8, May 1978.
12. K. S. Shanmugan, *Digital and Analogue Communication Systems*, John Wiley, 1979.
13. A. D. Whalen, *Detection of Signals in Noise*, Academic Press, 1971.
14. J. J. Spilker, *Digital Communications by Satellite*, Prentice-Hall, 1977.
15. A. J. Giger and W. T. Barnett, "Effects of Multipath Propagation on Digital Radio," *IEEE Trans. on Comms*, Vol. COM-29, No. 9, pp. 1345–1352, Sept. 1981.
16. W. D. Rummler, "A new selective fading model: Application to propagation data," *Bell Syst. Tech. J.*, pp. 1037–1071, May–June 1979.
17. R. P. Coutts and J. C. Campbell, "Mean Square Error Analysis of QAM Digital Radio Systems Subject to Frequency Selective Fading," *Australian Telecommunication Research Journal*, Vol. 15, No. 1, 1982.
18. M. T. Dudeck and J. M. Robinson, "A Decision Feedback Equalizer and Novel Carrier Recovery Circuit for Digital Radio-Relay Systems Operating at up to 5 bit/s/Hz," *Proc. of IEEE ICC'80*, pp. 41.5.1–41.5.6, Seattle, 1980.
19. A. T. Michaelides and G. Mackie, "2048 kbit/s Transmission over Analogue Radio Relay Systems," *Telecommunication Journal of Australia*, Vol. 31, No. 1, pp. 12–18, 1981.
20. J. G. Proakis, *Digital Communications*, McGraw-Hill, 1983.
21. V. Sargeant and J. Steele, "Planning Telephone Networks for Rural and Remote Areas using Digital Radio Concentrator Systems," *Proc. I.E.Aust. Engineering Conference*, Paper 314, pp. 138–45, 1981.
22. M. de Couesnongle and C. R. Garnier, "IRT 1500 Integrated Rural Telephony System," *Philips Telecom Review*, Vol. 41, No. 2, June 1983.
23. Telecom Australia, "Guidelines for the Application of Digital Radio Concentrator Systems," *Planning Application Bulletin*, No. 46, Sept. 1981.
24. J. C. Campbell and R. P. Coutts, "Outage Prediction of Digital Radio Systems," *IEE Electronics Letters*, Vol. 18, No. 25/26, pp. 1071–2, Dec. 1982.
25. I. Godier, "Application of the CCIR bit error ratio recommendation to real digital radio paths," *IEEE Trans. on Comms.*, Vol. COM-32, No. 9, pp. 1060-1, Sept. 1984.
26. G. Ungerboeck, "Channel coding with multilevel/phase signals," *IEEE Trans. on Inform. Theory*, Vol. IT-28, No. 1, pp. 55–67, Jan. 1982.
27. S. G. Wilson, H. A. Sleeper, H. P. J. Schottler and M. T. Lyons, "Rate 3/4 convolutional coding of 16PSK: Code design and performance study," *IEEE Trans. on Comms.*, Vol. COM-32, No. 12, pp. 308–1315, Dec. 1984.
28. G. D. Forney, R. G. Gallagher, G. R. Lang, F. M. Longstaff and S. U. Qureshi, "Efficient modulation for band-limited channels," *IEEE J. on Selected Areas in Comms.*, Vol. SAC-2, No. 5, pp. 632–647, Sept. 1984.

29. L-F Wei, "Rotationally invariant convolutional channel coding with expanded signal space—Parts I and II," *IEEE J. on Selected Areas in Comms.*, Vol. SAC-2, No. 5, pp. 659–686, Sept. 1984.
30. M. Miyake, T. Fujino, Y. Umeda and E. Yamazaki, "A study of a coded 8PSK modem with a Viterbi decoder using simplified metric calculation," *Proc. PTC'85*, pp. 279–286, Honolulu, Jan. 1985.
31. S. A. Rhodes, R. J. Fang and P. Y. Chang, "Coded octal phase shift keying in TDMA satellite communications," *COMSAT Tech. Review*, Vol. 13, pp. 221–258, Fall 1983.
32. T. Fujino, Y. Umeda and E. Yamazaki, "Coded octal PSK system performance disturbed by imperfect recovered phase reference," *Trans. IECE of Japan*, CS84-69, pp. 57–64, Aug. 1984.
33. J. C. Campbell, "Outage prediction for the route design of digital radio systems," *Australian Telecom Research (ATR)*. Vol. 18, No. 2, pp. 37–49, 1984.
34. K. Wu and M. J. Miller, "Point-to-multipoint digital subscriber radio systems in China," *Proc. PTC'85*, pp. 269–274, Honolulu, Jan. 1985.
35. S. V. Pizzi and S. G. Wilson, "Convolutional coding combined with continuous phase modulation," *IEEE Trans. on Comms.*, Vol. COM-33, No. 1, pp. 20–29, Jan. 1985.
36. C. Sundberg, "Continuous Phase Modulation," *IEEE Communications Magazine*, Vol. 24, No. 4, pp. 25–38, April 1986.
37. I. Korn, *Digital Communications*, Van Nostrand, 1985.

Chapter 2

DIGITAL TELEPHONE NETWORKS

2.1 DIGITAL SWITCHING

A digital telephone network consists of a combination of *digital transmission systems* based on pulse code modulation (PCM) or some other form of digital modulation such as adaptive differential pulse-code modulation (ADPCM), and *digital switching centers*, including a hierarchy of remote switching stages (concentrators), local exchanges, and trunk exchanges.

The remote switching stages and local exchanges serve as concentrating and distributing centers for telephone calls in a local area. They interconnect the subscriber local loop plant via shared interexchange digital transmission facilities. When calls involve transmission over distances greater than about 20 km, long-distance digital trunks are used. The trunk exchanges serve to interconnect the concentrated traffic to long-distance digital trunk circuits.

Where a group of lines are provided from a local exchange to serve a single customer premises, a private automatic branch exchange (PABX) is usually located at the customer's building. The PABX may provide for local telephone connections within the building and for interconnections to the local exchange. More recently, digital PABXs have become available that can provide for local area computer network connections and for telephone services.

In this section, the principles of digital switching will be reviewed. At the present time, almost all of the world's telephone networks rely on analogue switching techniques. That is, voice signals in analogue form are interconnected via electro-mechanical switches or crosspoints. These may be in the form of rotating switches known as uniselectors or rectangular switching matrices composed of crossbar switches or miniature reed relays.

Modern switching centers are stored-program control (SPC) exchanges where the supervision and control functions in the exchange are performed by digital computer equipment. Despite the use of digital computers for control, most existing SPC exchanges are *analogue* switching centers since they interconnect analogue signals via electro-mechanical switches. For more details on analogue switching, see Pearce (1) and Inose (3).

Digital telephone exchanges interconnect voice and signalling information in digital form (usually binary) using digital techniques based on solid-state memories and logic gate arrays. Time-division multiplexed signals flow between switching points with fixed-length frame formats as discussed in Chapter 1, Volume 1.

In analogue exchanges, the electro-mechanical switching devices require, by their nature, different forms of control signals and information (speech) signals. The signalling and speech signals must be kept separate in the switching center. This, in turn, requires reasonably complex interface hardware between the switching equipment and the transmission facilities. However, if the speech signals, as well as the control signals, are in digital form, the processing and routing through the digital exchange can be performed with common equipment in much the same way as in a digital computer. Modern reliable solid-state circuit technology can be used, switching times can be reduced, and the integration of voice, data, and other services is made feasible. The overall result is a switching center with significantly lower cost for a given degree of complexity and reliability.

A high proportion of the cost of PCM transmission systems lies in the terminal costs. The introduction of digital switching can substantially compensate for this cost since switching is performed directly on the digital bit stream and no costly analogue/digital conversion is needed. A combination of digital transmission and switching can, therefore, lead to lower overall costs.

As suggested by current trends, future developments will ultimately lead to completely digital transmission and switching networks. The development and penetration of digital switching techniques into the existing analogue network will be a gradual process over many years. Apart from technical and economic considerations, many problems of international standardization are still to be resolved.

In the following sections, we will examine the basic form a completely digital telephone switching system can be expected to take in the future, based on information from Ericsson (2), Inose (3), and Flood (4).

The principles of PCM switching systems, which have developed in recent years, have evolved from a concentrator/exchange model developed in the late 1950s and early 1960s. In 1959, Bell Laboratories began a research experiment known as ESSEX (Experimental Solid State Exchange). See Vaughan (5) for details. The ESSEX model has turned out to be the prototype for many current telephone exchange system designs.

2.1.1 Local Networks

We first examine networks interconnecting subscribers in a single local exchange area. It is instructive to compare local networks based on analogue and digital techniques, respectively. It is assumed that the subscriber telephone voice signals are analogue (although even this assumption is not likely to be valid in the future, if analogue-to-digital conversion is carried out within the subscriber's equipment). Consider the elemental local area networks shown in Figure 2.1(a) and (b). A conventional analogue scheme based on a local exchange is shown in Figure 2.1(a). Its digital counterpart is shown in Figure 2.1(b).

The digital local network of Figure 2.1(b) is based on a two-level exchange hierarchy. It consists of large local exchange control and switching equipment

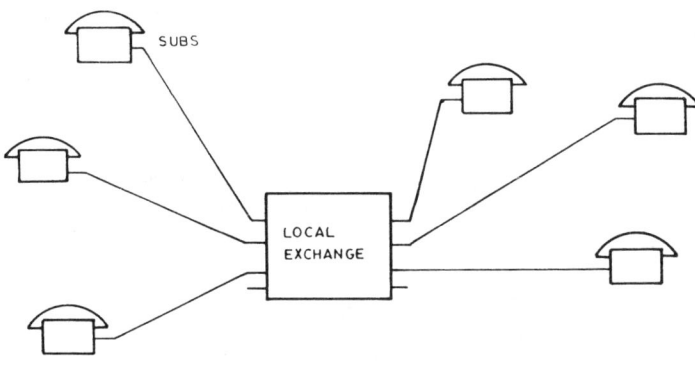

(a) ANALOGUE NETWORK WITH CONVENTIONAL LOCAL EXCHANGE

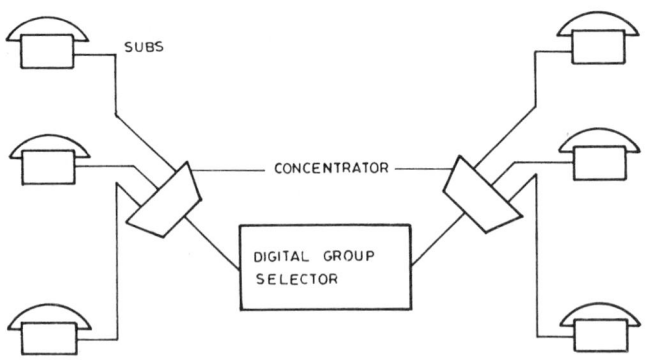

(b) DIGITAL NETWORK WITH DIGITAL GROUP SELECTOR AND CONCENTRATORS

Figure 2.1 Analogue and digital local area switching.

known as *digital group selectors*. These, in turn, control a number of remote *digital concentrators* or remote switching stages (RSSs).

The concentrators, together with their digital group selector, form a star structure. Transmission between concentrators and digital group selectors is digital, and is usually based on pulse code modulation (PCM) techniques. The voice signals between the customer's telephone and the RSS may be analogue. These signals will be converted into PCM form at the RSS. Alternatively, analogue-to-digital conversion may (in the future) be performed in the subscriber's telephone set.

In a local exchange area, a very significant component of the total capital cost is associated with the subscribers' line plant, sometimes referred to as "local loops." The total cost of the subscribers' line plant is very significant. It normally exceeds the cost of the exchange equipment.

2.1.2 Concentrators

A concentrator or RSS is a switching facility by which a number of subscribers can be connected efficiently to the local exchange. The subscribers share the use of a limited number of time-division muliplexed lines. This concentrator function can have the following benefits:

(1) provide opportunities to obtain savings in line costs
(2) decrease the central control cost
(3) provide higher flexibility in network planning

Concentrators become cost-effective partly as a result of digitalization of transmission and also because of the nature of the local exchange switching and control techniques. Recall that the signalling channel in PCM transmission systems provides high signalling capacity. For example, in a 30-channel system, time slot 16 provides a 64 kbit/s signalling facility between concentrator and digital group selector. This permits much of the concentrator control intelligence to be centralized in the digital group selector and used for a greater number of subscribers with resulting economic advantages. The most economic number of subscribers to be served by a single concentrator is typically somewhere between a few hundred and a few thousand subscribers, depending on geographical factors such as the subscriber population density.

A concentrator consists of two main parts, as illustrated in Figure 2.2. The first of these is the *remote part*, sometimes referred to as the "subscribers' unit," which performs the concentration function. This is connected by means of PCM transmission lines to the other component of the concentrator known as the *central part* or "exchange unit." The central part is located with the local exchange digital group selector. In some special circumstances, even the remote part of the concentrator may also be located with the digital group selector.

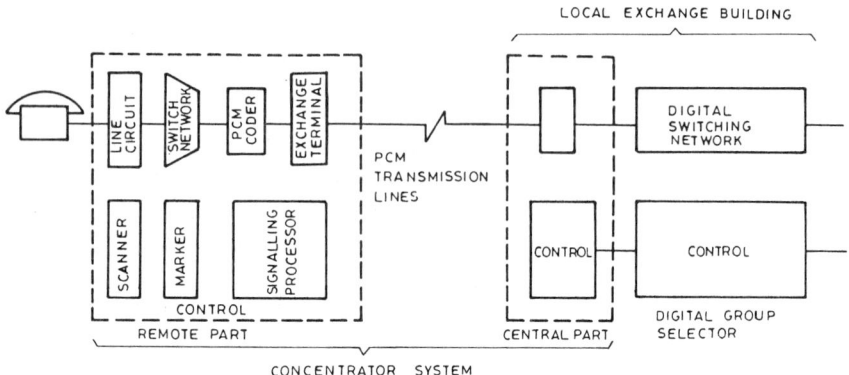

Figure 2.2 Principle of a concentrator.

The *remote part* of the concentrator consists of five subunits, as follows:

(1) *Line circuits* provide an interface between the subscribers' lines and the concentrator system. They transform the speech signals into a form suitable for the switching network unit and process the subscriber line signalling for compatibility with the signalling network.
(2) *Switch network* units perform the actual traffic concentration from a larger number of subscribers' lines to a smaller number of speech channels. Electromechanical or digital switching may be used depending on whether the subscriber telephone outputs are analogue or digital.
(3) *PCM convertors* convert the switch network output into a PCM time-division multiplexed signal if the voice signals in the switch network unit are not PCM.
(4) *Exchange terminal units* are the line terminal interface between the concentrator and PCM transmission line. The signals on the switch side and on the line side of the exchange terminal unit will be PCM signals.
(5) *Remote control units* work as slaves to the digital group selector and perform the RSS control functions which cannot be centralized. Such a unit consists of
   (a) a *scanner*, which repetitively examines the subscriber lines to detect "off hook" or "on hook" signals,
   (b) a *marker*, which selects a path and sets up a connection simultaneously through the switching network unit connecting the specified input and output terminals, and
   (c) a *signalling processor*, which receives commands from the central part via the PCM signalling channel. It performs error detection and if the

Digital Switching

command is found correct, forwards it to the appropriate subunit. For faulty commands, retransmission is requested.

The central part of the concentrator system located in the local exchanger consists of two subunits.

(1) *The exchange terminal unit*, which is a terminal unit for the PCM transmission line which interfaces with the digital group selector. In this terminal, the time slots coming from connected digital lines are brought into phase with the local exchange time slots. This will be discussed further in Section 2.2. Signalling information can be extracted in this unit.
(2) A *regional control unit* controls the operation of the remote part of the concentrator.

The exchange terminal units in the remote and central parts of the concentrator perform the functions of framing and clock extraction. They also carry out the injection and extraction of the signalling time slot into and from the line transmission sequence.

It is interesting to examine how the signalling and control functions are shared between the remote and the central parts of the concentrator. Typically, changes in subscriber line condition and dial pulses are detected in the remote part but push-button dial tone pulses are sent over the PCM channel to tone receivers in the digital group selector. This gives better utilization of the tone receivers than if they were situated in the remote part.

All speech signals must be routed via a digital group selector for switching. Selection of a free PCM channel from the remote switching stage is performed by a central control unit in the digital group selector. Then a command is sent to the remote part to connect the subscriber to the selected PCM channel.

Metering and call time supervision are carried out in the digital group selector. Ringing signals to the called subscriber and busy and ringing tones to the calling subscriber are generated within the remote part.

Although working in a digital environment, the concentrator does not necessarily use digital switching. Assuming that the subscriber local loop speech signal is analogue, the concentrator must perform an analogue-to-digital conversion function. In different concentrator designs, the analogue/digital conversion may take place at the input, in the middle or at the outlet of the remote part of the concentrator. Figure 2.3 illustrates some different arrangements. Figure 2.3(a) shows an arrangement where analogue switching techniques are used to perform the concentrator function. Then a PCM codec interfaces the switch to the digital line. In Figures 2.3(b) and (c) the analogue voice frequency signals are converted to pulse amplitude modulation (PAM) and delta modulation (DM) forms, respectively, before switching takes place. After switching, the signals

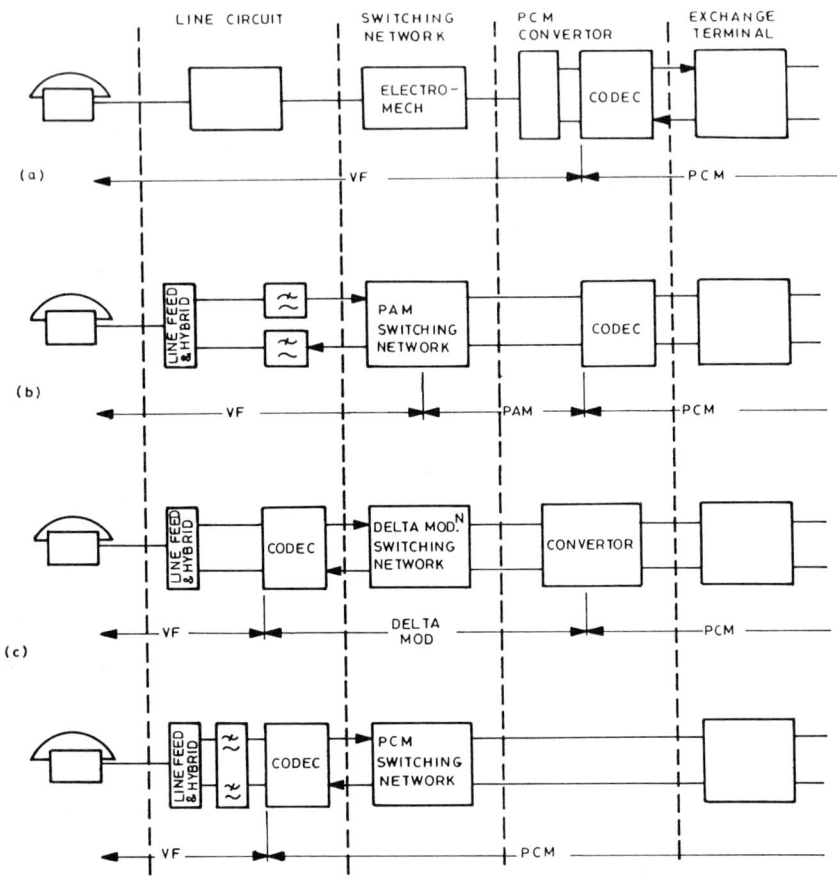

Figure 2.3 Possible configurations for the remote part of the concentrator.

are reconverted into PCM forms. In Figure 2.3(d), the speech signals are first converted into PCM forms before switching.

Depending on the choices made, the switching network may be analogue (electro-mechanical) or digital. We will examine techniques for digital switching of binary signals in Section 2.1.5.

Speech signals are sometimes converted into PAM forms for remote switching. This is achieved by sampling the speech waveforms, typically at a rate of 8000 samples per sec. The speech information is contained in the amplitudes of the sample sequence. However, after switching, the signals are converted into PCM form for transmission to the digital group selector.

Digital Switching

2.1.3 Digital Group Selectors

A digital group selector or digital switching center consists of a stored-program control (SPC) system and a switching network, as illustrated in Figure 2.4. Both the control and switching units are electronic. This is in contrast with more conventional SPC exchanges, which are semi-electronic. That is, the control devices consist of digital electronic components but the switching network is electro-mechanical. Although in principle, semiconductor switches could be used, metallic contacts are usually preferred because of their good transmission properties and ability to withstand high currents associated with ringing signals and lightning discharges on subscriber line circuits.

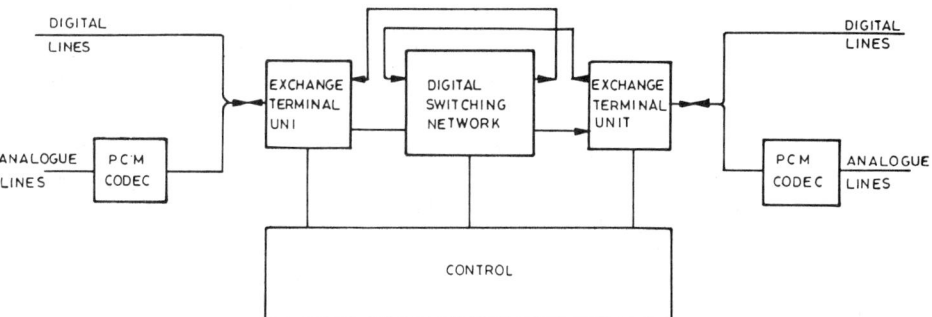

Figure 2.4 Digital Group Selector and associated control.

The digital group selector must often be compatible with a mixed analogue and digital transmission environment. It must, therefore, be able to interconnect PCM links as well as analogue links with frequency-division multiplexed (FDM) or voice frequency signals. The digital group selector consists of the following blocks:

(1) A *PCM multiplexer* and codec converts analogue signals to a PCM form and arranges for the multiplexing of speech and signalling elements.

(2) An *exchange terminal unit* acts as an interface for incoming PCM lines. This was discussed in relation to the concentrator in Section 2.1.2.

(3) A large *digital switching network* provides for switching between the incoming and outgoing time-division multiplexed digital busses. It consists of arrays of digital memory and gate components. This will be discussed in Section 2.1.5. In a digital group selector, four-wire switching (separate go and return paths) is necessary to obtain a bidirectional duplex connection.

(4) *Regional and central control units* provide for control of the routing, switching, and signalling functions. These control units use SPC computer control techniques as used for the control of analogue group selector exchanges.

Space does not permit a detailed discussion of conventional SPC techniques here. For further details see for example, Pierce (1) or Inose (3).

The control units for both analogue and digital exchanges have many common features. There are, however, minor differences, as indicated for the following.

(1) *Interfaces*—The digital group selector uses the same technology in both the central control processor and in the switching network. On the other hand, an electro-mechanical switching network (using reed relays or crossbar switches) must be controlled by an electronic processor using different circuit technology so that special equipment is needed as an interface.

(2) *Route selection*—The time-division multiplexed operation of the digital switching network results in higher traffic capacity for a given switch complexity. Simpler switch networks lead to easier searching for free paths through the network. This, in turn, reduces the load on the control system.

(3) *Continuity testing*—Galvanic through connection tests are not possible in a digital switch because of the digital circuits and time-division operation. Other methods such as parity checks on the bit streams must be used.

2.1.4 Advantages of Concentrators with Centralized Exchanges

The motives for the use of concentrator systems in digital telephone networks can be summarized as follows:

(1) *Saving of copper in the local cable plant*—This saving is achieved by placing the traffic concentration near the subscribers.

(2) *Network design flexibility*—A network with a central group selector and a number of subordinated concentrators can be very flexible in design. Rapid extensions to meet needs caused by changes in planning can be more readily achieved. Existing subscriber cables can be converted to PCM junctions and costly extensions of the cable plant avoided. Existing exchanges can be extended by concentrators, initially controlled from a distant group selector. After some time, a group selector can be installed replacing the old equipment.

(3) *Improved transmission quality to the subscribers*—Improved transmission quality is obtained by having shorter subscriber lines and by using digital transmission between the concentrator and its group selector.

(4) *Access to data channels for the subscribers*—By moving digital transmission closer to the subscriber, access to high-speed data transmission becomes simplified.

Digital Switching 77

(5) *Decrease of central control cost*—Taking a long-term view, the introduction of concentrators will raise the optimal size of the "local area" associated with a given digital group selector. This should distribute the fixed cost of the SPC system in the central group selector among a greater number of subscribers.

An essential factor for the realization of a concentrator system is the *high-capacity signalling* between the digital group selector and the concentrator. This is made possible through PCM transmission. This allows flexible control of the concentrator and permits concentration of control equipment to the efficiently-used central part of the system.

A further consideration is the trend in modern system design towards modular subsystems with standardized interfaces. This makes possible a clear division between subscriber stages and the group selector, which in turn, facilitates the development of remote concentrators.

2.1.5 Digital Switching Principles

In conventional analogue exchanges, the electro-mechanical switching techniques can be classified as *space-division switching*. That is, different calls are connected between different circuit paths separated in space. A switch contact provides a continuous speech path for the duration of the call. This remains in operation throughout, and handles no other call until the end.

In a digital telephone network with PCM transmission, the switching network must perform switching between two time-division multiplexed circuits or busses. It is necessary to perform switching between different time slots and between different circuits. As a result, both switching in space and switching in time are necessary. The latter is called *time-division switching*. In digital exchanges, time-division switching of PCM signals is achieved by buffer memories. Space-division switching is performed by logic gates in crosspoint matrices.

With time-division switching, the switch for a particular call need only be operating at the corresponding time slots; for the remainder of the time it can be available to switch calls in other time slots.

Space-division Switching

Space-division switching networks have traditionally been implemented in conventional analogue exchanges by networks of metallic contacts. The simplest representation of a switching structure is a rectangular array of metallic contacts known as crosspoints. This is illustrated in Figures 2.5(a) and (b). The network in Figure 2.5(a) is referred to as a *nonblocking* switching stage. It can connect any one of the N input lines to any one of the N output lines. Figure 2.5(b) is called a *graded* switching stage in which each inlet stage can only access a

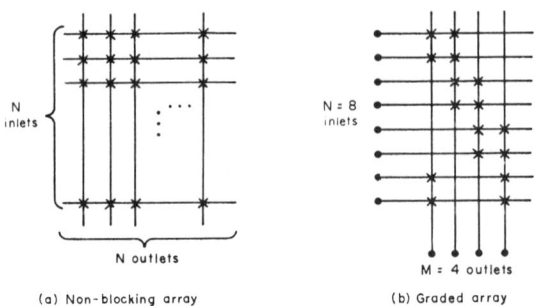

Figure 2.5 Rectangular switching matrices.

limited number of outputs. *Grading* is the process of allocating available outlet groups of connections for various inlet groups using overlapping.

In order to make the rectangular array nonblocking, N^2 crosspoints are required. For a large switching stage, N may be 10,000 lines say, then the number of crosspoints is 100 million! Furthermore, the large number of crosspoints on each inlet and outlet line can result in a large amount of capacitive loading.

On the other hand, a nonblocking network with reduced numbers of crosspoints can be designed by using a multi-stage switch as shown in Figure 2.6. The figure shows a three-stage switch in which the N inlet circuits are subdivided into N/n subgroups, each of n circuits. The inlet circuits of each subgroup are switched via $n \times k$ first-stage rectangular crosspoint arrays. Each one of the k outlets is connected to one of the k second stage arrays (known as junctors). Then the

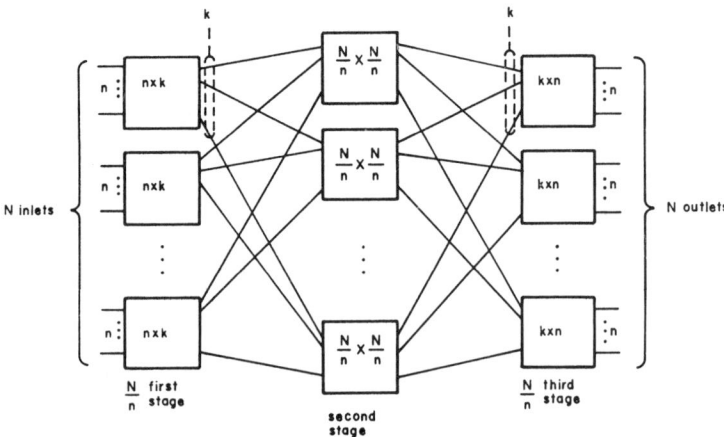

Figure 2.6 Three-stage switching matrix.

circuits are connected to the third-stage outlet switches, such that the switch is symmetrical about the junctors.

Note that each of the k paths from a given inlet switch is connected to a separate second stage array. This multiple stage structure provides protection against failures because of the availability of alternate paths.

For the three-stage switch to be nonblocking, it is necessary to make the number of second-stage switches $k = 2n - 1$. This follows from considering the worst case situation for blocking as follows. Let an inlet i require connection to outlet j. The worst case arises if each of the $n - 1$ other inlets connected to the same inlet switch as i are busy and furthermore, the $n - 1$ remaining outlets connected to the same switch as j are busy. To ensure that one additional center stage exists for connecting i to j, we require that

$$k = (n - 1) + (n - 1) + 1 = 2n - 1.$$

For the three-stage array with $k = 2n - 1$, the total number of crosspoints is

$$N_x = (2n - 1)\{2N + (N/n)^2\}.$$

To see that this is considerably less than the number of crosspoints (N^2) for a rectangular array, consider the case where $N = 10{,}000$, $n = 100$ and $k = 199$. Then we obtain $N_x = 5.97 \times 10^6$ (compared to $N^2 = 100 \times 10^6$).

Digital PCM Switching

Figure 2.7 illustrates a digital PCM switching system configuration. To illustrate the principles involved, the diagram shows only two incoming PCM systems (numbered 1 and 2) and two outgoing systems (numbered 3 and 4).

Suppose a channel in time slot x of system 1 is to be connected through to a channel in time slot y of system 3. The 8-bit information words arriving on system 1 in each frame is first stored in a time-shift switch. The central control unit selects paths to pass the 8-bit words from system 1 to system 3 through the electronic space switch. These paths are known as "highways." Suppose that highways labelled A and C are selected. Then each 8-bit word from time slot x in system 1 is read out from the time-shift switch at a time when highways A and C are both free. Crosspoint P is operated at this time and the information passes into the time switch associated with system 3. From here, it is read out at time slot y for forward transmission.

By similar operations, a simultaneous call can be connected from, say, time slot z on system 2 via crosspoint Q to time slot z on system 3. Other connections may be similarly established. Note that in these examples, only signals in one direction have been considered. In practice, this must be duplicated by switching circuits for the signals flowing in the opposite direction. That is, four-wire (full duplex) switching is required.

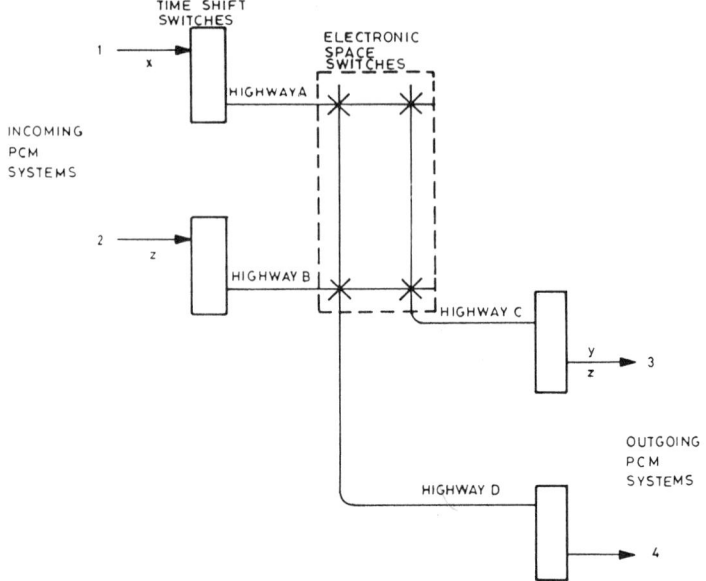

Figure 2.7 Digital PCM switching (T-S-T switching).

The switching system shown in Figure 2.7 uses a space-division switch at the center of the connection and time-division switches at the ends. This arrangement is called *time-space-time* (T-S-T) trunking.

The time-shift switching stages are known as time slot interchange (TSI) memories. Each consists of two random access memories (RAMs): one for storing the incoming bytes and one for storing the memory read address list. The space switching crosspoint arrays consist of logic gates and associated control circuits. See for example, Bellamy (16) or Ronayne (17) for details.

It is also possible to use a *space-time-space* (S-T-S) configuration. A central time-shift switching unit is flanked by input and output space-shift switches. A very simple S-T-S configuration is shown in Figure 2.8. Consider the switching of a channel in time slot x of system 1 to time slot y in system 2. During a time slot x in system 1, the crosspoints marked P are activated and the PCM 8-bit word from input 1 is written into a specific cell in the time switch. This PCM word is stored there until time slot y arrives. When time slot y arrives, the crosspoints marked Q are activated and the system 1 input word is read out towards the system 2 output. At the same time, when Q is operated, the PCM word from the system 2 input is written into memory cell. During the next x time slot this word from 2 is read out towards outlet 1.

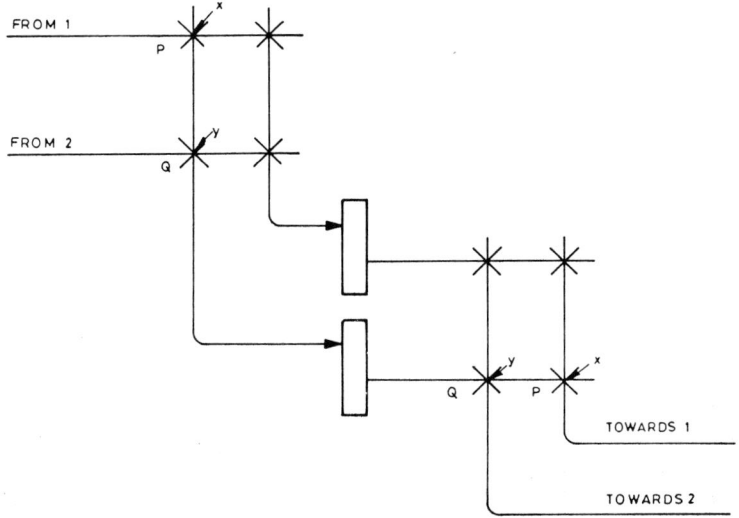

Figure 2.8 S-T-S switching.

When switching exchanges become very large, the T-S-T and S-T-S switching structures can be modified by the addition of further switching stages. For example, this may lead to SSTSS or TSSST structures. Further details on these techniques are given in Ericsson (2), Inose (3), and Ronayne (17).

In some of the switching literature, T-S-T switching is referred to as M-G-M type since it consists of memory stages (M) and a gate stage switching matrix (G). Likewise, S-T-S may be referred to as G-M-G.

2.1.6 Exchange Congestion

In the design of any telephone switching network, it is necessary to estimate, using statistical methods, the number of circuits and the amount of switching and control equipment that must be provided. Sufficient speech paths and control equipment have to be provided through appropriate parts of the network to carry the expected number of calls simultaneously present under peak-traffic conditions.

If during the process of establishing a call, it occurs that there are more calls than the necessary equipment to handle them, then the call attempt may not succeed. It is referred to as blocked or lost.

The *blocking probability* that a customers' incoming call will be blocked under specified conditions of traffic is used as a basis for estimating how much switching and transmission equipment must be provided. In some countries, an alternative measure used is the *grade of service*. This is defined as the percentage of calls lost during the busiest hour of the day.

Statistical measures used to characterize a call attempt include the *time of arrival* and the *service time* of the call. The average load on the system is referred to as the *traffic*. This is defined as the product of the *average arrival rate* C (calls per hour) and the *average service time* H (hours). It is given the unit of *erlangs* (E).

That is, we can write the traffic as

$$A = C H \qquad (E). \qquad (2.1)$$

The traffic (A) is measured by multiplying the number of calls which the circuits carry in one hour by the average duration in hours of the calls. It is obvious that, in practice, $A < 1$ (E) for a single circuit.

For example, if the average call duration is $H = 0.05$ hours (three minutes) then 1 erlang of traffic represents $C = 20$ calls per hour. The calling rate of a subscriber's line typically varies from 0.02 (E) for domestic areas to 0.2 (E) for heavily used business lines.

Within the switching network, the amount of equipment provided is typically designed to provide a grade of service of 1 in 100. That is, there will be a blocking probability of about 0.01 that call attempts will be blocked. Various design procedures have been developed to estimate the amount of equipment required to ensure this grade of service. See, for example, Inose (3).

Digital switching techniques allow the economic design of switching systems with extremely low blocking probability. Because time-division switching is involved, the switches for a particular call need only be operated for the corresponding incoming and out-going time slots. For the remainder of the time, they can be available to switch calls in other time slots.

In T-S-T digital switching networks, the blocking probability is dependent on whether a pair of vacant time slots can be found for the transmission between the two time switches. In practice, it is possible to obtain complete freedom from blocking by such methods as doubling the internal bit rate or building the S-switch in two parallel planes. See, for example, Ericsson (2) or Inose (3) for details.

The blocking probability of the S-T-S switching network depends on the probability of finding a free memory cell in one of the centralized time switches. The more time-switching memory elements there are, the lower the blocking. Zero blocking can be achieved by ensuring that the number of time switches equals double the number of inlet/outlets minus one. See Ericsson (2) for details.

Very large switching exchanges can be more economically constructed using digital techniques than conventional analogue techniques. For example, a 100,000 line switching network (50,000 incoming and 50,000 outgoing lines) can be readily accommodated by one digital group selector and associated concentrators.

2.2 NETWORK SYNCHRONIZATION

In a digital telephone network, the digital exchanges inject digital PCM signals into the transmission network. These are transmitted through the digital TDM systems, may be multiplexed into higher-order systems, and may be switched onto new routes in one or more other exchanges before the signals reach the called subscribers.

Each exchange has its own clock which determines the rate at which bits are stored into receiving buffers, switched, and sent out into other routes. Each exchange may receive incoming bit streams at a variety of symbol rates from other exchanges. It is evident that it is necessary to ensure that the incoming bit rates to a given exchange and the rate of the exchange clock must have the same long-term mean value. Otherwise, bits will be lost or bits will be duplicated.

In the following sections, the problems associated with a network with geographically separated clocks is examined. Several network synchronization methods will be reviewed. *Network synchronization* is a collective expression for those measures that aim at bringing about and maintaining a common bit rate for all digital exchanges.

2.2.1 Synchronization Requirements—Slips

If each exchange clock operated independently, then for any practical clock design, their bit rates would ultimately differ. This fact, combined with other problems such as jitter and variations in transmission delays between terminals can give rise to events known as *slips*. A slip is defined as the irretrievable loss or gain of a digit or a set of consecutive digits in a digital signal.

Consider an exchange with its own internal clock. Signals are being continuously received from other exchanges with their own clocks. Incoming pulses from other exchanges are typically written into an input buffer having the capacity of one PCM frame. The signals are read into the buffers under the control of the clock signals which are recovered from the incoming signals by the terminal regenerators. The contents of the buffers are read out under the control of the local exchange clock. This is illustrated in Figure 2.9.

Two cases that cause slips can be considered. The first is the case where the incoming bit rate r_i (the write-in clock) is slightly higher than the exchange clock rate r_o and the second is the case where the reverse is true.

Case (1) $r_i > r_o$

Figure 2.10(a) illustrates the case where the write-in clock pulses from the remote exchange occur at a rate slightly exceeding the read-out clock pulses of the local exchange. The read clock eventually falls so far behind that two input writes occur before the next read interval as illustrated for input symbol 5. As a result,

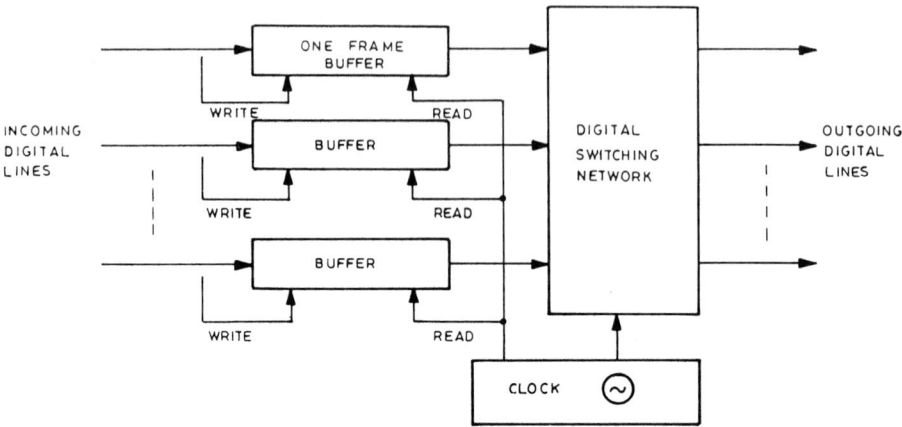

Figure 2.9 Switching exchange interfaces.

one bit extra is accumulated in the buffer. Over a long period, this effect will accumulate so that eventually buffer overflow occurs and one or more bits are lost.

Case (2) $r_i < r_o$

Figure 2.10(b) illustrates the case when the external write clock rate frequency is slightly less than the internal read clock frequency. As shown in the diagram,

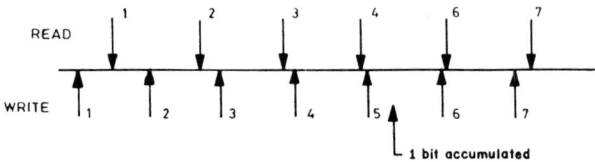

(a) WRITE CLOCK RATE > READ CLOCK RATE

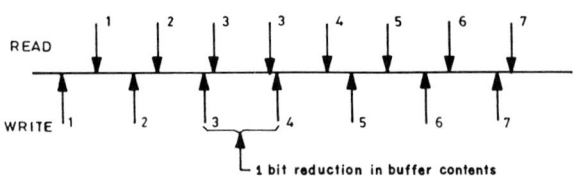

(b) WRITE CLOCK RATE < READ CLOCK RATE

Figure 2.10 Effect of different clocks.

Network Synchronization

after an input symbol is read into the buffer, two read pulses may occur before the next input pulse. This is illustrated by the two read pulses occurring after input symbol 3 in Figure 2.10(b). When this occurs, the buffer contents are reduced by one bit. Eventually the buffer may become empty.

When the buffer overflows or when it becomes empty, bits may be lost or duplicated. These events are referred to as slips. The input buffers are also referred to as *elastic stores*. It is the function of the elastic stores to avoid the occurrence of slips.

Slips that do occur because of either buffer overflow or underflow can result in signal distortion. The effects may be, for example:

(1) clicks in PCM transmission (slips are found to give rise to audible clicks on about 1 occasion in 25)
(2) misrouting if an error is caused to the signalling bit stream (parity-check error detection techniques usually protect against this)
(3) errors in data transmission (the severity being dependent on the transmission protocol error-handling procedures.)

The CCITT recommends (Rec. G.822) minimum objectives for network synchronization in terms of slip intervals. The end-to-end slip rate of an international digital circuit of 27500 km in length should be less than 5 slips in 24 hours for more than 98.9% of the time. (This represents a mean slip interval of at least 4.8 hours.) Suggested slip rate limits for various services are as follows:

(1) for *speech* channels—0.01 hour slip interval
(2) for *data transmission* channels—for fixed block lengths (packet networks) the minimum objective is a one-hour slip interval. For variable block length systems, three-hour slip interval.
(3) for *facsimile transmission* without error control, six-hour slip intervals.

See CCITT Rec. G.822 (15) and Okima (7) for details.

2.2.2 Causes of Slips

The three most significant causes of slips are:

(1) *Imperfect clocks* of limited frequency accuracy and stability: The *accuracy* of a clock is the degree to which the frequency corresponds to the frequency of a primary standard. This can be expressed in terms of a frequency error. *Stability* is a measure of the degree to which a clock will produce the same frequency over a period of time. This can be described in terms of *short-term* and *long-term* drift rate as illustrated in Figure 2.11. Let $f_r(t)$ and $f_c(t)$

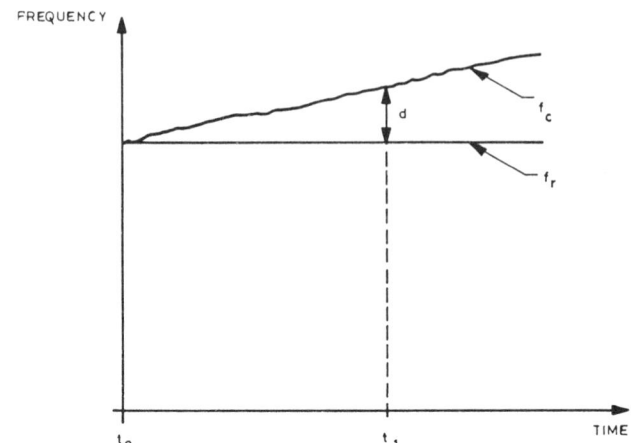

Figure 2.11 Clock accuracy and stability.

represent the variation of frequency with time for a primary reference standard and exchange clock, respectively. In the figure, it is assumed that $f_r(t)$ is constant, that is, the reference clock is ideal. Then we can define the *error* $A(t)$ of the exchange clock at any time t_1 as

$$A(t) = \frac{d}{f_r(t_1)} \qquad (2.2)$$

where $\quad d = |f_r(t_1) - f_c(t_1)|$.

Likewise, we can define the *long-term drift rate* (S) as

$$S = \frac{d}{f_r(t_1)} \cdot \frac{1}{t_1 - t_0} \qquad (2.3)$$

Typical values of long-term drift rate for various types of clocks are as follows:

Atomic clocks
(a) Cesium beam standard—most stable and expensive of all clocks $S \approx 10^{-13}$ per day.
(b) Rubidium vapour cell—less costly and smaller than cesium clocks $S \approx 10^{-12}$ per day.

Quartz crystal oscillators (including digital watches)
Simple, reliable, and economical clocks. Long-term stability $S \approx 10^{-9}$ per day. These are popular for networks because they are inexpensive and the clock frequency can be adjusted by controlling the current flowing through the crystal.

Exercise 2.1

Consider two switching centers exchanging bits at a rate of 2 Mbit/s. Assume the use of 8-bit input buffers to attempt to eliminate slips. Further, assume that the buffers are initially set to half full. If a slip occurs because of buffer overflow or emptying, the buffer contents are reset to half full by appropriate inhibiting of read clocks. Determine the frequency of occurrence of slips for two cases. In one case, consider the use of atomic clocks and assume a fixed frequency error of 10^{-12}. As a second case, consider the use of crystal oscillators with an assumed error of 10^{-7}.

Solution. The timing differences t_d, expressed in bits accumulated during one day of operation are calculated as follows:

Atomic clock

$$t_d = 2 \times 10^6 \times 10^{-12} \text{ bit/s}$$
$$= 2 \times 10^{-6} \times 60 \times 60 \times 24 = 0.17 \text{ bit/day}.$$

Crystal clock

$$t_d = 2 \times 10^6 \times 10^{-7} \times 60 \times 60 \times 24 = 17{,}280 \text{ bit/day}.$$

Since an 8-bit buffer is used, the number of slips per day will be $t_d/4$. That is, the number of slips will be approximately 0.04 and 4320 (slips per day) for the atomic and crystal clocks respectively.

(2) *Transmission channel delay variations*—The pulse propagation characteristics of cable, terrestrial radio, and satellite systems are subject to change with time. For example, an increase in ambient temperature may result in longer propagation times in a cable system because the cable has a longer electrical length.

As the temperature increases, the phase delay (expressed in bits) increases. Even if the terminal clocks are in synchronism, this delay may become large enough to cause a slip because the receiver buffer empties. As an example, for a 500 km cable link with 20°C temperature variation, the resultant transmission delays may be approximately as follows.

(a) 60 bits for a 2 Mbit/s system over paper-insulated cable pairs.

(b) 240 bits for an 8 Mbit/s system over paper-insulated cable pairs.

(c) 20 bits for an 8 Mbit/s system over coaxial cable.

(3) *Jitter*—Jitter or fluctuations in bit arrival times caused by regenerators and other devices in the link can also give rise to significant delay variations. As discussed in Chapters 3 and 4 of Volume 1, a very low frequency form

of jitter known as *baseline wander* can occur in digital transmission systems, typically at rates below 0.01 Hz.

The effects of channel delay variations and jitter can be taken care of in the input buffers (elastic stores.) However, the problem of different rate clocks requires additional remedial measures. We will examine the use of elastic stores and clock synchronization in the next section.

2.2.3 Approaches to Network Synchronization

Different network synchronization approaches can be taken to achieve slip control. The most appropriate technique depends on the state of development of the network, its geographical configuration, and its complexity. In most countries, digital networks primarily intended for voice communication are developing in stages. As the network develops, so the synchronization techniques may need to change. A typical scenario might be as follows:

Stage 1: Consider the introduction of primary-PCM junction links between exchanges in an otherwise analogue network as illustrated in Figure 2.12(a). In this case, the transmit end of the PCM terminals can use its own clock and point-to-point synchronization is used. At the receiving end, the clock is recovered from the received line waveform as discussed in Chapter 4, Volume 1. The clock at the receiver terminal is locked to the sending end and slips do not occur. This is known as independent point-to-point synchronization. For independent point-to-point synchronization, quartz oscillator clocks at each transmitting terminal can provide sufficient accuracy.

Stage 2: When the network develops to the point when second or higher-order digital multiplexing is required, several lower-rate tributary bit streams from different sources may be multiplexed into a higher rate composite signal. For example, four 2 Mbit/s primary PCM streams may be multiplexed into an 8-Mbit/s secondary stream as illustrated in Figure 2.12(b). Then a *pulse stuffing* technique in conjunction with elastic stores is effective for multiplexing the four asynchronous tributary signals. This is also known as *justification*, a term deriving from printing where spaces are incorporated in or removed from text to ensure right and left margin alignment.

Figure 2.13(a) illustrates second-order multiplexing. Figure 2.13(b) shows the second-order frame structure recommended by CCITT and Figure 2.13(c) shows how pulse stuffing is achieved.

Consider the multiplexer of Figure 2.13(a) with inputs from four 2 Mbit/s regenerators. The functions can be summarized as follows:

Network Synchronization

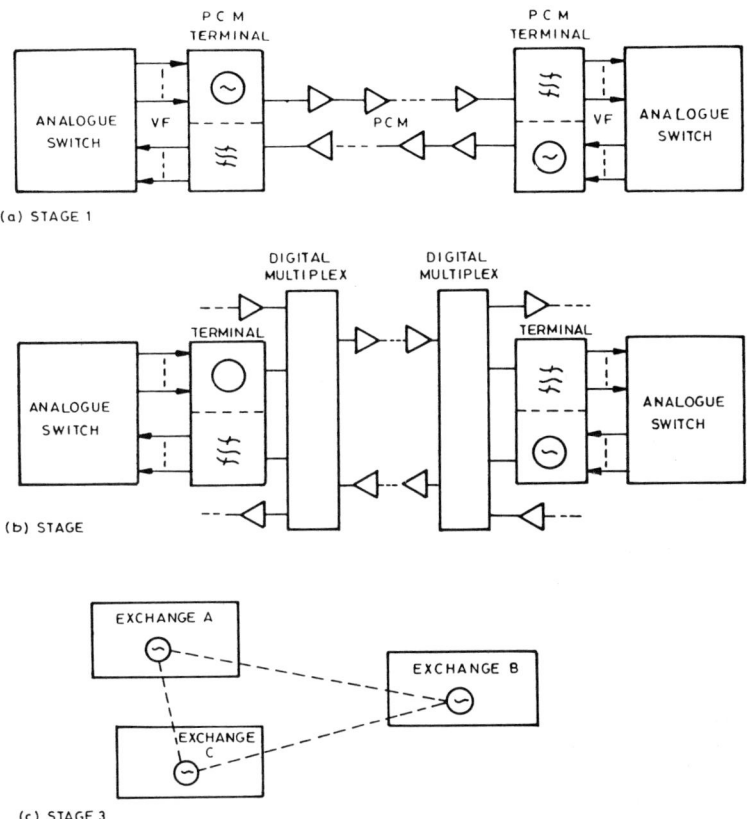

Figure 2.12 Evolution of digital network synchronization.

(1) The *code translator* converts the HDB3 incoming signal into binary format and extracts the timing information from the receiving terminal.

(2) Each binary signal is read into the elastic store. This is arranged as a first-in first-out (FIFO) buffer. The bit stream is read in at the rate of the clock recovered from the received bit stream. This is illustrated in Figure 2.13(c).

(3) The common control unit reads a bit out of each of the four elastic stores in turn. Each store is read with a "read" clock of frequency f_r. This frequency is arranged to be nominally 8448 kHz but with

$$f_r > \sum_{i=1}^{4} f_w(i)$$

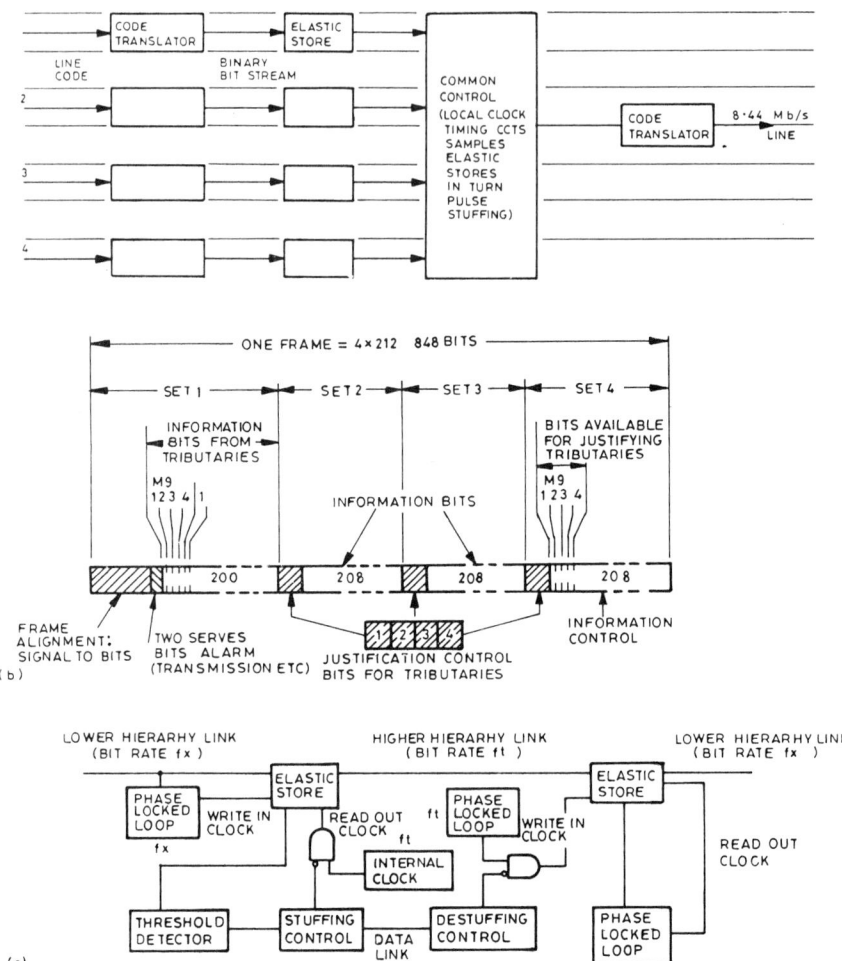

Figure 2.13 Pulse stuffing in second-order multiplexing.

where $f_w(i)$ represents the rate of the ith tributary write clock. The inequality is arranged to have a difference margin sufficient to accommodate transmission delay variations and jitter.

Because of the frequency difference between write and read rates, the number of stored bits in the elastic store may be gradually reduced. The threshold detector in Figure 2.11(c) supervises this. When the number of stored pulses is reduced to a predetermined value, this is indicated to the control circuit. The control circuit then inhibits the readout clock for a one-bit period at the next instant at which it is possible to send a stuffing bit in the higher-order multiplexed sequence. This allows the one-bit period for the elastic store contents to be increased by one bit.

In the corresponding bit location in the multiplexed sequence, a justifying bit is inserted instead of an information bit. This location is in set 4 as shown in Figure 2.13(b). The secondary pulse frame of 848 bits is made up of four sets of 212 bits each. In set 4, there are four successive bit locations available for stuffing a bit (if necessary) for each of the four tributaries. If a justifying bit is to be inserted for the first tributary say, but not for the the other three, then an extra dummy 1 is inserted (stuffed) into the fifth bit location in set 4. No other bits are stuffed here so the information bits from tributaries 1–4 follow immediately after bit 5.

Identification of this justified bit slot (dummy bit) is required at the received terminal so that it can be removed. To indicate whether or not dummy bits are stuffed, justification control bits are used. There are 12 bits for this purpose, 4 in each of sets 2, 3, and 4.

The first bit in each of the three groups of justification control bits indicate whether the first bit is stuffed for tributary 1. If the pulse is stuffed, the three associated control bits are set to 111. Otherwise they are set to 000. Likewise the second bits in each of the three groups of justification control bits indicate whether or not the second stuffing bit is present and so on for the third and fourth stuffing bits. The 111 or 000 control sequences are code words in an elementary (3,1) repetition code chosen to provide error protection against transmission errors. More details of other justification procedures are given in (8) and (9).

Stage 3: When the network develops to the point where multiple digital exchanges (nodes) are interconnected by digital transmission lines, incoming signals on each line must be synchronized symbol-by-symbol and frame-by-frame. Then it is necessary to ensure that the clocks of all switching exchanges are synchronized sufficiently accurately to ensure slip rates are maintained within the specified limits. This requires much more sophisticated clock synchronization techniques.

As illustrated in Figure 2.14, there are three classes of technique commonly used to synchronize the clocks of nodes in a network.

(1) A *plesiochronous* (or stable clock) system in which each exchange or switching node has a clock, and the clocks are sufficiently stable that they can be operated independently of each other.
(2) A *mutual synchronization* system in which each exchange has a variable frequency oscillator VFO which is phase locked to the average or some other function of the incoming clock phases.

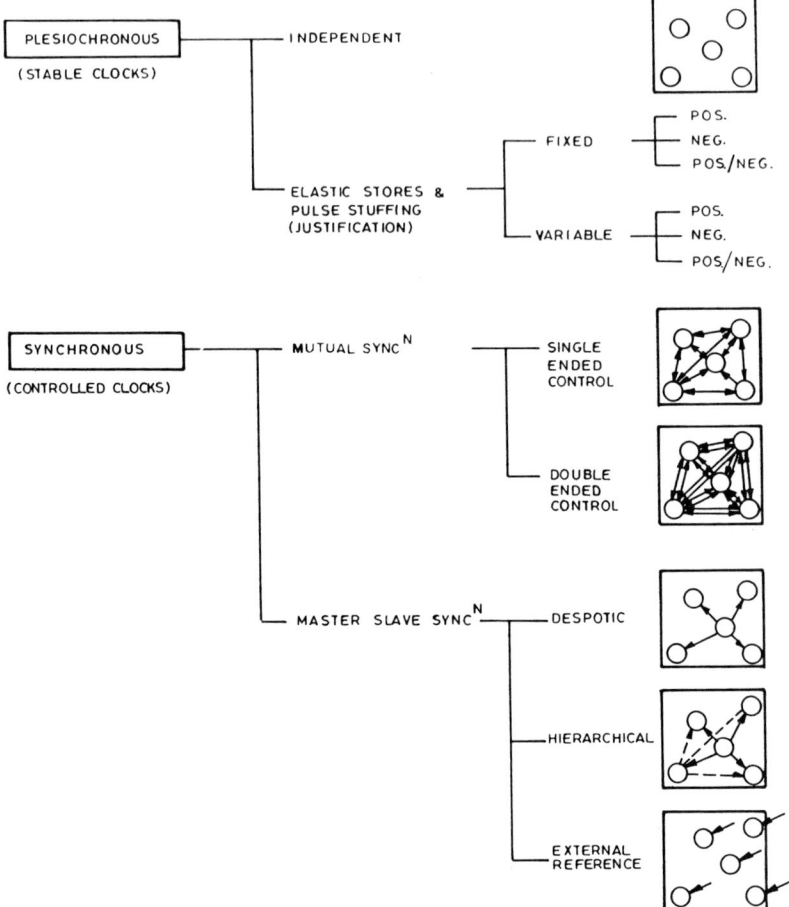

Figure 2.14 Network synchronization.

(3) A *master-slave* system in which a reference or master clock signal is distributed to each slave office via a timing distribution network with a tree-type topological structure.

The latter two systems are classed as *synchronous* networks techniques. They are essentially different from the plesiochronous type. Next we consider each of these schemes in more detail.

2.2.4 Plesiochronous Networks

The CCITT recommendation G.702 (6) describes two signals as plesiochronous if their corresponding "significant instants" occur at nominally the same rate,

Network Synchronization

any variation in rate being constrained within specified limits. The term *plesiochronous* is based on the Greek root "plesio," meaning "near."

Plesiochronous networks of independent stable clocks require the use of elastic buffers and pulse stuffing to minimize the impact of the occurrence of slips. In addition, clocks in the network have to be checked from time to time against some external reference. This is to compensate for limitations in long-term stability that increase slip rate with time.

For switching purposes, any stuffing pulses included in a frame must be removed before switching is performed. This is necessary, for example, in the case of trunk switching of second-order and higher-rate PCM signals where the pulse stuffing has been used in multiplexing lower rate PCM signals. In some cases, it may be simplest to perform switching exclusively on the primary group signals and to demultiplex higher-order signals before switching.

The allowed maximum error for each clock in a tandem connection of plesiochronous links (7) is given by (3)

$$A(t) = \frac{3}{2} \cdot \frac{T_f}{T_s} \cdot \frac{1}{N-1} \qquad (2.2)$$

where

T_f = frame interval
T_s = mean interval between slips
N = number of nodes.

Equation (2.2) is based on the premise that the average frequency error is approximately two-thirds of the maximum possible error summed over all $N-1$ links.

Exercise 2.2

If the maximum number of nodes envisaged in a digital connection with international links is 14, find the required frequency accuracy for each clock in a plesiochronous network. Assume a second order (8448 kbit/s) system is to be used for transmission.

Solution. CCITT Rec. G.822 (see Section 2.2.1) suggests the slip interval should satisfy

$$T \geq 0.5 \text{ hours.}$$

From Figure 2.13(b), the second-order multiplex frame size is 848 bits. Hence, the frame interval is

$$T_f = \frac{848}{8448 \times 10^3}$$
$$= 100.4 \ (\mu\text{sec}).$$

Then using Equation (2.2), the required accuracy for each clock must satisfy

$$A(t) = \leq \frac{3 \times 100.4 \times 10^{-6}}{2 \times 4.8 \times 60 \times 60 \times 13}$$
$$= 3.2 \times 10^{-9}.$$

For international links, CCITT Rec. G.811 suggests that the clock accuracy in plesiochronous operation should be better than 10^{-11}. See (14) for details. Such highly stable performance requires the use of atomic clocks.

2.2.5 Master-Slave Synchronization

Less expensive phase locked crystal oscillators can be used in a master-slave configuration sometimes also called *despotic* or *subjective* synchronization. Each slave exchange clock is phase-locked to the master clock. The level of the elastic buffer at the slave terminal can be used to indicate the phase difference between master and slave clocks. In master-slave synchronization systems, provision must be made for:

(1) a transmission network for distribution of a master reference clock to each slave node

(2) a control protocol for maintenance of timing in the event of transmission link failure.

According to how these requirements are implemented, various types of master-slave synchronization systems can be envisaged:

(1) *Hierarchical (self-control) system*—This method may be adopted for a meshed network of nodes. All exchange clocks are assigned a rank in a hierarchy. Each node receives the clock signal from one or more nodes of equal or higher ranking. This ranking assigns priorities to them and synchronizes with the first priority node if it is available. If the clock distribution system becomes faulty, the node automatically changes over to synchronization with the node with the second highest priority, and so on. There is no need to transfer control signals between nodes. In the worst case, if some nodes cannot receive any clock signal, then they have to operate in a plesiochronous mode.

(2) *Self-organizing system*—The mechanism of this system is to transmit control signals between nodes in order to automatically rearrange the network in

Network Synchronization

case of failure. A typical example is the Canadian Dataroute system. Three kinds of control data are continuously transmitted, namely:

(a) the destination of the node (D say), from which the clock is originally coming.

(b) the number of links between the node D, and the local node, and

(c) the destination of the previous intermediate node.

(3) *Hierarchical (self-protected) system*—This method is illustrated in Figure 2.15(a). It consists of two or more master clocks (Caesium oscillators) each

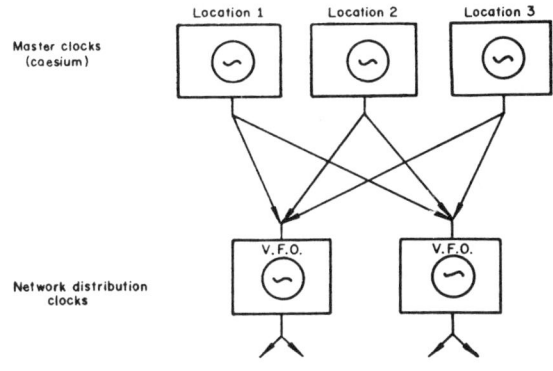

(a) SELF-PROTECTED SYSTEM

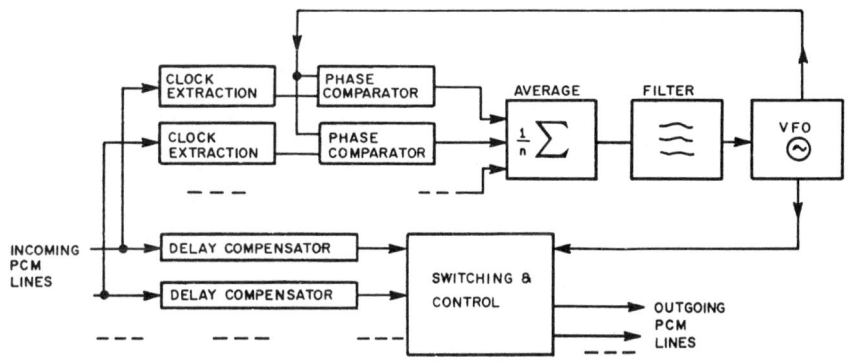

(b) AVERAGE PHASE LOCKED OSCILLATOR

Figure 2.15 Hierarchical synchronization.

connected to the next levels, called network distribution clocks. Each of the latter is a phase locked oscillator synchronized with the average phase of the masters.

Figure 2.15(b) illustrates a typical configuration for an average locked phase oscillator. Phase comparators are provided for each clock signal incoming from other connected nodes. The outputs of the phase comparators are averaged in a summer and passed through a low-pass filter. The output is used to control the frequency of the local variable-frequency oscillator (VFO). If a transmission path to a network distribution clock fails, the clock is still controlled by the remaining incoming master clock reference signals.

2.2.6 Mutual Synchronization

Methods have been developed for achieving a synchronous network without having to use a single master clock as in the master-slave systems. In master-slave synchronization systems with a single master, the major concerns are the reliability hazards associated with loss of the master clock or its transmission bearers to the slave stations. *Mutual synchronization* systems avoid this problem. Each exchange clock is mutually synchronized *to the average* of every other clock signal incoming to that exchange. The aim of mutual synchronization is to establish a unique clock frequency throughout the network through the interdependence of phase locked variable-frequency oscillators (VFO) in each of the exchanges. Each node clock frequency is dependent on each other node clock in the network.

The reference phase at each node can be determined by averaging all incoming clock phases using the average-phase VFO configuration of Figure 2.15(b). The system is a multiple feedback system with considerably increased complexity over simple master-slave systems.

One problem with this synchronization method is the difficulty brought about by the effects of delay variations on incoming transmission lines because of temperature changes. A *double-ended control* scheme of bilaterally connected nodes can be used to remedy this temperature dependence of the system clock frequency. Here the input to the VFO control circuit at node i say, is made up of the difference between:

(1) the single-ended average phase difference signal measured at node i
(2) the phase differences measured at other cooperating nodes and sent to node i

This double-ended control procedure is indicated by double arrows as shown in Figure 2.14. This method can ensure that the system operates in a node in which the network frequency is insensitive to delay variations. Further details are given in Inose (3).

2.2.7 Comparison of Synchronization Methods

Each of the network synchronization techniques has its own specific advantages and its own shortcomings. In any synchronous system of clock control, the network can be regarded as a large control system with mutual synchronization techniques being generally more complex than master-slave techniques. We now compare the synchronization techniques.

Plesiochronous synchronization systems:

(1) do not suffer from any stability constraints,
(2) are independent of the network topology, but
(3) for low slip rates, require expensive stable clocks.

For a synchronous network, the complexity of master slave synchronization is less than that of mutual synchronization with double-ended control.

The *master-slave* method:

(1) is easy to implement,
(2) has no stability problems,
(3) may involve reliability hazards because of dependence on a single master clock
(4) requires relatively high stability clocks so the system can survive (without serious slip rate) if the master fails, and
(5) is attractive for a network having few alternative routes, that is, a star network topology.

The *mutual synchronization* method:

(1) can provide improved reliability
(2) allows the use of clocks with lower stability than those required in master-slave networks
(3) requires relatively complex double-ended control for system stabilty and may, therefore, involve higher costs than master-slave systems
(4) requires system clock realignment methods to be used when a new clock is added to the network. Otherwise, any phase difference of the new clock propagates through the network and may cause some buffers to slip.

An optimum synchronization scheme may involve a combination of the different synchronization methods and takes advantage of the benefits offered by each. For example, it is normal for local concentrators to be controlled from their parent exchanges using the master-slave principle. In turn, the parent exchanges may be synchronized with other local exchanges in the individual networks using mutual synchronization. Alternatively, a hierarchical self-protected system could be used, as illustrated in Figure 2.15(a). In turn, national networks must be synchronized with international gateway exchanges. The latter are equipped with highly stable atomic clocks and operate plesiochronously.

2.3 FRAME SYNCHRONIZATION

2.3.1 Introduction

In the previous section, we examined one aspect of network synchronization, namely the procedures for ensuring that the exchange clocks are adjusted to ensure that symbol synchronization can be achieved without slips. Methods for recovering symbol synchronization from received baseband digital waveforms were discussed in Chapter 4, Volume 1.

Another level of synchronization is frame synchronization. This determines the end of one frame and the beginning of another. Frame synchronization techniques are required in most forms of digital transmission systems as in the following examples:

(1) in *30-channel PCM systems*, described in Chapter 1, a frame consists of 32 eight-bit time slots (Figure 2.16(a)). The first of these time slots (TS0) is allocated for framing purposes. The eight bits in this time slot are

        X0011011   for even frames, and
        X1AYYYYY   for odd frames

where the bits labelled X and Y are usually set to 1. (CCITT has not decided on the use of these bits and recommends that they be set at 1 until their use is agreed internationally. Bits Y are available for use for any national purpose but should be set to 1 for PCM signals crossing national boundaries.) The bit labelled A can be used as an alarm bit that is set when frame alignment is lost in the other direction. For details see CCITT Recommendation G.732 (13).

(2) In *24-channel* PCM D1 channel banks used in North America and Japan, a framing bit is inserted once in every frame in bit position 193 (the last bit position in each frame). The framing sequence is an alternating 1,0 pattern. Note that when the T1 digital line system is carrying only voice traffic, no information bits can sustain an alternating 1,0 pattern. Such a pattern represents a 4 kHz signal component which is rejected by the filtering in the PCM coder.

Frame Synchronization

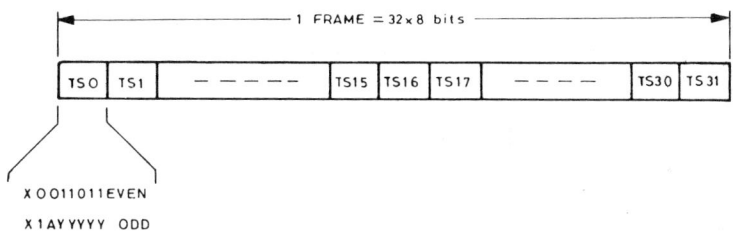

(a)   30-PCM FRAME ALIGNMENT SIGNAL

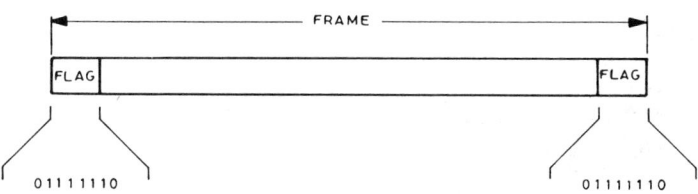

(b)   X.25 INFORMATION FRAME FLAG USED IN PACKET NETWORKS

Figure 2.16   Frame synchronization.

(3) In *British 24-channel PCM systems*, synchronization is achieved by combining four of the basic frames (192 bits each) into a multiframe of 768 bits. In the first and third frames, bit 1 of each word is used for channel signalling. In the second frame, its use is unspecified. In the fourth frame, bit 1 of words 9–24 contains the frame alignment pattern

$$1101010101010101$$

(4) In *higher-order multiplexed PCM signals*, a frame structure is used which contains bits allocated specifically for frame synchronization. For example, the 8.448 Mbit/s 120-channel secondary PCM multiplex frame illustrated in Figure 2.13(b) contains 848 bits. The first 10 bits are set to the pattern

$$1111010000$$

for frame synchronization purposes.

(5) In *packet-switched data networks*, a frame may be of variable length, as discussed in Chapter 7. Each frame starts and ends with flag sequence patterns

$$01111110$$

known as the opening and closing flags. These flags permit the receiver to establish frame synchronization and also to determine the number of bits present in the frame. This is illustrated in Figure 2.16(b).

2.3.2 Frame Alignment Systems

As can be seen from the examples above, most frame synchronization systems rely on the use of a specified digital sequence. Such a sequence is called a *frame alignment signal* or *flag*. The frame alignment signal can be one of two types, namely

(1) *bunched* when the signal elements occupy consecutive digit time slots, or
(2) *distributed*, when a specified pattern of symbols occupy nonconsecutive digit time slots.

The 30-channel PCM frame alignment signal described in (1) above is an example of a bunched frame alignment signal. The 24-channel PCM scheme of (2) uses a distributed frame alignment signal.

In the following, we will consider frame alignment techniques using bunched framing. Similar principles apply to distributed framing techniques. Consider a frame of length n bits, as illustrated in Figure 2.17(a). Let the first n bits of the frame be the bunched frame alignment signal or flag.

The fraction of time occupied by the m-bit flag is $\Delta = m/n$. In general, the design objective of a frame alignment scheme is to achieve a reliable system with minimum Δ.

A simple frame alignment system is illustrated in Figure 2.17(b). It consists of an n-bit frame store, an m-bit comparator as a flag detector and associated control circuitry.

Assume that the frame synchronization system is initially correct. That is, n bits of the received sequence can be shifted in to the frame store such that the bits appearing in the first m locations are the received flag bits. Then the m-bit comparator flag detector checks whether there is a match between the known flag sequence and the received m bits. Assuming the match is correct, the next $n - m$ bits are delivered to the output. The next n bits are shifted in to the frame store and the process repeats itself. When operating in this manner, the frame alignment system is said to be in the *LOCK mode* of operation.

If for some reason, frame synchronization is lost, then the frame alignment signals will not be received in their predicted positions. This will be detected by the comparator. The frame alignment system cannot deliver the $m - n$ information bits to the output but must begin to attempt to regain synchronization. It is then said to be in the *SEARCH mode*.

There are many possible methods of operating in the SEARCH mode. For example, the frame alignment system may move all bits one position to the left

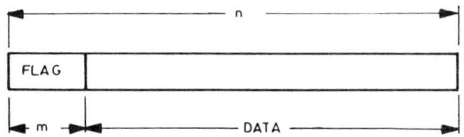

(a) GENERAL FRAME STRUCTURE

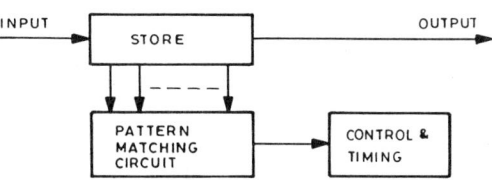

(b) ELEMENTARY FRAME SYNCHRONIZATION SYSTEM

Figure 2.17 Frame alignment system.

or right and recheck whether the first m bits in the frame store match the flag. If not, a further 1-bit sequence shift may be implemented and the process continued until a match is obtained.

When a match is found, the system returns to the LOCK mode. It delivers the last $m-n$ bits in the frame store to the output and then it proceeds to enter the next n bits into the frame store in the received sequence and to check the flag. If a match is found the process repeats itself. If not, the system proceeds to the SEARCH mode.

In summary, frame alignment systems must be designed to operate in either the LOCK or the SEARCH mode as determined by the flag detector output:

(1) In the *LOCK mode*, the receiver assumes it is in alignment. For each n new bits shifted in, $m-n$ bits are shifted out, providing a flag detector test succeeds. If it fails, then the system does not necessarily exit immediately from the LOCK mode, since a single channel error may have occurred. However, if the flag test fails consecutively a preset number of times, the system proceeds to the SEARCH mode.

(2) In the *SEARCH mode*, the receiver assumes it has lost frame alignment. It then initiates some search strategy until the first m bits in the frame store agree with the expected flag sequence.

There are three major problems that frame synchronization systems must be designed to deal with. They are as follows:

(1) *Channel errors* may corrupt the received frame alignment signal. If the frame synchronization system is in the LOCK mode, these errors may cause a transition to the SEARCH mode.
(2) *Slips* may result in a loss of bits or the addition of bits to the sequence. If the system is in the LOCK mode, slips will result in repetitive errors in the frame alignment detector tests, and hence, necessitate a transition to the SEARCH mode.
(3) *Information sequences* may happen to match the frame alignment signal. These will have no effect if the system is in the LOCK mode. However, if it is in the SEARCH mode, it may be caused to go into the LOCK mode but still remain out of alignment.

Note that in some systems, such as with CCITT X.25 packet network protocols, it is necessary to ensure that no information sequence matches the flag sequence. This is because the frame length is not fixed but may vary with the number of information bits to be transmitted. Simple methods are available for ensuring that the flag sequence pattern never occurs in the information bits. For example, if the flag is 01111110, this can be accomplished by inserting a zero bit after all information sequences of five consecutive ones at the transmit end, and by discarding any zero bit which directly follows five consecutive ones at the receiving end.

Frame alignment systems of the forward-acting and closed-loop types can be used (3). These systems are illustrated in Figure 2.18. Forward-acting framing systems, as in Figure 2.18(a), may be used when the frame alignment signal precedes the information sequences and steps are taken to preclude any occurrence of information bit patterns simulating the flag sequence.

A closed-loop framing system is illustrated in Figure 2.18(b). It consists of a pattern matching flag detector circuit, a protection circuit, and a frame counter. The protection circuit is required to achieve two functions:

(1) to prevent an erroneous transition from the LOCK to the SEARCH mode when individual channel errors occur in a frame alignment signal. For example, in second-order 8.448 Mbit/s PCM systems, loss of frame alignment is assumed to have taken place when four consecutive frame alignment signals have been incorrectly received in their predicted positions.
(2) to detect erroneous synchronization caused by an occasional occurrence of the frame alignment signal in the information sequence.

Frame Synchronization

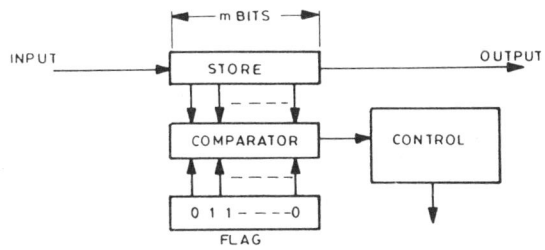

(a) OPEN LOOP SYSTEM

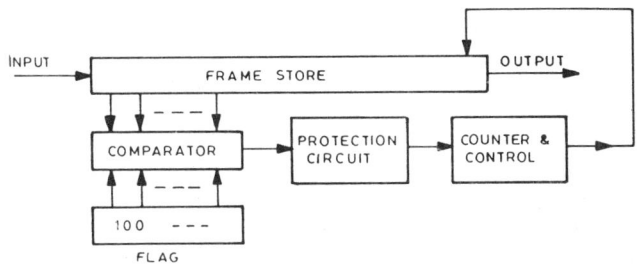

(b) CLOSED LOOP SYSTEM

Figure 2.18 Frame synchronization systems.

Exercise 2.3

Propose a scheme to carry out function (2) in the protection circuit.

Solution. This is left as an exercise for the reader. See for example Bylanski (12).

2.3.3 State Diagrams and Design Principles

State diagrams can be used to provide a diagrammatic description of frame alignment systems. These can be useful in establishing design principles or in comparing the performance of different frame alignment circuits. Figure 2.19 shows a state diagram for a typical frame synchronization scheme. It consists of four *states* S_1, S_2, S_3 and S_4. These states represent the possible conditions of the system at each time interval following a flag detection test:

(1) The system is in state S_1 if it is in alignment and in the LOCK mode.
(2) The system is in state S_2 if it is in alignment but in the SEARCH mode. This may have occurred because of channel errors corrupting the frame alignment signal.

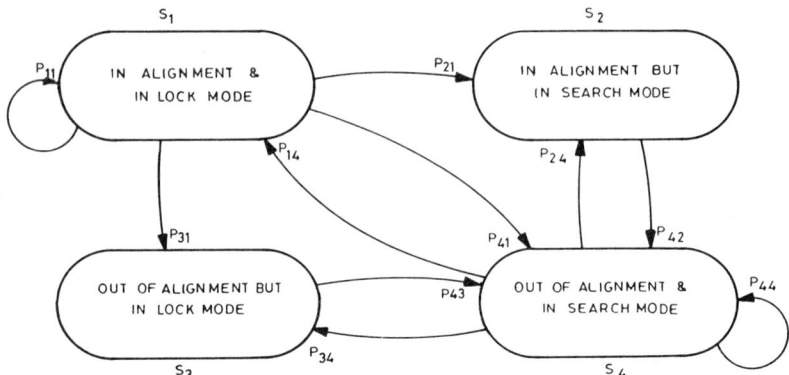

Figure 2.19 State diagram for a frame synchronizer.

(3) The system is in state S_3 if it is out of alignment but in the LOCK mode. This may have occurred as a result of an information sequence pattern simulating a flag.

(4) The system is in state S_4 if it is out of alignment and in the SEARCH mode. This may have been brought about because of a slip.

The state diagram also shows by means of arrows the possible transitions that can occur between states. Each has an associated transition probability p_{ij} which is the probability of that transition occurring between each flag detection test. That is, p_{ij} is the probability of the transition during each frame period where

p_{ij} = Prob (the system goes to state S_i given that it was in state S_j during the previous frame period).

These probabilities are functions of the design of the synchronization scheme and the nature of the sources of synchronization error. For example, the following interpretations may apply:

p_{21} = Prob (Channel errors in the flag | system previously in alignment and in LOCK)

p_{42} = Prob (the system goes looking for a flag | previously in state S_2)

p_{14} = Prob (found the flag and now back in alignment | previously in state S_4)

p_{34} = Prob (found a flag and changed to the LOCK mode but the flag was a spurious simulation by information bits | previously in state S_4)

p_{43} = Prob (subsequent checks on the flag showed an error so changed back to SEARCH mode | previously in state S_3)

Frame Synchronization

It is possible to analyze the operation of the synchronization system in terms of the transition probabilities. For example, assume the initial state of the system to be in alignment and in LOCK (state S_1). Let $P_k(i)$ represent the probability of the system being in state S_i just before the kth flag test.

The probabilities that the system will be in the various states at the $(k+1)$th test can be found using the matrix equation

$$\begin{bmatrix} P_{k+1}(1) \\ P_{k+1}(2) \\ P_{k+1}(3) \\ P_{k+1}(4) \end{bmatrix} = \begin{bmatrix} p_{11} & p_{12} & \cdots & p_{14} \\ p_{21} & & & \\ & & \cdots & \\ p_{41} & & \cdots & p_{44} \end{bmatrix} \begin{bmatrix} P_k(1) \\ P_k(2) \\ P_k(3) \\ P_k(4) \end{bmatrix} \quad (2.3)$$

with the initial condition

$$\begin{bmatrix} P_1(1) \\ P_1(2) \\ P_1(3) \\ P_1(4) \end{bmatrix} = \begin{bmatrix} 1 \\ 0 \\ 0 \\ 0 \end{bmatrix}. \quad (2.4)$$

It is possible to use these iterative relationships to analyze the system. We will not proceed further with this analysis. For details, see Bylanski (12).

However, it is easy to establish certain *design criteria* for a frame synchronization system. These would include the following:

(1) For *rapid realignment* maximize p_{14}, and minimize p_{34} to ensure that when the system is out of alignment and in the SEARCH mode, it will arrive back at state S_1 rapidly. This implies the use of a relatively long flag to reduce the probability of it being simulated by random information bits.
(2) To minimize the probability that alignment will be lost as a result of channel errors causing unnecessary SEARCH node operations, minimize p_{21} (and p_{24}). This could be achieved by using a *relatively short flag* so as to minimize the chance of errors occurring within it.

Obviously, there is a conflict between these two requirements. This conflict is often overcome in practice by the use of more complex strategies. These may include the following:

(1) the use of a more tolerant frame alignment test—for example, in 30 channel PCM systems, it is usually assumed that frame alignment has been lost when four consecutive frame alignment signals have been found to contain an error.
(2) the introduction of a new CHECK mode of operation—during the search for realignment, the flag test in (1) would be unnecessarily harsh and require testing over an excessively lengthy period. A CHECK mode can be used

such that if a frame alignment signal is located, it is provisionally assumed correct. However, if there is a failure in a flag test in the next three tests, for example, then this results in a resumption of the SEARCH mode.

Exercise 2.4

Draw a state diagram for a frame alignment system which incorporates the CHECK mode in addition to the LOCK and SEARCH modes of operation.

Solution. Two extra states are required, namely:

S_5: System in alignment and in the CHECK mode
S_6: System out of alignment and in the CHECK mode.

A state diagram is given in Figure 2.20.

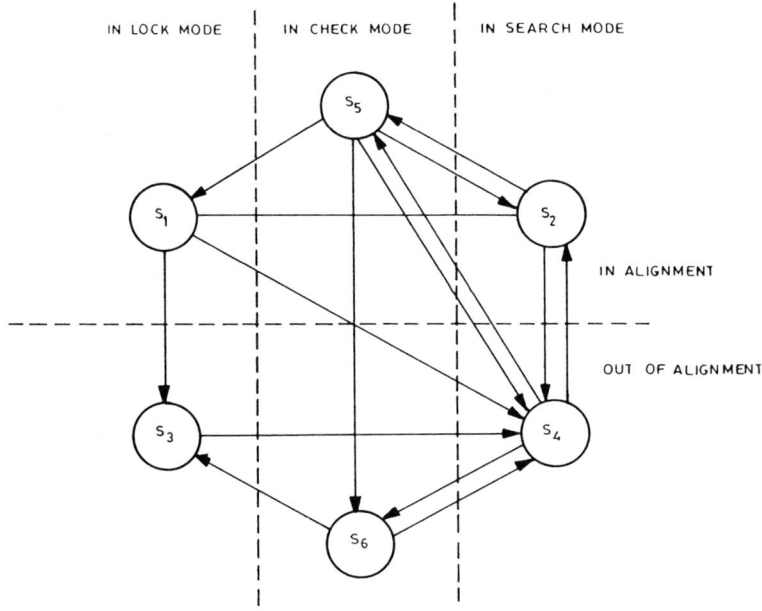

Figure 2.20 Six-state diagram for more complex frame alignment strategy.

Once the state diagrams for frame alignment systems are established, it is possible to analyze the system behavior in terms of assigned transition probabilities. For example, Bylanski (12) provides results for the two schemes outlined above. Results are computed in respect to the comparative performance of various misalignment detection systems and various search/realignment schemes.

Frame Synchronization

Other more sophisticated SEARCH mode schemes have been developed that can reduce the mean time to realignment. One method proceeds as follows. On detection of misalignment, the synchronization system examines the next n digits in the sequence where n is the length of one frame. Then the system records in a store, the locations where a match to the flag occurs. This is then repeated for the next n digits until successive comparsions reveal the location of the true flag.

2.3.4 Choice of Frame Alignment Signal

The choice of a suitable frame alignment signal or flag involves the selection of the flag length (m) and sequence pattern to be used. We previously discussed the factors which affect the choice of flag length. The choice of flag sequence pattern should be such that the probability of random data causing spurious simulation of a flag is minimized.

Consider the situation where the system has slipped out of alignment by only a small number of bits less than the flag length m. That is, the flag search test window overlaps part of the true flag. As a specific example, assume that synchronization is lost to the extent of one bit slip and an 8-bit flag is being used. This is illustrated in Figure 2.21(a) for the case where the flag pattern is

$$11111111 \ .$$

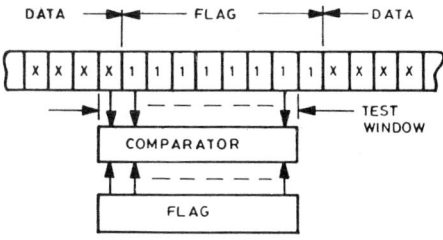

(a) THE USE OF AN UNSUITABLE FLAG 11111111

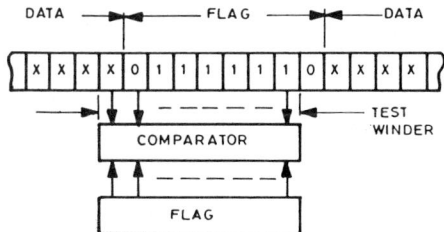

(b)   USE OF FLAG 01111110 AVOIDS SPURIOUS ALIGNMENT FOR $\leq$ m-1 SLIPS

Figure 2.21   Choice of frame alignment signal (flag).

For this case, it is easy to see that the probability of the adjacent data bit causing spurious simulation of the flag is 0.5. This will also be the case for the all-zero flag. Clearly, these are unsuitable patterns. However, if the pattern

01111110

is used, the probability of spurious flag simulation by random data is zero for up to $m-1$ time slips. This is illustrated in Figure 2.21(b).

Considerable research effort has been spent on the search for good flag patterns. The following patterns known as Barker codes have been found to have a high tolerance to spurious simulation that results in data and transmission errors.

110
1110010
11100010010 .

Other more extensive lists of good patterns are given in Bylanski (12).

2.4 PROBLEMS

2.1 Consider an hierarchical telephone switching network as illustrated in Figure 2.1(b). The network consists of remote switching stages (Concentrators) connected to local exchanges (Digital Group Selectors) by two Mbit/s primary PCM lines systems. It is argued that most of the "intelligence" in these systems can be located economically in the local exchange because of the capacity of the primary PCM systems to carry signalling information. Determine the average bit rate available for signalling using time slot 16 in each 2 Mbit/s primary PCM signal.

2.2 Explain why the switching exchange equipment for PCM digital telephony signals must perform a combination of "space" and "time" switching.

2.3 Explain with appropriate examples how rectangular crosspoint space switching arrays can be designed to be "non-blocking" or "graded," respectively.

2.4 Determine the number of memory elements N_M and the number of crosspoints N_x required for a TST digital switch with 32 input and output two Mbit/s lines. A nonblocking single stage space switch is to be used.

2.5 Consider the three-stage space switching network of Figure 2.5 consisting of N inlets, N outlets, N/n first and third-stage $n \times k$ switching arrays, and k second stage arrays.

(1) Show that for nonblocking operating, the number of second-stage switching arrays must be
$$k = 2n - 1$$

(2) Show that the total crosspoint count is given by
$$N_x = (2n-1)(2N + N^2/n^2)$$
(3) Show that the optimum value for n to minimize N_x is given by
$$n = (N/2)^{1/2}$$
for large N.

2.6 Determine the number of memory elements N_M and crosspoints N_x required for a TSSST digital telephony switch with the following characteristics:

Number of input and output PCM 2 Mbit/s lines = 5000 each
Number of channels per PCM frame = 30
Nonblocking 3-stage space switch to be used with $k = 10$ second-stage 100×100 switching arrays.

(Use the information given in Problem 2.5).

2.7 Consider two switching centers communicating via 2 Mbit/s PCM links. If the centers do not use elastic stores in their receiving terminals, determine the average number of slips per day for each PCM link if both centers use crystal clocks with an accuracy of 10^{-7}.

2.8 A certain copper pair has a coefficient of expansion of 17×10^{-6}/°C and a velocity of propagation of 2×10^8 M/s. Consider a 200 km long 2 Mbit/s primary PCM transmission system using a copper pair. Determine the change in the number of bits in the path if the temperature changes by 20°C over a period of two hours.

2.9 A method of pulse stuffing used in second order multiplexing is illustrated in Figure 2.13. If the 12 justification control bits are received as follows
$$1\ 0\ 0\ 1 \quad 1\ 0\ 0\ 1 \quad 1\ 0\ 0\ 0$$
explain how this pattern should be interpreted.

2.10 The 8-bit pattern used for frame synchronization in a certain PCM system is
$$1\ 0\ 0\ 1\ 1\ 0\ 1\ 1$$
(1) Determine the probability that a particular 8-bit segment of the random information signal may match the above frame alignment signal.
(2) Determine the probability that two successive specified 8-bit information segments will both match the frame alignment signal.

2.5 REFERENCES

1. J.G. Pierce, *Telecommunications Switching*, Plenum Press, 1981.
2. O. Borgstrom, B. Andersson, A. Marlevi, J. Anas and S. Braugenhart, *Digital telephony: An introduction*, L.M. Ericsson Telephone Company, Sweden, 1977.

3. H. Inose, *An introduction to digital integrated communication systems*, University of Tokyo Press, 1979.
4. J.E. Flood, *Telecommunication Networks*, Peter Peregrinus, 1977.
5. H.E. Vaughan, "Research model for time-separation integrated communication," *Bell System Tech. J.*, Vol. 38, No. 4, pp 909–932, July 1959.
6. CCITT, Recommendation No. G.702, "Vocabulary of Pulse Code Modulation (PCM) and Digital Transmission Terms," Geneva, 1972 and 1976.
7. K. Okimi and H. Fukinuki, "Master-Slave synchronization techniques," *IEEE Communications Magazine*, pp 12–21, May 1981.
8. A. Brinin and P. Frueh, *Transmission Systems Technology—PCM*, Telecom Australia Engineer Development Programme, Publication ED0027, Melbourne, 1980.
9. Bell Telephone Laboratories, *Transmission Systems for Communications*, Bell Telephone Laboratories, 4th edition, 1970.
10. R.D. Ramsay, "Network Synchronization Techniques—A Review," *Proc. Radio Research Board Symposium on Digital Communications*, Sydney, Dec. 1977.
11. H.S. Wragge, "The Effect of digital switching techniques on the form of telecommunication networks," *Proceedings of the I.R.E.E. Australia*, Nov. 1970.
12. P.B. Bylanski and D.G.W. Ingram, *Digital Transmission Systems*, Peter Peregrinus Ltd., 1980.
13. CCITT, Recommendation No. G.732, "Characteristics of Primary PCM Multiplex Equipment operating at 2048 kbit/s," Geneva 1972, amended 1976, 1980.
14. CCITT, Recommendation No. G.811, "Performance of Clocks suitable for Plesiochronous Operation of International Digital Links," Geneva 1976, amended 1980.
15. CCITT, Recommendation No. 822, "Controlled Slip Rate Objectives on an International Digital Connection," Geneva, 1980.
16. J. Bellamy, *Digital Telephony*, John Wiley, 1982.
17. J. Ronayne, *Introduction to Digital Communications Switching*, Pitman, 1986.

Chapter 3

COMPUTER NETWORKS

Prepared by Dr. Teresa Buczkowska, N.S.W. Inst. of Tech.

3.1 INTRODUCTION

The merging of computers and communications in recent years has had a significant effect on the way in which computer systems are organized and on the characteristics of communication networks. The term "computer networks" has come to be used to represent a system consisting of one or more computers with associated communication lines and terminals interconnected to provide service to a set of users.

This chapter provides an introduction to the principles of computer networks including network classification and models for network architecture. The characteristics of some public computer networks will also be discussed.

In view of the lack of any generally accepted interpretation of the terms

(1) computer networks
(2) computer communications network
(3) data communications

some clarification of these concepts will first be discussed.

A computer network may be viewed as an interconnection of computers and peripherals. The term *computer network* is defined very often as an interconnected collection of autonomous computers capable of exchanging information, or more precisely, as an interconnected group of independent computer systems which communicate with one another and share resources such as programs, data, hardware, or software. Both definitions exclude master/slave types of systems and computers used for handling communications or controlling terminals.

The term data communication was historically associated with a terminal-oriented network. A *data communication network* is a system which consists of at least one computer with remote terminals.

In the early stages of the development of computer networks, data communication systems were referred to as teleprocessing systems. In stages, the networks became more complex and included devices such as concentrators for handling communications and equipment such as terminal control units for controlling terminals. These systems with some distributed intelligence are sometimes referred to as decentralized computer networks.

The term *computer communication* refers to the means by which computers are able to exchange information. Primarily it has come to refer to networks in which a packet switching technique is used.

There is another term which is used in asssociation with computer networks, namely a *distributed system*. There is considerable confusion in the literature about the distinction between a computer network and a distributed system. One of the best known definitions of a distributed system was suggested by Enslow in 1978 (1), namely, a system-wide operating system with services requested by name and not by location. In the same year, however, Liebowitz and Carson (2) gave a much more general interpretation of a distributed system as one in which the computing functions are dispersed among several physical computing elements.

In this chapter the term *computer network* will be used exclusively since the purpose is to emphasize the principles of network architectures and structures, and the communications procedures.

3.2 CLASSIFICATION OF COMPUTER NETWORKS

3.2.1 Terminology

As discussed above, a computer network is a set of one or more computers, communication lines, and terminals interconnected to provide service to a set of users. A *node* is a computer system that is attached to, or part of, a computer network. A *switching node* is a node whose primary function is switching data in the network. A *host* is a node which has primary functions other than switching data. A *homogeneous network* is one consisting of physically and logically identical processors which are capable of executing copies of the same software. A *heterogeneous network* is one that is not homogeneous. *Terminals* are interfaces between the users and the computer or network.

3.2.2 Network Classification

The networks used for computer communication may be classified in many ways. Perhaps the best taxonomy of data networks was given by Chou (3). This results from deriving all possible subsets of networks by examining their architectures. It is illustrated in Figure 3.1. As shown, this classification scheme subdivides networks into *architectural* groups according to their

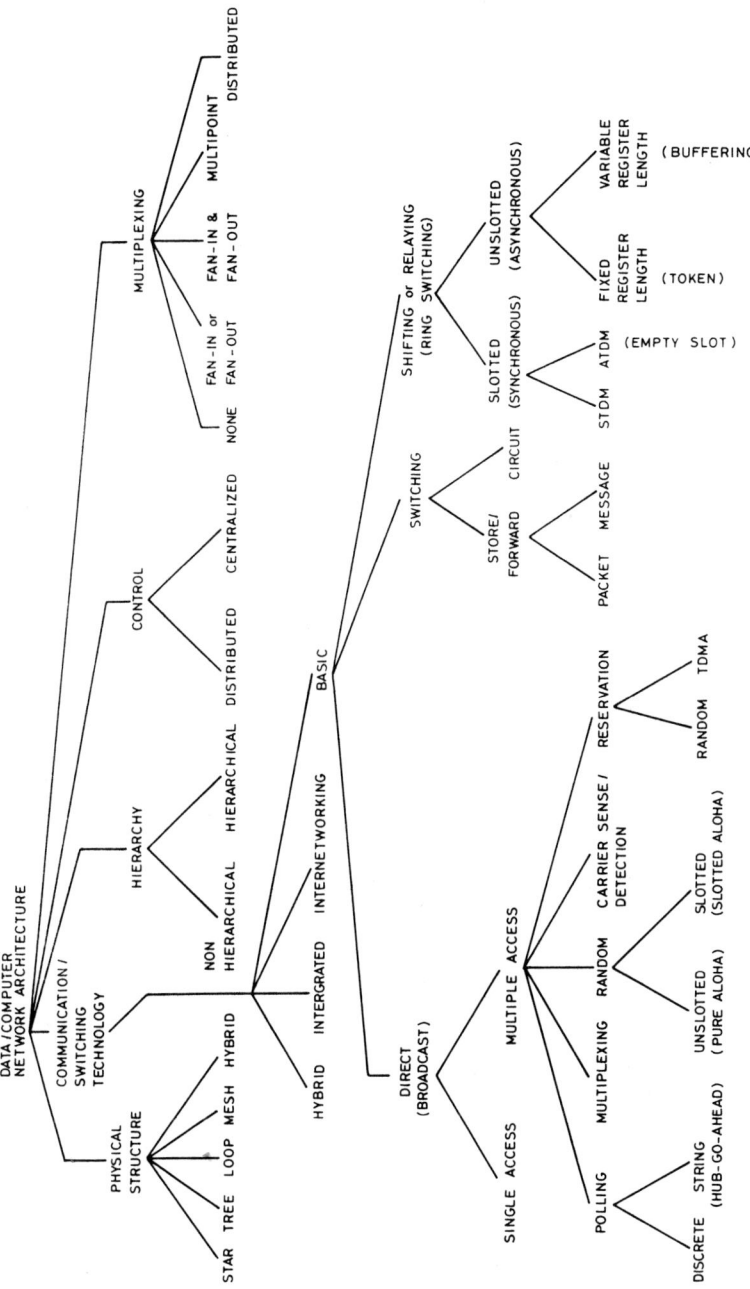

Figure 3.1 Taxonomy of data networks.

(1) physical structure
(2) communication/switching technology
(3) hierarchy
(4) control
(5) multiplexing arrangements

The classification shown does not provide for the division of data networks according to their functions and operations.

From the *functional* point of view networks could be classified as:

(1) terminal oriented remote-access networks
(2) value-added networks
(3) problem oriented networks (business oriented, process control, scientific, engineering, time sharing, information management networks and so on.)

From the *operational* point of view, networks could be classified according to whether they use

(1) a deterministic algorithm (flooding, fixed, split traffic or ideal observer network)
(2) a stochastic algorithm (random, isolated, or distributed network)
(3) a flow control algorithm (isarithmic, buffer allocation or special route assignment).

Many of these terms will be explained further in this chapter. However it would be impractical to discuss all possible subsets of networks in this limited discussion. We will restrict ourselves to a description of the four basic subsets indicated in Table 3.1. Then special attention will be given to a *layered architecture* description based on an International Standards Organization (ISO) model.

**Table 3.1
Computer Network Classifications**

Physical Structure	Switching Techniques	Access Techniques	Control
Star	Circuit	Single	Centralized
Tree	Message		
		Multiple	Distributed
Loop	Packet		
Mesh			
Hybrid			

3.3 COMPUTER NETWORK STRUCTURES

3.3.1 Data Networks

We preface our study of computer networks by reviewing their predecessors, data communication networks.

The simplest connection of terminals to a computer is by point-to-point lines forming a structure known as a *star configuration*. This is illustrated in Figure 3.2(a). This type of structure may be expensive in terms of cost of lines, line interfacing devices, and input/output ports of the computer.

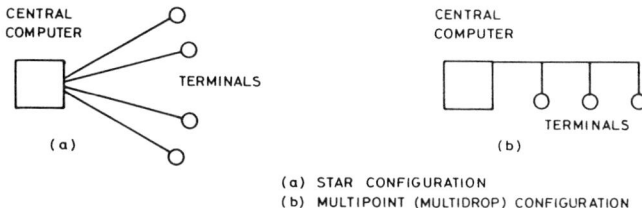

(a) STAR CONFIGURATION
(b) MULTIPOINT (MULTIDROP) CONFIGURATION

Figure 3.2 Point-to-point network structures.

Cost considerations resulted in the concept of sharing one line amongst a number of terminals. One implementation of this concept results in a structure referred to as a *multidrop* or *multipoint* configuration. It is illustrated in Figure 3.2(b). This configuration reduces the number of line interfacing devices and the number of input/output ports of the computer, but lowers the reliability and requires special control of the flow of data. The sharing of one line may also be achieved by connecting the terminals to the computer through a multiplexer.

Both structures illustrated in Figure 3.2 belong to the group of non-hierarchical centrally controlled data communication networks. This is in contrast with an hierarchical architecture where terminals are connected to concentrators. These in turn are connected point-to-point or multipoint to the computer. In hierarchical architectures, the concentrators are controlled by the computer, but the terminals are controlled by concentrators. The concentrator may perform many functions including:

(1) polling (selecting terminals)
(2) error checking
(3) code and/or speed conversion
(4) local data base accessing
(5) local processing
(6) local switching

The structural characteristics of such centrally controlled hierarchical networks are illustrated in Figure 3.3. The networks in both Figures 3.2 and 3.3 are examples of data communication networks.

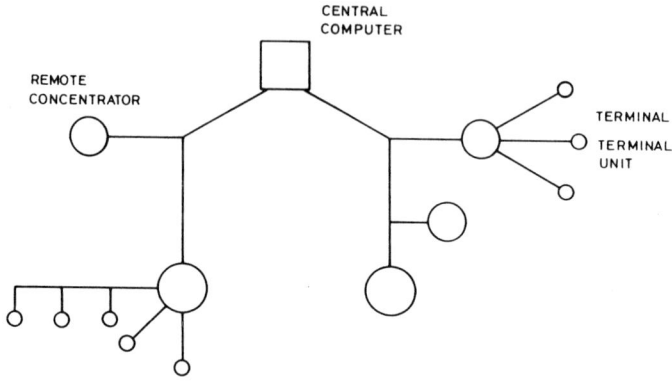

Figure 3.3 Centrally controlled network—tree structure.

3.3.2 Computer Networks

Towards the end of the 1960s a new concept in network topologies was introduced. This occurred as a result of an urgent need for access to the many distributed data bases and different processing services residing in geographically scattered computers. Networks were developed in which a number of interacting computers were interconnected. We refer to these as computer networks.

Computer networks can have many different forms. They may have the physical structure of either a fully or a not-fully connected mesh. They may exhibit layered structural characteristics, the number and function of each layer differing from network to network. Terminals may be connected to computers by point-to-point or by multipoint techniques, through multiplexers or through concentrators.

A computer network is illustrated in Figure 3.4. It consists of a number of host computers and associated terminals and concentrators. The hosts are interconnected via a communication subnetwork consisting of a number of switching nodes and transmission links.

3.3.3 Circuit, Message, and Packet Switching

Communication between any arbitrary pair of end-users of a computer network is performed by establishing a communication path through the switching nodes of the communications subnetwork. The task of this subnetwork is to carry

Computer Network Structures

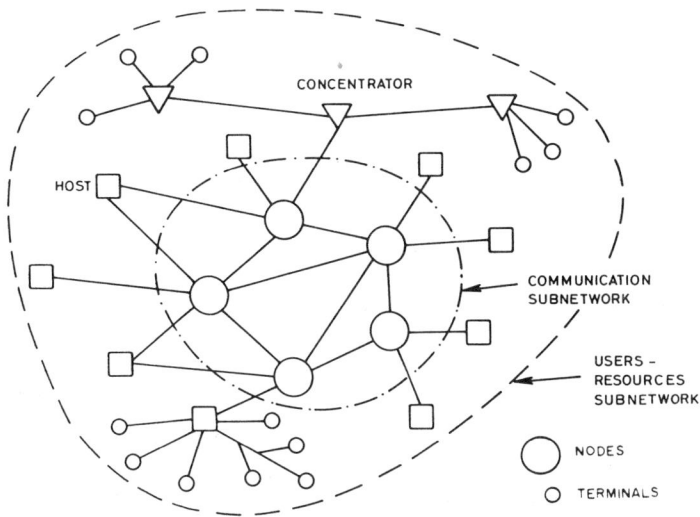

Figure 3.4 Layered structure computer network.

messages form host computer to host computer. For this purpose, computer networks may use alternative types of switching techniques, namely

(1) circuit switching, or
(2) store-and-forward switching, which includes
   message switching
   packet switching

In circuit switching, a dedicated communication path is established between two hosts through the nodes of the communication subnetwork for the duration of the data communication. An end-to-end path must be set up before any data can be sent. The most common example of the use of circuit switching is in the conventional telephone system. In a circuit switched computer network, each node is an electronic or electromechanical switching device which passes on message bits as fast as it receives them.

Store-and-forward techniques represent a very different approach. They take advantage of the fact that it is not necessary to maintain a dedicated path between the two communicating hosts for the duration of the message. In store-and-forward switching, the transmitting station appends an address to each message or to portions of a message that it wishes to transmit. The message or portion of a message is then sent to the first switching node, and then later passed from node to node through the communication subnetwork, one hop at a time. Each node is typically a mini- or microcomputer system with sufficient storage to store messages as they come in and subsequently, it transmits them to the next node.

Store-and-forward switching can be performed by one of two techniques, known as message switching and packet switching. With message switching, there is no limit on the size of the block of data or message being sent from node to node. A single message requires the establishment of a node-to-node link for the duration of the message.

In packet switching, data is formed into blocks known as packets. A tight upper limit is placed on packet size so that the data information can be readily stored in the switching computers (switching nodes) and efficiently routed to the next node as soon as a suitable path is available.

Store-and-forward switching techniques have a number of advantages as compared to circuit switching. These may be summarized as follows:

(1) A message/packet input from one terminal or host computer can be broadcast to a number of other hosts
(2) A message can be accepted by the switching and transmission subnetwork irrespective of whether the addressed station is available or not at that time
(3) Switching nodes can perform many functions including error detection and correction, editing, message duplication, message protection, code and speed conversion, handling of different communications protocols, flow control and routing, monitoring the network's status, and more.

In addition to having the above advantages, packet switching is more suitable for interactive real-time applications then message switching. The use of packet-switching offers the possibility of shorter network transmission delays and more effective channel utilization for bursty data requiring low or moderate transfer throughput. For these reasons, computer networks usually use packet switching.

3.4 THE ISO ARCHITECTURAL MODEL FOR OPEN SYSTEMS INTERCONNECTION (OSI)

Computer network design involves a complex amalgamation of protocols, network structures and interfaces. To ensure a logical and comprehensive design approach, it has become a standard procedure to organize the network description in terms of a series of layers or levels. The function and structure of these layers may vary from network to network. The set of layers and protocols is called the *network architecture*. The description of the network architecture is sometimes referred to as an architectural model.

The ISO Architectural Model for Open Systems Interconnection (OSI)

The International Standards Organization (ISO) has developed a layered computer network model which has become widely accepted as a standard. It is called the *Reference Model for Open Systems Interconnection (OSI)*.

The Open Systems model is viewed as logically composed of a succession of layers, each layer embedding the lower layers and isolating them from the higher layers. The basic principles of the structuring technique referred to as layering are:

(1) each layer adds value to services provided by the set of lower layers
(2) each layer defines services provided to the next higher layer, independent of how these services are performed between the layers

The ISO model defines seven layers (or levels) as illustrated in Figure 3.5. These are:

The Physical Layer (lowest layer)
The Data Link Layer
The Network layer
The Transport Layer
The Session Layer
The Presentation Layer
The Application Layer (highest layer)

We will examine the characteristics of each of these layers to examine the functions envisaged for them. A given layer on one device in the network carries

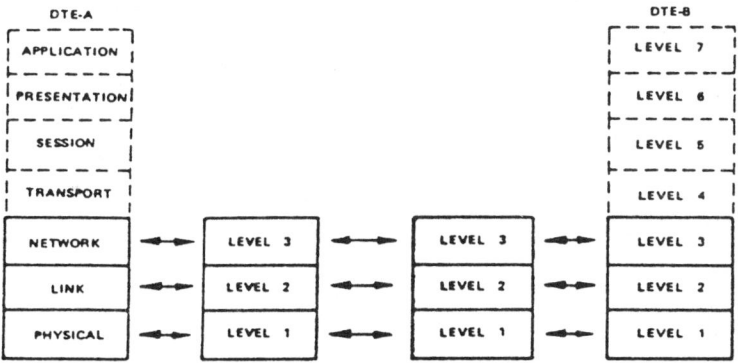

Figure 3.5 Seven-layer ISO Model.

on a conversation with the same layer on another device. The rules and conventions used in this exchange are collectively known as the layer *protocol*. In the following sections, we will examine the main features of each of the layers.

3.5 THE PHYSICAL LAYER

Level 1 of the OSI model specifies the physical interface between the devices. It covers the rules by which bits are passed from one device to another across physical connections. A physical connection may be permanent or dynamically established and released, and may allow half-duplex or full-duplex transmission of a bit stream in serial or parallel synchronous or asynchronous modes. Therefore, the physical level provides the set of mechanical, electrical, functional, and procedural services to establish, maintain, and disconnect circuits between data terminal equipments (DTE), data circuit-terminating equipment (DCE), and data switching exchanges (DSE) or nodes. As discussed in Chapter 1, Volume 1, the DTE (in CCITT nomenclature) represents the users computer terminal equipment. The DCE represents the user's interface to the communication subnetwork.

The various interfaces and their characteristics are described by various standards published by the International Standard Organization (ISO), by the International Telegraph and Telephone Consultative Committee (CCITT) and by the Electronic Industries Association (EIA), the latter body representing the U.S. electronic industry.

The *mechanical* aspects of an interface include the specifications of the physical connector, (typically a plug-in type) including the assignment of interchange circuits to pins of the connector, the description of its dimensions, and the mounting arrangements at the point of demarcation. There are five connectors that have been standardized by ISO. They are a 25 pin connector (described by ISO document 2110), 34 pin connector (ISO 2593), 37 and 9 pin connectors (ISO 4902), and a 15 pin connector (ISO 4903).

The *electrical characteristics* of the physical layer have been standardized by CCITT. The three most familiar standards are:

V.28 (EIA RS-232-C),

V.10/X.26 (EIA RS-423-A) and

V.11/X.27 (EIA RS-422-A).

The V.28 standard specifies the electrical characteristics at the point of demarcation, while more recent standards, V.10/X.26 and V.11/X.27, specify the

The Physical Layer

characteristics of the generators and receivers. The V.10/X.26 standard was designed for discrete component technology, and the V.11/X.27 for integrated circuit technology. The V.28 interface standard describes an unbalanced interface. This uses one conductor per circuit with a common signal ground (return). The maximum signal rate is 20 kbit/s over distances less than 15 meters.

The more recent unbalanced interface standard, V.10/X.26 is designed for signalling over distances up to 1 km with speeds up to 3 kbit/s, and for signalling up to 10 meters with speeds up to 300 kbit/s. It generates much reduced crosstalk as compared to the earlier standard.

The CCITT V.11/X.27 standard describes a balanced interface requiring a pair of wires per circuit. The maximum data signalling rate over cable distances up to 1000m is 100 kbit/s. The maximum rate is 10 Mbit/s over distances below 10m. The interface described by V.10/X.26 permits interoperation with V.28 and V.11/X.27.

The *functional* characteristics of a physical interface may be described as the functions assigned to the connecting circuits. Figure 3.6 illustrates the basic functions. These may be classified as:

(1) control
(2) data transmission
(3) timing
(4) earthing

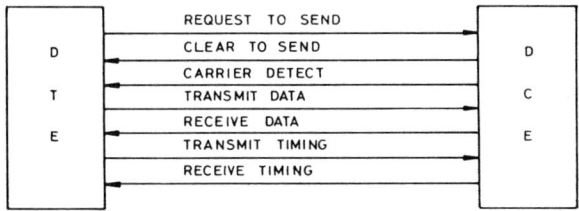

Figure 3.6 An example of DTE-DCE functional signals.

The functional characteristics are specified in the CCITT recommendations V.24 and the EIA RS-232-C and RS-449. The 1978 version of Recommendation V.24 defines 41 interchange circuits for use in various DTE/DCE interfaces and 12 interchange circuits for the DTE/ACE interfaces.

Recommendation X.24 was developed to describe the functions of a physical interface for use in public data networks (described in CCITT Recommendations X.20, X.21 and X.22). While V.24 employs the concept of one function per interchange circuit, X.24 defines a smaller number of interchange circuits because some of the control function circuits and data circuits are multiplexed.

The *procedural characteristics* of the physical layer form a set of procedures for using the interchange circuits. They describe the procedures to be performed that enable the transmission of bits so that the higher level functions can take place. Space does not permit a more detailed discussion here.

3.6 THE DATA LINK LAYER

The purpose of the data link layer is to provide the functional and procedural means to establish, maintain and release data links between network entities. The functions of this layer are:

(1) to provide error free transmission (error checking)
(2) to break the input data into frames (note that the ISO term for a frame is a physical layer service data unit)
(3) to transmit the frames sequentially
(4) to process the acknowledgement frames sent by the receiver
(5) to recognize frame boundaries
(6) to retransmit an erroneous frame
(7) to resolve the problem of relatively fast transmitters and slow receivers.

Data link control protocols may be divided into two groups:

(1) character-oriented protocols
(2) bit-oriented protocols

In the following sections, we will examine some of the more commonly used data link control protocols.

3.6.1 Character-Oriented Protocols

The character-oriented protocols are described in several different standards. They are:

(1) the American National Standard Institute (ANSI)—Standard ANSI X 3.28
(2) the International Standard Organization (ISO)—Standard ISO 1745
(3) the European Computer Manufacturers Association (ECMA)—Standard ECMA-16
(4) the IBM Binary Synchronous Communication (BSC) Protocol
(5) the International Air Transport Association (IATA) Synchronous Link Control (SLC) Standard
(6) the Digital Equipment Corporation (DEC) Digital Data Communications Message Protocol (DDCMP)

The Data Link Layer

The common feature of character-oriented protocols is that they are based on the use of characters defined from a given code. Some characters are used to convey information and others to control and supervise the link.

The most common communication character set in use is the American National Standard Code for Information Interchange known as ASCII (Standard ANSI X3.4-1968). This is illustrated in Table 3.2. ASCII is essentially identical to the CCITT Alphabet 5 and the ISO Standard 646.

Another well-known character set is the Extended Binary Coded Decimal Interchange Code (EBCDIC), illustrated in Table 3.3.

Table 3.2
ASCII Standard Code for Information Exchange

ROW	COLUMN	0	1	2	3	4	5	6	7
	BITS 4321 ⇩	765⇨ 000	001	010	011	100	101	110	111
0	0000	NUL	DLE	SP	0	@	P	'	p
1	0001	SOH	DC1	!	1	A	Q	a	q
2	0010	STX	DC2	"	2	B	R	b	r
3	0011	ETX	DC3	#	3	C	S	c	s
4	0100	EOT	DC4	$	4	D	T	d	t
5	0101	ENQ	NAK	%	5	E	U	e	u
6	0110	ACK	SYN	&	6	F	V	f	v
7	0111	BEL	ETB	'	7	G	W	g	w
8	1000	BS	CAN	(	8	H	X	h	x
9	1001	HT	EM	)	9	I	Y	i	y
10	1010	LF	SUB	*	:	J	Z	j	z
11	1011	VT	ESC	+	;	K	]	k	{
12	1100	FF	FS	,	<	L	\	l	∫
13	1101	CR	GS	−	=	M	[	m	}
14	1110	SO	RS	.	>	N	∧	n	~
15	1111	SI	US	/	?	O	_	o	DEL

Table 3.3
Extended Binary Coded Decimal Interchange Code (EBCDIC)

COLUMN (HEX) →	0	1	2	3	4	5	6	7	8	9	A	B	C	D	E	F
BITS 4567 → / ROW (HEX) ↓ BITS 0123 ↓	0000	0001	0010	0011	0100	0101	0110	0111	1000	1001	1010	1011	1100	1101	1110	1111
0 0000	NUL	DLE	DS		SP	&	-						{	}	\	0
1 0001	SOH	DC1	SOS						a	j	~		A	J	/	1
2 0010	STX	DC2	FS	SYN					b	k	s		B	K	S	2
3 0011	ETX	DC3							c	l	t		C	L	T	3
4 0100	PF	RES	BYP	PN					d	m	u		D	M	U	4
5 0101	HT	NL	LF	RS					e	n	v		E	N	V	5
6 0110	LC	BS	EOB/ETB	UC					f	o	w		F	O	W	6
7 0111	DEL	IL	PRE/ESC	EOT					g	p	x		G	P	X	7
8 1000		CAN							h	q	y		H	Q	Y	8
9 1001		EM							i	r	z		I	R	Z	9
A 1010	SMM	CC	SM		¢	!	|	:								
B 1011	VT				.	$	,	#								
C 1100	FF	IFS		DC4	<	*	%	@								
D 1101	CR	IGS	ENQ	NAK	(	)	_	'								
E 1110	SO	IRS	ACK		+	;	>	=								
F 1111	SI	IUS	BEL	SUB	|	¬	?	"								

The Data Link Layer

In character-oriented protocols, ten transmission control characters are used, namely:

(1) ACK (acknowledgement). This is an affirmative reply from a receiver to a sender.
(2) NAK (negative acknowledgement). Indicates a negative reply from a receiver to a sender.
(3) SOH (start of heading). The first character in a heading of an information message or text.
(4) STX (start of text). This character indicates the start of the text and the end of the heading.
(5) ETX (end of text). A character used to terminate a text.
(6) ETB (end of transmission block). This indicates the end of transmission block of data where data is divided into such blocks for transmission purposes.
(7) EOT (end of transmission). This indicates the conclusion of transmission of one or more texts.
(8) ENQ (enquiry). A character used as a request for a response from a remote station. It may be used to ask a station to identify itself.
(9) SYN (synchronous idle). This character is used to establish or maintain synchronization in a synchronous transmission system.
(10) DLE (data link escape). A character which will change the meaning of a limited number of the characters which follow immediately behind it. It is used exclusively to provide supplementary data transmission control functions.

3.6.2 The Binary Synchronous Communications Protocol

The phases of link establishment, information transfer and termination, error control, and transparency in a character oriented protocol can be illustrated by the IBM BSC protocol (sometimes also called the BISYNC protocol).

The operation of the data link depends on its configuration. The data link can be designed to operate either point-to-point (two stations) or multipoint/multidrop (two or more stations). In point-to-point operation, both stations can attempt to use the communications line simultaneously resulting in a contention situation. A station bids for the line by sending the ENQ control character preceded by two SYN characters. If simultaneous bidding occurs, one station must persist in its bidding attempt to resolve the contention situation.

The station which gains control of the line becomes a "master" station and starts the phase of information transfer (transmit status). A station which is in a receive status is referred to as a "slave" station.

The text transmitted from the master is usually divided into blocks of a fixed length. This is illustrated in Figure 3.7. Each block is preceded by two contiguous

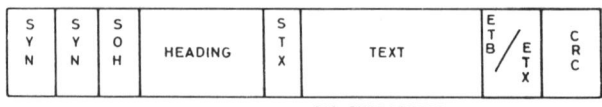

Figure 3.7 BSC Message Format.

SYN characters and is identified by a start of heading (SOH) character if it is the first block of a message. It is identified by a start of text (STX) character if it is an intermediate block. Each block of text, except the last, is immediately followed by an end of transmission block (ETB) character. The last block of text in a message is terminated by an end of text (ETX) character.

As each message block is completed, it is checked for transmission errors at the receiving station before the next block is transmitted.

Error Checking: Each block of data transmitted is error checked at the receiving end by using a method known as vertical redundancy checking (VRC) and longitudinal redundancy checking (LRC). Alternatively, a more powerful method known as cyclic redundancy checking (CRC) may be used. This will be discussed in Chapter 4.

If the received block is found to contain no detectable errors, the receiving station sends a positive acknowledgement (ACK) which indicates that the transmission may continue. When they are different, the receiving station sends a negative acknowledgement (NAK) character to indicate that the received block was erroneous and that it must be retransmitted.

If the first block is received correctly and the slave station is ready to receive the next block, it transmits the positive sequence ACK1. Then the ACK0 sequence is used as a positive acknowledgement of the second block and next "even" blocks. That is, ACK1, ACK0 are sent alternately. It should be noticed that the ACK0 sequence is also used as a positive reply to selection (discussed later) or a line bid indicating a "slave" status.

The character sequences used to represent ACK0 and ACK1 are shown in Table 3.4.

**Table 3.4
BSC Character Conversion**

DATA LINK CHARACTER	CHARACTER SEQUENCE	
	ASCII	EBCDIC
ACK0	DLE0	DLE'70'
ACK1	DLE1	DLE/
WACK	DLE;	DLE,
RVI	DLE<	DLE@

The Data Link Layer   127

Other Control Characters and Control Segments: There are two other groups of control characters and sequences, namely:

(1) backward control characters/sequences
(2) forward control characters/sequences

The backward control characters or sequences of characters are issued by a slave station. Apart from the sequences ACK0, ACK1, and NAK discussed previously, there are other sequences.

(1) WACK (Wait-Before-Transmit Positive Acknowledgement) allows a receiving station to indicate a "temporarily not ready to receive" condition. The character sequences representing this data link character are shown in Table 3.4.
(2) RVI (Reverse Interrupt) is a positive response used in place of the positive acknowledgement. It is transmitted by a receiving station to request a termination of the current transmission. The RVI character sequences are given in Table 3.4.

The forward control characters/sequences are ENQ and TTD. The ENQ character is used to bid for the line (point-to-point configuration), to obtain a repeat transmission of the response to a message block, and to terminate transmission.

The TTD (Temporary Text Delay) control sequence (STXENQ), is sent by a master station when it wishes to retain the line but is not ready to transmit.

The reader should now be in a position to follow the typical BSC communication sequence illustrated in Figure 3.8.

In the sequence illustrated, the transmitter first sends SYN SYN ENQ requesting a response from the receiver. Note that in Figure 3.8, the transmitter blocks are shown with characters to be sent in order from right to left. The receiver responses are sent with characters shown in order from left to right. On receipt of the first block, the receiver replies with ACK0 indicative that the block was received without detectable error. The transmitter then sends a block of text. The receiver then replies with a WACK indicating that it received the block without detectable error but that it is temporarily not ready to receive a further block.

The transmitter subsequently makes a further enquiry about the receiver's readiness. An ACK1 indicates that the receiver is now ready to receive, and so the transmitter proceeds to send a further segment of the text, commenced earlier.

The significance of the remainder of the sequence in Figure 3.8 is left for the reader to determine.

Transparency: In character-oriented procedures the data link control characters would not normally be permitted for use as data characters. To enable all data

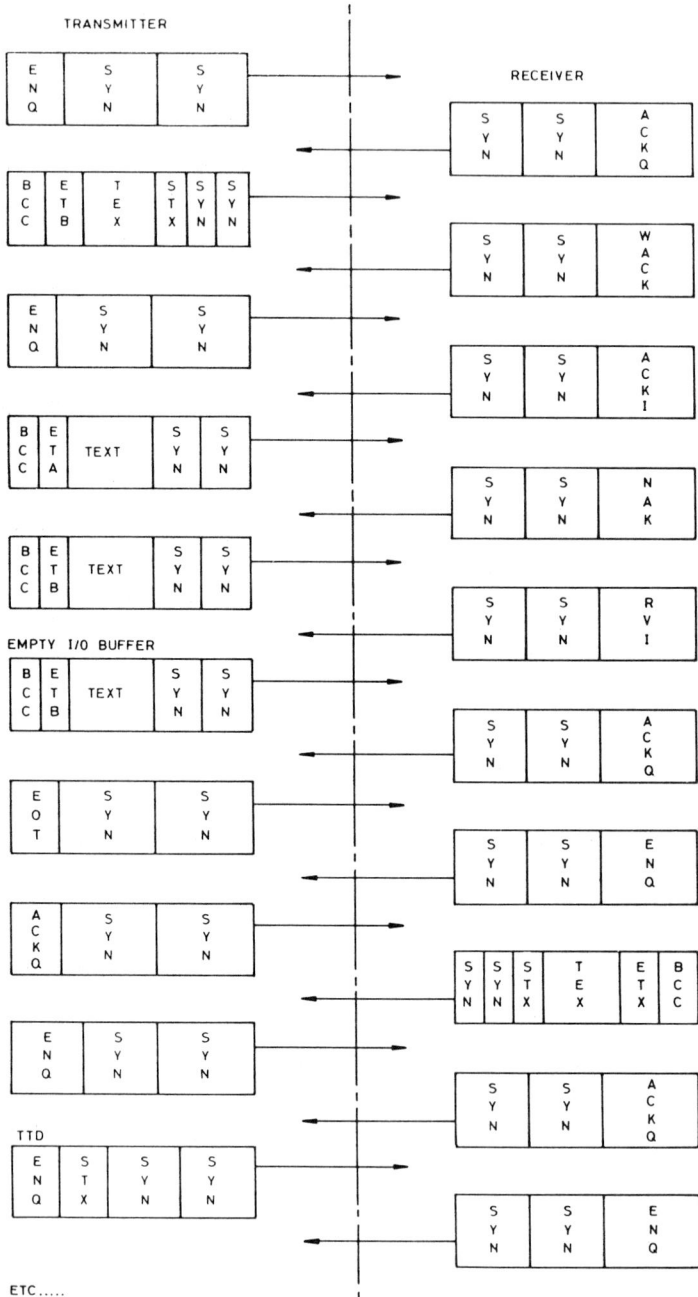

Figure 3.8 A typical BSC communication procedure.

link control characters to be transmitted as transparent data without taking on control meaning, a transparent mode of operation has to be used.

To initiate the transparent mode for text transmission, the sequence DLE STX is used. To terminate this mode, DLE ETB or DLE ETX is used. This illustrated in Figure 3.9. If the DLE character is to be sent as a data character, it has to be preceded by another DLE character.

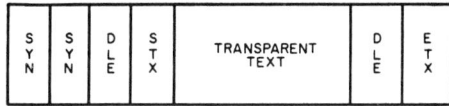

Figure 3.9 Transparent mode sequence in BSC.

Multipoint Transmission: In a multipoint (multidrop) configuration, one station is always designated as a control (master, primary) station, and the receiving stations are designated as tributary (slave, secondary) stations. The master station controls the information flow by polling and selection.

Polling is "an invitation" to transmit, issued by the control station to each tributary station. The polling sequence consists of the control character ENQ preceded by the poll address of the desired station and by SYN characters. This is illustrated in Figure 3.10.

| PAD | SYN | SYN | POLLING OR SELECTION | ENQ | PAD |

Note: PAD - leading 55(Hex), trailing FF (Hex).

Figure 3.10 Polling message format.

The possible replies from a polled tributary station are:

(1) Heading data—SYN SYN SOH . . .
(2) Text data—SYN SYN STX . . .
(3) Transparent text data—SYN SYN DLE STX . . .
(4) Negative reply (no message to be sent)—SYN SYN EOT . . .
(5) Temporary text delay—SYN SYN STX ENQ . . . .

Selection is a procedure which allows the master station to send a message to a tributary station. The possible replies from a selected tributary stations are:

(1) Positive reply indicating that the station is ready to receive—SYN SYN ACK0

(2) Negative reply indicating that the station is not ready to receive—SYN SYN NAK

(3) Reply indicating that the station is temporarily busy—SYN SYN WACK.

Synchronization: Synchronization between the transmitting and receiving stations is established at bit and character level by using a flag or "sync pattern." The sync pattern used to precede a transmission consists of at least two contiguous SYN characters. This establishes initial character synchronization. It has to be re-established for each block of data.

Bit synchronization is established either by the self-clocked modem or by a special bit-phase sync pattern which must precede the character phase. Usually the bit synchronizing pattern consists of two consecutive hexadecimal '55' characters.

In some protocols, special characters known as pad characters are used to assure that the first and last characters of a transmission are properly transmitted. At the start of a data stream, the leading pad prepares the receiver for searching for the sync pattern. At the end of the data stream, the trailing pad assures that the last character has been transmitted by the associated modem before it turns off for line turnaround. The leading pad is usually defined as a SYN character while the trailing pad as a sequence of 1 bits (hexadecimal 'FF').

3.6.3 Byte Count Protocols

Some character-oriented protocols do not utilize escape sequences, based on DLE, to provide transparency. Instead, they maintain a character count (byte count) in each block. The character count is usually transmitted after the SYN characters and the receiver counts characters instead of searching for a control character identifying the end of the block. One widely used byte count protocol is Digital Equipment Corporation's DDCMP protocol.

3.6.4 Bit-Oriented Protocols

Bit-oriented data link control protocols are based on a bit organization and format. The most common bit-oriented procedures are:

(1) HDLC (ISO standard IS 4335, 1977)—High-level Data Link Control
(2) ADCCP (ANSI standard X 3.66, 1979)—Advanced Data Communication Control Procedure
(3) SDLC, IBM's Synchronous Data Link Control
(4) BDLC—Burroughs Data Link Control
(5) UDLC—Univac Data Link Control

The Data Link Layer

All bit-oriented procedures are based on the basic transmission unit called a frame. All transmissions (data, control, or both) are by means of frames as illustrated in Figure 3.11. Each frame may have one of the two following formats:

(1) If there is an information field to transmit: Flag, Address, Control, Information, Frame, Check Sequence, Flag. This is illustrated in Figure 3.11(a).

(2) If there are only control sequences to transmit: Flag, Address, Control, Frame Check Sequence, Flag. This is illustrated in Figure 3.11(b).

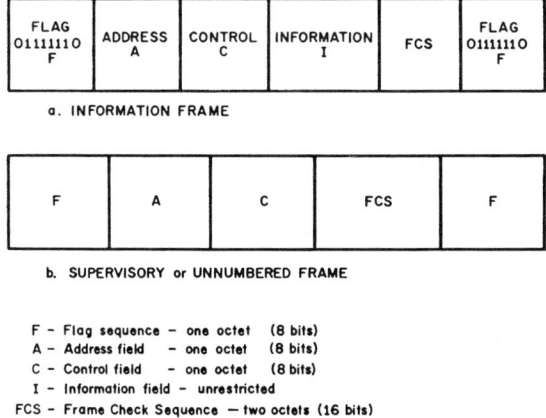

Figure 3.11 Frame structures for bit-oriented protocols.

The *flag field* is a unique 8-bit pattern 01111110. It identifies the start of a frame as well as the end of the frame. It synchronizes the receiver with the incoming frame, and is also used by the sending station to fill time between multiple frames. To achieve transparency (this unique 8-bit pattern cannot occur in other fields), a "bit stuffing" method is used as discussed in Chapter 2. When the transmitter encounters five consecutive 1's anywhere before sending the closing flag it automatically inserts a 0 bit following the 1's. This 0 bit is removed automatically by the receiver.

The *address field* identifies the secondary station on that link which is to receive the frame. The exception occurs when the frame carries the response to a command frame. Then the address field identifies the address of the sending station.

The *control field* formats are shown in Figure 3.12. In general, the control field identifies the function and purpose of the frame. There are three different control field formats:

(1) information frame (I)

(2) supervisory frame (S)
(3) unnumbered frame (unsequenced) (U).

CONTROL FIELD

CONTROL FIELD BITS	1	2	3	4	5	6	7	8
I frame	0	N(S)			P/F	N(R)		
S frame	1	0	S		P/F	N(R)		
U frame	1	1	M		P/F	M		

N(S) — Transmitter Send sequence count
N(R) — Receive sequence count
(the expected sequence number of the next received frame)
S — Supervisory function bits
M — Modifier function bits
P/F — Poll bit when issued by Primary
Final bit when issued by Secondary

1 = Poll/Final

Figure 3.12 Control field formats for bit-oriented protocols.

The first bit of the control field is set to 0 for the information frame and to 1 otherwise. The first bit set to 1 with the second bit set to 0 identifies the S frame. A 01 sequence identifies the U frame.

In the I frame, bits 2, 3, and 4, carry the sequence number N(S) of frames sent and the bits 6, 7, and 8 carry the number N(R) of the next frame expected to be received.

Bit 5, denoted by P/F in all three types of frames, is considered to be the P bit if the frame is a command (poll bit) and the F bit if the frame is a response (final bit). For example, in the Normal Response Mode, the P bit is set to 1 when the control station polls the addressed tributary station. The F bit is set to 1 in the final frame in a series of frames which is sent by the tributary station in response to polling. In the asynchronous response mode, the frame received with the P bit set to 1 causes the secondary station to set F bit to 1 in the next appropriate frame transmitted.

Bits 3 and 4 in the S frame define the four supervisory functions, namely:

(1) Receive Ready (RR)—bits 3 and 4 are set to 00. This frame is used as an acknowledgement frame and indicates that the next frame is expected.
(2) Reject (REJ)—bits 3 and 4 set to 01, respectively. This frame is a negative acknowledgement, requesting retransmission of all I frames starting from a designated point in the numbering cycle.

(3) Receive Not Ready (RNR)—bits 3 and 4 set to 10, respectively. This frame acknowledges all frames up to, but not including, the next frame, that is, up to $(N - 1)$ frames. It also requests the transmitter to suspend transmission temporarily.

(4) Selective Reject (SREJ)—bits 3 and 4 set to 11, respectively. SREJ frames are used to request retransmission of a single designated I frame previously transmitted.

Bits 3, 4, 6, 7, and 8 in the U frame are used to establish the particular mode of operation to be used. Since five bits are available for this purpose there can be 32 possible modes, but normally not all 32 are used.

Some of the commands used in the U frame are:

SNRM — Set Normal Response Mode
SARM — Set Asynchronous Response Mode
SABM — Set Asynchronous Balanced Mode
SIM — Set Initialization Mode
DISC — Disconnect Command
RSET — Reset Command
UP   — Unnumbered Poll
XID — Exchange Identification
RD   — Request Disconnect
DM   — Disconnect Response

All frames include a 16-bit frame check sequence (FCS) field prior to the terminating flag sequence for error detection. The error detection procedure, using the CCITT V.41 generator polynomial

$$g(X) = X^{16} + X^{12} + X^5 + 1$$

will be discussed in Chapter 4.

3.7 NETWORK LAYER

The *network layer* is also known as the communication subnet layer. It defines the characteristics of the interface between the user's computing equipment (*host*) and the packet switching equipment (*node*) in the communication subnetwork. The functions performed by the network layer are:

(1) to accept messages from the source host
(2) to convert them into packets

(3) to direct packets to the destination
(4) to determine a transmission route
(5) to prevent congestion and deadlocks

In the following sections we examine some of these functions in detail.

3.7.1 Virtual Circuits and Datagrams

One mode of operation of the network layer is in support of the *virtual circuit* model, illustrated in Figure 3.13. In this model a logical path is established between the transmitting and receiving stations. The network layer must ensure that packets are delivered in the same order as they were sent. All packets belonging to the same message are transmitted along the logical path in the same order as they were sent by the sending node.

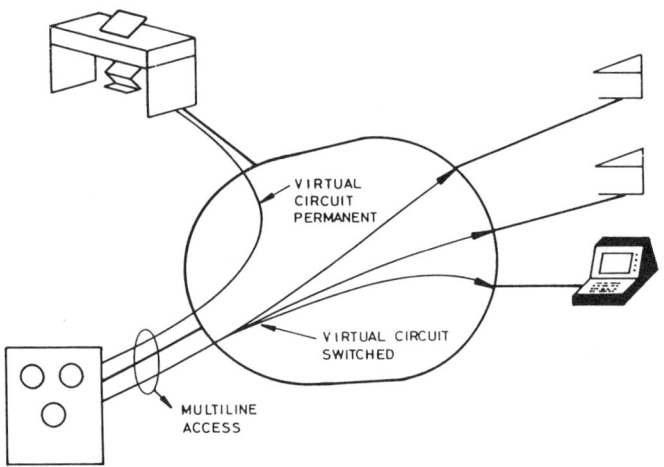

Figure 3.13 Permanent and switched types of virtual circuit.

On the other hand, the network layer may operate in another mode, associated with a so-called *datagram model*. In datagram operation, the network layer accepts messages from the transport layer, packetizes them, and then sends packets as isolated entities. That is, each packet may be sent to the destination via a different route.

Figure 3.14 illustrates the differences between virtual circuit and datagram operation. In datagram operation, the concept of sending packets as separate entities via different routes leads to the possibility of packets arriving at their destinations out of order, of losing packets, and of packet duplication. The datagram service is, however, very efficient when messages are short, preferably

Network Layer

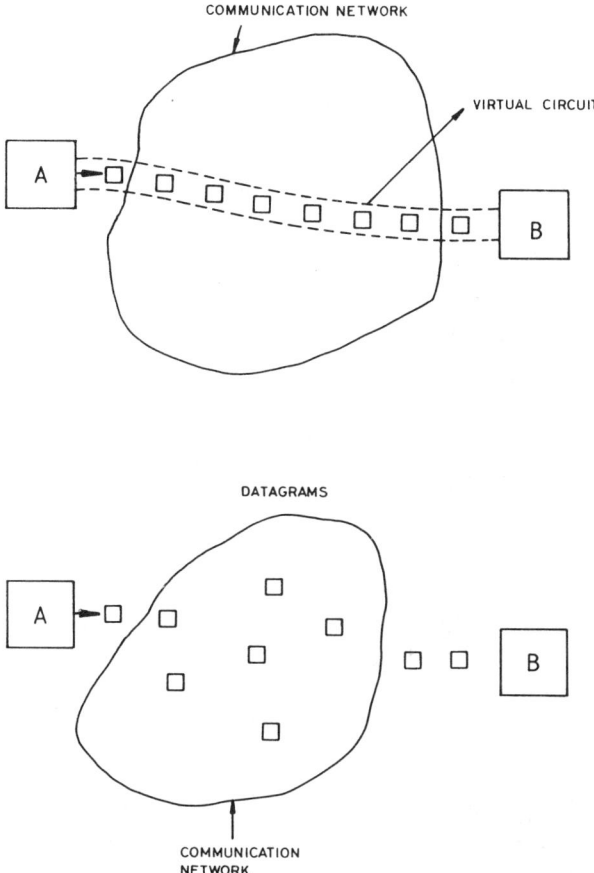

Figure 3.14 A comparison between virtual circuit and datagram connection.

one packet long, and for this reason is sometimes used in networks offering electronic mail as a service.

The virtual circuit service and datagram service also differ in the link set up and clear procedures. In the virtual circuit service, the path is set up between sending and receiving ends before the packets are exchanged. The destination address is needed only during setup procedure and when end-to-end flow control is provided. In the datagram service, a destination address must be included in every packet, since it is carried within the subnetwork independently. The end-to-end error and sequence control are performed by the host only.

The virtual circuit service is provided by many public data networks, for example, the French TRANSPAC and the Canadian DATANET. The IBM Systems Network Architecture, SNA, also uses virtual circuits being provided by the network layer to the transport layer.

3.7.2 Routing

Routing is the process of establishing a path in a network between any pair of source and destination nodes, along which message transfer is to be performed. The routing process is handled by the network layer software.

There are many different routing techniques currently being implemented in computer networks. Generally, they may be divided into two major groups:

(1) deterministic routing techniques
(2) random routing techniques

The algorithms which belong to the first group are:

(1) a least time delay algorithm
(2) the shortest path algorithm
(3) flooding

The *least time delay* algorithm is based on minimization of the overall delay time for a given pair of nodes. A so-called traffic matrix for the network is established, from which the least time path is found. Least time delay paths are stored in routing tables. Since the source-destination traffic may change, the routing tables have to be updated from time to time.

The *shortest path* algorithm may be derived taking into account the lengths of each of the possible links in the source-destination path. Alternatively, the shortest path may be estimated on the basis of certain additional "weights" attached to each link, such as, for example, link cost and link performance. The path with the minimum overall weight (or penalties) is chosen to be the optimal (shortest) path.

In the technique known as *flooding*, a packet is sent by the originating node to each of its neighboring nodes. Each node which receives a packet, determines whether it has been received already and if it has, discards it. A packet received for the first time is forwarded to all the neighboring nodes except the packet from which it was sent. The procedure continues until all packets are passed through the network.

The first two algorithms require a look-up table, either a traffic matrix table, or a so-called directory routing table. Consider for example, the network topology shown in Figure 3.15. The network is shown with six nodes A, B, C, D, E, and F. The links between nodes are numbered 1 to 10. Each node uses a routing table provided for the network. Table 3.4 shows the routing table for the network in Figure 3.15.

Network Layer

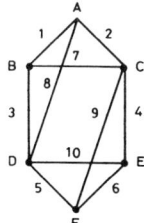

Figure 3.15 A network topology example.

Table 3.4
Routing Table for the Network of Figure 3.15

Source	A	B	C	D	E	F
Destination						
A	X	1	2	8	4-2	5-8/9-2
B	1	X	7	3	10-3/4-7	5-3
C	2	7	X	3-7/10-4	4	6-4
D	8	3	7-3/4-10	X	10	5
E	2-4	7-4/3-10	4	10	X	6
F	8-5/2-9	3-5	9	5	6	X

In general, deterministic routing techniques may be divided into two categories, namely "fixed indefinite," and "fixed for session."

The characteristic feature of the technique referred to as the *fixed indefinite* method is that there is always at least one alternative route apart from the primary route to every destination in the routing tables at each node. These routing tables may be changed several times a year as traffic and/or network topology vary.

The routing technique referred to as the *fixed for session* method is used primarily in timesharing networks (for example, TYMNET). When the user logs in, the control node calculates the best route at that time. It selects the route with the least cost penalty. The cost penalty is usually the lowest for high-speed links. Some additional penalty is usually added if nodes in the source-destination path report overload. A high penalty is added for a link with large delay such as a satellite link. All messages for a given session follow the same route (assuming no failures). The next session between the same nodes may be assigned a different route.

The best known *random (stochastic) routing* algorithm is an adaptive random routing strategy. In this algorithm each node carries out least-time estimates (every second or so) and decides which outgoing link to use to minimize the estimated time delay.

In the American network ARPANET for example, the estimate is determined by each node as follows. Note that in the ARPANET terminology, nodes are referred to as IMPs (Interface Message Processors).

(1) Each IMP advises all neighbors of the delay it is experiencing to all other destination IMPs

(2) This delay information is provided by "routing information packets" exchanged by adjacent IMPs at defined intervals of time (for example, every half second or every second)

(3) The "best" outgoing link for a given destination is then determined

To illustrate this, consider the simple network shown in Figure 3.16. Let us examine the calculations carried out by IMP A if IMP B is the destination. The calculations performed by node A are as follows:

delay via B (link L_1) = delay to B + estimate by B of
the delay from B to D.

delay via C (link L_2) = delay to C + estimate by C of
the delay from C to D.

Therefore, the minimum delay is

$$\text{delay min.} = \min \begin{cases} \text{A's calc. to B} + \text{B's calc. to D} \\ \text{A's calc. to C} + \text{C's calc. to D} \end{cases}$$

This minimum delay will be placed in the routing table of A.

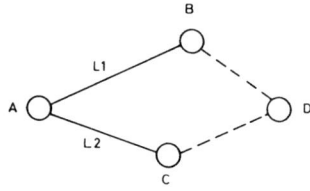

Figure 3.16 Network for delay calculations.

3.7.3 Congestion Control in Networks

Congestion may occur when the fluctuations in traffic are above the average level for which the routing algorithms were designed. This may occur when the traffic increases unexpectedly at certain nodes, or when the traffic builds up after routing decisions have been made and routing tables updated.

Network Layer

Congestion results in excessive delays in transmitting information. Different congestion control mechanisms are depicted in Figure 3.17. They are summarized as follows:

(1) *Bottleneck:* congestion on a link between two nodes of the network. If, for example, congestion occurs on link A-B, a remedy is to increase the capacity of the A-B link, or to provide alternative routes.
(2) *Reassembly lockup (deadlock):* occurring if node A is full of packets to B and node B is full of packets destined for A. No remaining buffer space is available.
(3) *Information trapped in network:* if there is no path available to the destination, or if the destination is out of service or not known.

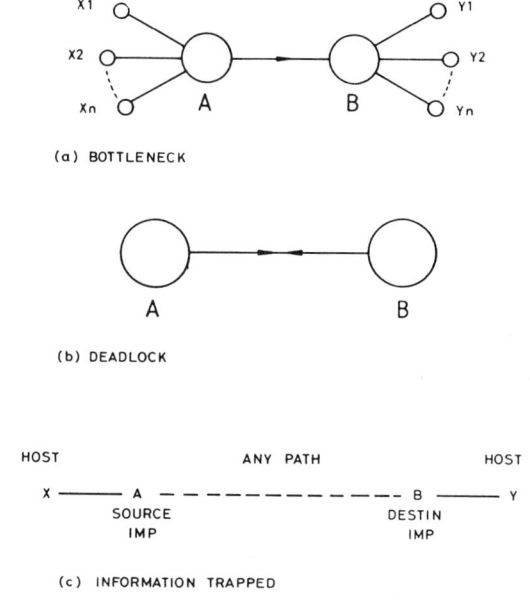

Figure 3.17 Congestion mechanisms.

In order to deal with these congestion problems, it is necessary to provide *end-to-end control*. To illustrate, end-to-end control in the ARPANET system includes the use of the following strategy between two nodes A and B:

(1) A sends a message to B—"allocate buffers"
(2) B replies with an "allocate" message
(3) A sends eight packets, B reassembles, and then
(4) B sends RFNM (Ready For Next Message) to A.

3.8 THE X.25 INTERFACE

A standard network access protocol has been proposed by the CCITT. Known as Recommendation X.25, it encompasses levels 1, 2, and 3 of the seven-layer ISO architectural model for networks. In a short time, X.25 has come to dominate the design of all national data networks. X.25 defines the interface between a user's computer terminal equipment or DTE (Data Terminal Equipment) and the telecommunications circuit provider's equipment or DCE (Data Circuit-terminating Equipment). X.25 describes the format and meaning of the information exchanged across the DTE-DCE interface for the layer 1, 2, and 3 protocols.

3.8.1 The X.25 Packet Characteristics

The first version of X.25 (1976) made provisions for two basic virtual circuit services in packet-switched public data networks. One, the *Permanent Virtual Circuit* (PVC) provides an end-to-end connection equivalent to a fixed route via a leased line. The second, the *Virtual Call* (VC), known also as the *Switched Virtual Circuit* provides for an establishment of a switched connection equivalent to a telephone connection between a calling and called subscriber. Consequently, it must also provide for a clearing procedure to release a connection. Later versions of the X.25 Recommendation introduced two new services, *Datagram* and *Fast Select*. These will be discussed in the next sections.

Many different types of packets are possible in such networks. Packet formats are described in terms of a stack of octets (a term used by CCITT to describe eight-bit bytes). One such packet structure is shown in Figure 3.18. In the packet

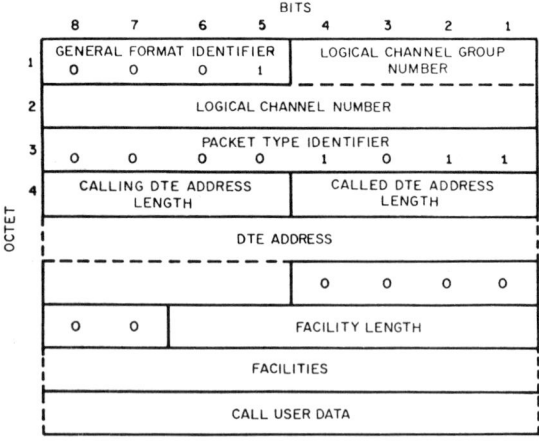

Figure 3.18   Packet structure of CALL REQUEST and INCOMING CALL packets.

The X.25 Interface 141

structure, the bits of an octet are numbered 8 to 1 where 1 is the lower order bit and is transmitted first. The octets in a packet are consecutively numbered starting from 1 and transmitted in order.

Each packet contains a header of three or more octets. This may contain address, routing, and control information. We will first examine the packet header structure. Next, the different types of packets will be outlined. The packet header structure is as follows:

(1) The header includes a General Format Identifier (GFI). The GFI field is a four-bit binary coded field located in bit positions 8, 7, 6, and 5 of octet 1. It is provided to indicate the general format of the rest of the header.
(2) In bit positions 4, 3, 2, and 1 of octet 1 a logical channel group number is located (except in restart packets). The logical channel number appears also in every packet (except in restart packets) in all bit positions of octet 2.
(3) The third octet's content depends on the type of packet. For the CALL REQUEST and INCOMING CALL packet structure of Figure 3.18, the third octet carries information on packet type. An example of some packet type identifiers is shown in Table 3.5.
(4) Octet 4 consists of field length indicators for the called and calling DTE addresses. Octet 5 and the following octets consist of the called DTE address and the calling DTE address.
(5) Then follow the *facilities length* field octets (if present) which indicate how many octets of facility field follow. The facilities field is used to request certain features for the virtual circuit.

Following the facility field, user data may be present.

There are several different types of packets used for various control and data transmission functions. We will briefly survey some of the more important types of packets.

CALL ACCEPTED and CALL CONNECTED packets consist of only three octets, as illustrated in Figure 3.19. The first octet includes a General format identifier (GFI) and logical channel group number. The second octet contains a logical channel number, and the third octet contains the packet type identifier.

CLEAR REQUEST and CLEAR INDICATION packets are illustrated in Figure 3.20. These include an octet indicating a reason for clearing the connection. This is termed a clearing cause. A clearing cause in a CLEAR INDICATION packet may be one of the following:

(1) Number busy
(2) Out of order
(3) Remote procedure error
(4) Number refuses reverse charges

(5) Invalid call
(6) Access barred
(7) Local procedure error
(8) Network congestion
(9) Not obtainable

Table 3.5
Packet Type Identifiers

Packet type		Octet 3 Bits
From DCE to DTE	From DTE to DCE	8 7 6 5 4 3 2 1
Call set-up and clearing		
Incoming call	Call request	0 0 0 0 1 0 1 1
Call connected	Call accepted	0 0 0 0 1 1 1 1
Clear indication	Clear request	0 0 0 1 0 0 1 1
DCE clear confirmation	DTE Clear confirmation	0 0 0 1 0 1 1 1
Data and interrupt		
DCE Data	DTE Data	X X X X X X X 0
DCE Interrupt	DTE Interrupt	0 0 1 0 0 0 1 1
DCE Interrupt confirmation	DTE Interrupt confirmation	0 0 1 0 0 1 1 1
Flow control and reset		
DCE RR	DTE RR	X X X 0 0 0 0 1
DCE RNR	DTE RNR	X X X 0 0 1 0 1
	DTE REJ	X X X 0 1 0 0 1
Reset indication	Reset request	0 0 0 1 1 0 1 1
DCE Reset indication	DTE Reset confirmation	0 0 0 1 1 1 1 1
Restart		
Restart indication	Restart request	1 1 1 1 1 0 1 1
DCE Restart confirmation	DTE Restart confirmation	1 1 1 1 1 1 1 1

Note—A bit which is indicated as "X" may be set to either 0 or 1.

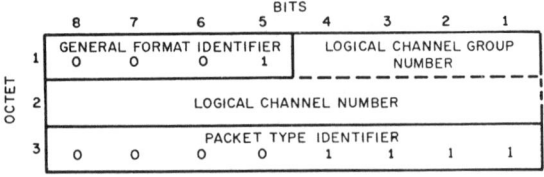

Figure 3.19 CALL ACCEPTED and CALL CONNECTED packets.

The X.25 Interface 143

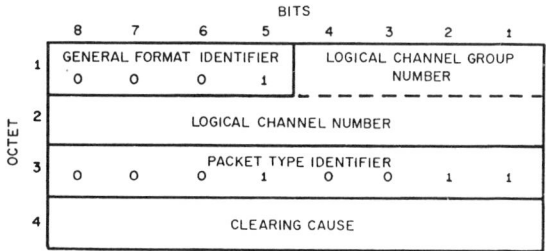

Figure 3.20 CLEAR REQUEST and CLEAR INDICATION packets.

The bits of the cause field in CLEAR REQUEST packets are set to 0.

RESET REQUEST and RESET INDICATION packets are illustrated in Figure 3.21. In these packets, octet 4 is the resetting cause field, while octet 5 is the diagnostic code field. This contains additional information on the reason for the RESET when the cause field indicates a local procedure error. The resetting cause may be one of the following:

(1) Out of order
(2) Remote procedure error
(3) Local procedure error
(4) Network congestion

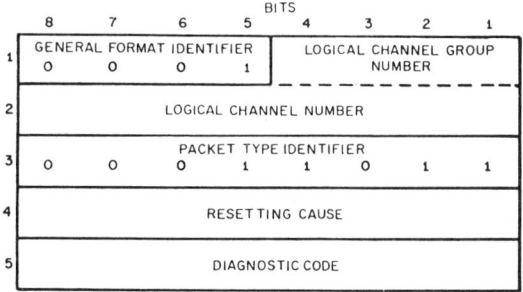

Figure 3.21 RESET REQUEST and RESET INDICATION packets.

CLEAR CONFIRMATION packets are illustrated in Figure 3.22. As in the CALL RESTART and CALL ACCEPTED packets, they consist of three octets.

The data transmission process uses DATA packets as illustrated in Figure 3.23. They consist of a header and a data field. These packets may be of variable length depending on the length of the user data field.

RESTART REQUEST and RESTART INDICATION packets are illustrated in Figure 3.24. In these packets, bits 4, 3, 2 and 1 of the first octet and all bits

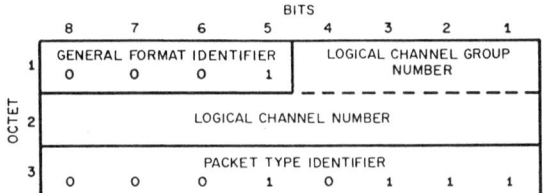

Figure 3.22 CLEAR CONFIRMATION packet.

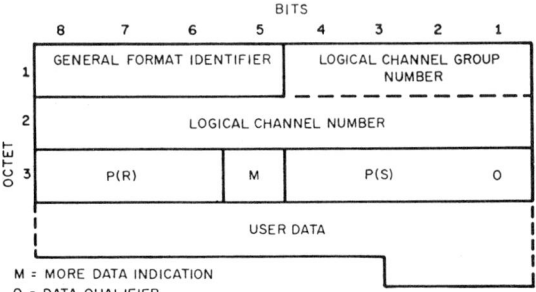

Figure 3.23 DATA packet.

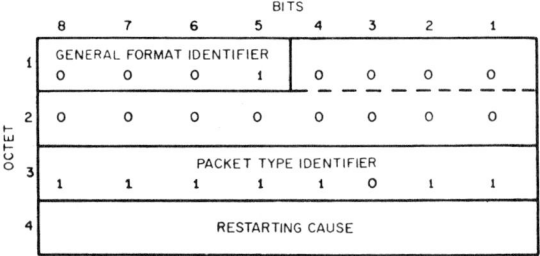

Figure 3.24 RESTART REQUEST and RESTART INDICATION packets.

of the second octet are set to 0. Octet 4 is the restarting cause field and contains the reason for the restart. The coding of the restarting cause field in the RESTART INDICATION packets may be used to indicate a local procedure error, or network congestion. The bits of the restarting cause field in RESTART REQUEST packets are set to 0.

3.8.2 Services Provided by X.25

Recommendation X.25 defines procedures used across the interface between the packet-mode terminal (DTE) and the common-carrier equipment (DCE). It provides for the following services on public data networks:

(1) switched virtual circuits (SVCs) also called virtual calls
(2) permanent virtual circuits (PVCs)
(3) datagrams

The characteristics of the X.25 end-to-end service (DTE-to-DTE) provide for the establishment and clearing of a virtual call and for data transfer. They also provide for error recovery including the reset procedure, restart procedure, and error handling.

The sequence of events in the process of *establishing a virtual circuit and clearing it* is as follows:

A DTE initiates a call by sending a CALL REQUEST packet, then either—

(1) the calling DTE will receive a CALL CONNECTED packet as a response indicating that the called DTE has accepted the call, or
(2) if the call is refused by the called DTE or if the attempt fails, the calling DTE will receive a CLEAR INDICATION.

Once the call is in a process of the data transfer, *Call clearing* may be initiated by either DTE (or by the network in the case of failure). This is done using the CLEAR REQUEST packet (in switched virtual circuits). It is followed by the transfer of a CLEAR CONFIRMATION packet.

The *data transfer* process proceeds using the packet structures illustrated in Figure 3.23. Each DATA TRANSFER packet includes a packet receive sequence number P(R) and the packet send sequence number P(S). Bit M set to 1 indicates that data in the following packet is associated with data in this packet. The P(R) numbers sent across a logical channel ensure that a sending DTE does not transmit data at an average rate greater than that at which the receiving DTE can accept that data (flow control).

The *error recovery* process is as follows:

(1) A *reset procedure* is used to reinitialize the flow control procedure. RESET REQUEST and CONFIRMATION packets are used in this procedure. All sequence numbers are set to zero.
(2) A *restart procedure* provides a mechanism to recover from major failures.

The *error handling* process is as follows:

(1) procedural errors during the establishment and clearing are reported to a DTE by *clearing* the call.
(2) procedural errors during the data transfer phase are reported to the DTE by *resetting* the virtual circuit.

(3) a diagnostic field is included in the reset packet to provide additional information to the DTE.
(4) *timers* are essential in resolving some deadlock conditions.

Timeouts are used where necessary in case the called DTE fails to respond. The minimum length of time the DTE has to respond to an incoming call is three minutes. The actions of the DCE when no confirmation has been received to an indication packet (that is during resetting, clearing and restarting) are as follows:

(1) On expiry of a 60-second timer, after issuing a RESET INDICATION, the DCE clears the call, indicating the reason for clearing via the diagnostic code and call progress signal. On a permanent virtual circuit, a DIAGNOSTIC packet is sent.
(2) On expiry of a 60 second timer after issuing a CLEAR INDICATION, the DCE will issue a DIAGNOSTIC packet and assume the interface has entered the ready state. (Packets sent by the DTE are not ignored).
(3) On expiry of a 60 second timer after a RESTART INDICATION has been issued, the DCE will use a DIAGNOSTIC packet. The DCE considers this condition to be serious and will stay in this state indefinitely.

3.8.3 The Transaction-Oriented Features of X.25

The CCITT Recommendation X.25 for a packet switched service in public data networks was amended in recent years to accommodate two services suitable for small amounts of data. These new services are:

(1) a *datagram service* for the transport of packets each containing an independent message
(2) the *fast select* facility which provides for the inclusion of 128 octets of user data in the call establishment packet for virtual circuits

The packet that is unique to the datagram service is the DATAGRAM SERVICE SIGNAL packet. It is illustrated in Figure 3.25. The *datagram* packet format has one unique field referred to as the Datagram Identification. This field follows the Facilities field and is a part of the User Data field. If this packet is generated by the network relative to a specific datagram issued by a DTE then it is known as the DATAGRAM SERVICE SIGNAL-SPECIFIC packet. There are three classes of the Datagram Service Signal-Specific, namely:

(1) Datagram Rejected
(2) Nondelivery Indication
(3) Delivery Confirmation

The X.25 Interface

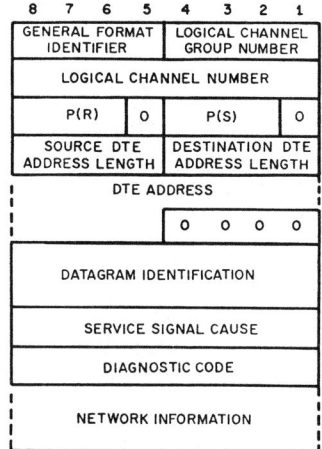

Figure 3.25 DATAGRAM SERVICE SIGNAL packet.

On the other hand, if the datagram packet is generated by the network relative to the overall datagram operation of the DTE or network, it is called a DATAGRAM SERVICE SIGNAL-GENERAL packet.

The *Fast Select* service allows a calling packet-mode terminal to transmit up to 128 octets of data in a Call Request packet and to receive up to 128 octets of data in the Clear Indication packet from the calling terminal. This is illustrated in Figure 3.26.

Some public data networks offer either the Datagram or the Fast Select but not both.

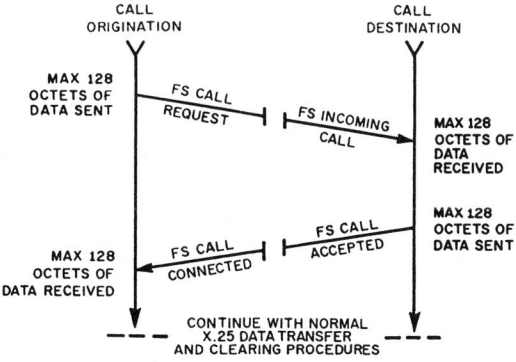

Figure 3.26 FAST SELECT with CALL ACCEPTED response.

3.9 THE X.75 INTER-NETWORK PROTOCOL

It is often desirable to be able to arrange the interconnection of individual computer networks. This may be required for data and program exchange and sharing or to provide remote access to resources. Often this involves the international connection of national data networks.

The interconnecting of different networks calls for the standardization of protocols. CCITT Recommendation X.75 has been developed for this purpose. It specifies the interface for the interconnection of public data networks that conform to the X.25 Recommendation.

X.75 provides terminal and transit control procedures and data transfer systems on international circuits between packet-switched data networks. The interface between two data networks specifies the operation between a Signalling Terminal (STE) in one network and an STE in another. These are known as *gateways* and are controlled by each of the national data networks.

The interconnection of data networks via X.75 interfaces results in a series of virtual circuits, each being a distinct entity with separate error recovery, flow control, and so forth. The packet structure used in X.75 is illustrated in Figure 3.27.

Standard *addressing* techniques become important considerations when networks are interconnected. In X.75 the address field is up to 15 binary coded decimal digits long. The maximum address is then 60 bits.

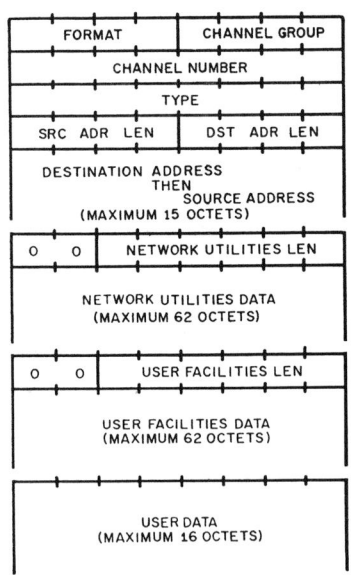

Figure 3.27 X.75 CALL REQUEST and DATA packets.

For *routing* purposes, each virtual circuit has independent flow control specific to the data network which is responsible for its establishment. The flow controls associated with the STE-STE connection are designed to be adapted on a "per call basis." This is necessary in order to provide for variations between network types. However, the drawbacks of this approach include the possibility of longer delays. Also, the use of different types of flow control in data networks may lead to errors.

Each portion of the end-to-end path has a local *acknowledgement* procedure. This indicates that the receiving data network has accepted the message for transmission, not that it has arrived at the destination.

The X.75 Recommendation does not specify how the communicating data networks deal with errors internally. The networks signal a RESET if unrecoverable errors occur. During RESET the virtual circuit still exists but the flow control is reset. This means that messages may have been lost. More serious errors result in the call clearing. More details are given in Tanenbaum (4).

3.10 HIGHER LEVELS OF THE ISO REFERENCE MODEL

The X.25 protocol, discussed in Section 3.5, is a node to packet network protocol. It does not, however, provide end-to-end path management. The management of the access path from end to end is provided by higher level protocols such as the SNA (System Network Architecture) of IBM and DECNET (the computer network design offered by Digital Equipment Corporation).

We will briefly describe here only the functions of the SNA protocol layers. The layers and their functions can be summarized as follows:

(1) The *Data Link Control (DLC)* constructs frames and provides for the transfer of them across a potentially noisy transmission facility (detecting and recovering from transmission errors). This layer corresponds to the Open System Interconnection (OSI) second level. The protocol of this layer is known as the *Synchronous Data Link Control* (SDLC).

(2) *Path Control*, layer 3 of SNA provides routing and congestion control within the subnet.

(3) *Transmission Control* creates and manages session establishment/release. It controls the rate of flow for an individual session, allocating buffers, resolving priorities, and handling multiplexing/demultiplexing of data and control messages. It may also perform encryption and decryption.

(4) The *Data Flow Control* layer is not concerned with the data flow control concepts as they were described previously. It maintains the order of the messages exchanged between two end users, providing such functions as chaining of requests and management of session sequence numbers.

(5) The *Presentation Layer* is also referred to as the Network Addressable Unit (NAU). It provides three classes of services, namely presentation services, session services, and network services. Presentation services include text compression, encryption/decryption, virtual terminal protocol, and file transfer protocol. The other two classes of services include provision for setting up the session connection, management of the configuration, maintenance (provision of diagnostics and collection of statistics), and interfacing with the operator.

For details on the relationships and relative correspondence between the SNA layers and those of DECNET and the ISO model, see Tanenbaum (4).

3.11 SOME ASPECTS OF DATA NETWORK DESIGN

The goal of network design is to achieve a specified performance at minimum cost. The factors that have the major influence on network design are the geographical locations of sources of data and destinations of data, the type and quantity of data, messages or codes to be transferred to and from each location, and the response time required.

Design considerations arising from the above factors may be classified into three groups. They include:

(1) *Constants:* including the locations of terminals and hosts, the traffic generated (traffic matrix), and the costs (cost matrix)
(2) *Variables:* including the line capacities, network topology, and flow assignment
(3) *Performance constraints:* such as reliability and delay/throughput (system response time)

A very important step in the design process is to include a comprehensive study of the environment in which the network will be operating and the understanding of the services it has to offer to the end users.

As an illustration of sets of different objectives, let us compare the information services required by a hospital information system and a large airline reservation system, respectively.

The objectives of a hospital information system could be:

(1) to provide on-line diagnostic services to all participating hospitals and medical centers, with response time of 10 minutes or less
(2) to maintain an updated medical history file for purpose of increasing reliability of the diagnostic service

(3) to maintain complete records of all patients admitted to the major hospitals and provide response to the enquiries of patients withing three seconds

On the other hand, the primary objectives of the large airline reservation system might be:

(1) to provide a central file to control the seat inventory for all scheduled flights and make this file available to all reservations clerks
(2) to provide responses to reservations clerks' enquiries in three seconds or less, regardless of their geographical location.

Once the network environment and required services are understood, the network design process can begin. The sequence of steps required in the design process are as follows:

(1) Determine the major objectives of the system
(2) Determine the sources and destinations of data
(3) Determine the quantity and characteristics of the data such as the distribution of the length of messages, message rates, codes to be used, and traffic volume
(4) Establish the response criteria
(5) Determine the characteristics, type, and number of terminals per location
(6) Determine the possible network structure (multiplexers, concentrators, multipoint, point-to-point)
(7) Determine possible line control procedures (polling sequences, priorities, half-duplex, full-duplex)
(8) Select the modems and other line equipment types
(9) Develop analytical models (employ queueing theory and simulation)
(10) Establish a minimum-cost geographic layout of leased lines
(11) Make refinements to the design

Step (9) gives an indication of the expected delays within a network as a function of throughput. This delay/throughput analysis is performed by employing the basic principles of queueing theory. We will summarize the principles involved for a single communication path. See for example, Kobayashi (5) for details.

Consider the basic structure of a queueing model that consists of arriving messages, a queue, and one or more servers. The behavior of the queueing system is analyzed in terms of assumed statistical properties for the message arrival patterns, the queueing discipline, and the service mechanism. There are many alternative assumptions that can be made about such elements of the queueing system.

Often the arrival of items is assumed to have a Poisson distribution. That is, items are assumed to arrive at the queue at random from an infinite source population. In that case, the interarrival time is exponentially distributed. Under such conditions, the probability of exactly n arrivals in time t is given by the Poisson distribution

$$P_n(t) = \frac{(\lambda t)^n}{n!} e^{-\lambda t} \qquad n = 0, 1, 2, \ldots \qquad (3.1)$$

where λ is a constant known as the traffic rate or mean arrival rate, and λt is the mean number of arrivals in the interval t.

If the service time is assumed to follow an exponential distribution then the probability density function for the service time is given by

$$f_S(t) = \mu e^{-\mu t} \qquad (3.2)$$

where $1/\mu$ is the mean service time in seconds per customer.

The notation commonly used to describe queueing models is of the form $A/B/M$ where A represents the interarrival time probability density function, B represents the service time density function, and M represents the number of servers. For example, a queueing system designated $M/M/1$ represents a system with random (Poisson) arrivals, exponential service times and a single server. Alternatively, an $M/G/1$ model represents a system with random arrivals and with a general (unspecified) service distribution.

The symbol M stands for Markov (exponential probability density). G stands for general, that is, an arbitrary probability density. Queueing system models range from $M/M/1$, about which all the statistics are known, to $G/G/M$, for which very few properties can be determined.

We can summarize some of the properties of the more important queueing models used in analyzing computer networks. Consider those cases where the input messages must queue in a single line if one or more arriving messages find the server (node) busy. Each message must wait until all earlier messages have been processed. This is referred to as a single server model. Let

$E(n)$ represent the expected (average) arrival rate
$E(t_s)$ be the average service time
$E(q)$ be the average numer of items in the system
$E(w)$ be the average number of items waiting to be served
$E(t_w)$ be the average waiting time
$E(t_q)$ be the average queueing time including waiting and service time.

Then the following relationships are assumed:

$$t_q = t_w + t_s \qquad (3.3)$$

Some Aspects of Data Network Design

$$E(t_q) = E(t_w) + E(t_s) \tag{3.4}$$

$$E(w) = E(n)E(t_w) \tag{3.5}$$

$$E(q) = E(n)E(t_q). \tag{3.6}$$

Equation (3.6) is referred to as Little's formula. Defining the server's *utilization factor* ρ as

$$\rho = E(n)E(t_s) \tag{3.7}$$

we obtain from Little's formula that the total number of items in the system on average is

$$E(q) = E(n)E(t_w) + E(n)E(t_s)$$

and, hence, that

$$E(q) = E(w) + \rho. \tag{3.8}$$

Consider an $M/G/1$ queueing model. The average number of waiting items for any $M/G/1$ model is given by the Polloczek-Khinchine formula as

$$E(w) = \frac{0.5\rho^2}{1-\rho}[1 + (\sigma_{ts}/E(t_s))^2] \tag{3.9}$$

where σ_{ts} is the standard deviation of the service times. If the service time is constant then the parameter $\sigma_{ts} = 0$ and the mean number of items in the queue is

$$E(w) = \frac{\rho^2}{2(1-\rho)}. \tag{3.10}$$

It follows that the mean number of items queueing or being served is

$$E(q) = \rho + \frac{\rho^2}{2(1-\rho)} \tag{3.11}$$

and the average queueing time including waiting and being served is

$$E(t_q) = E(t_s)\left\{1 + \frac{\rho}{2(1-\rho)}\right\}. \tag{3.12}$$

On the other hand, if the service time is exponentially distributed as described in Equation (3.2), and has average service time $E(t_s) = 1/\mu$, then the utilization factor can be written

$$\rho = \lambda/\mu. \tag{3.13}$$

Furthermore, for an exponential distribution, it is easy to show that

$$\sigma_{ts} = E(t_s). \tag{3.14}$$

It follows from Equation (3.9) that

$$E(w) = \frac{\rho^2}{1-\rho}. \tag{3.15}$$

Then the mean number of items in the system is

$$E(q) = \frac{\rho}{1-\rho} \tag{3.16}$$

and

$$E(t_q) = \frac{1/\mu}{1-\rho}. \tag{3.17}$$

The mean queue size, mean queueing time, and their standard deviations can be plotted against utilization for different service time distributions as shown in Figures 3.28 and 3.29.

To evaluate the delay experienced by a packet on each link and the average total delay within a communication subnetwork, it is important to determine the total time T spent in the system (the delay). If

$$T = E(t_q)$$

then an important relationship obtained from Little's formulae is

$$E(t_q) = E(q)/E(n). \tag{3.18}$$

For an exponential service time, $E(q)$ is given by Equaton (3.16). For Poisson arrivals, the mean arrival rate is $E(n) = \lambda$. In that case, the total delay experienced by a packet on each link is

$$T = \frac{\rho}{\lambda(1-\rho)}$$

and since $\rho = \lambda/\mu$ we obtain

$$T = \frac{1}{\mu - \lambda}. \tag{3.19}$$

Some Aspects of Data Network Design

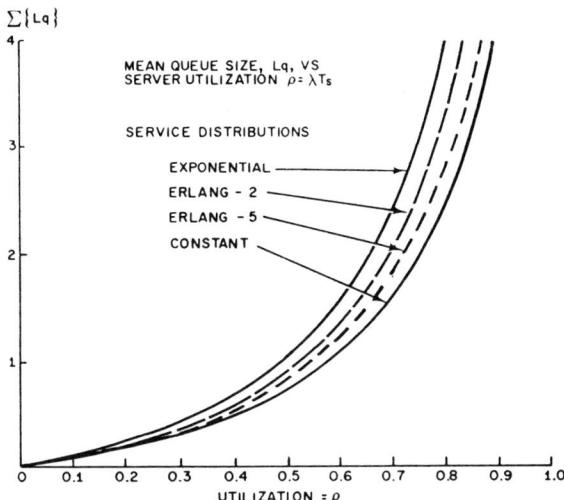

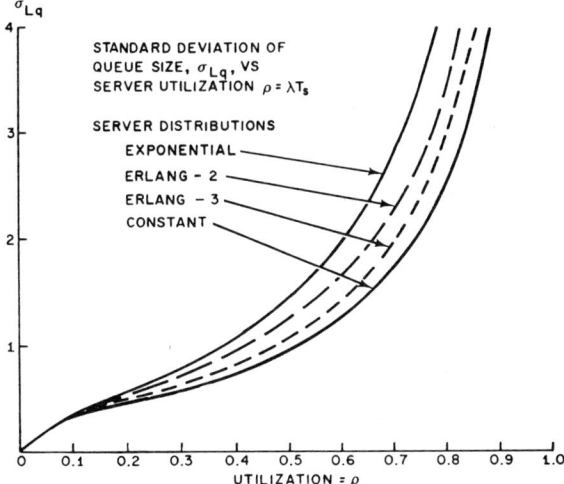

Figure 3.28 Queue size for single server—from IBM Systems Reference Library "Analysis of Some Queueing Models in Real Time Systems," GF20-0007-1 with permission.

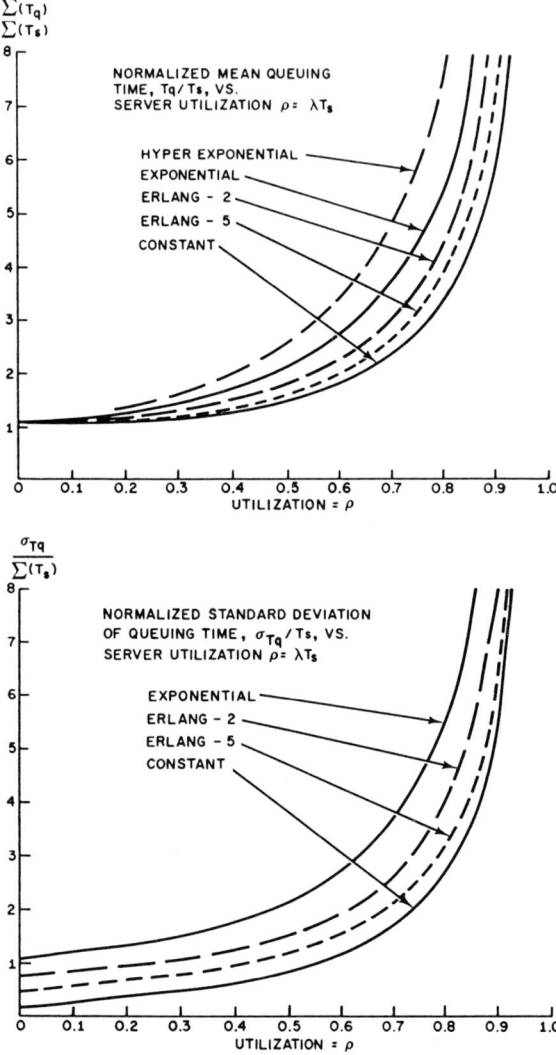

Figure 3.29 Queue time for single—From IBM Systems Reference Library "Analysis of Some Queuing Models in Real Time Systems," GF20-0007-1, with permission.

Some Aspects of Data Network Design

Exercise 3.1

Consider a node at which packets arrive according to a Poisson distribution. The mean arrival rate is 300 packets/min. The service time for each packet is exponentially distributed with a mean of 100 msec/packet. On the average, how long does a packet have to queue at the node?

Solution. The mean arrival rate is $\lambda = 5$ packets/sec. The mean service rate is $\mu = 10$ packets/sec.

From Equation (3.19), the total time (queueing plus service) is $T = 200$ msec. Since the mean service time $(1/\mu)$ is 100 msec the mean queueing time is $E(t_q) = 200 - 100 = 100$ msec.

Let the probability density of the packet size be

$$f(x) = \mu' e^{-\mu' x} \tag{3.20}$$

with mean packet size $1/\mu'$ bits per packet. Also, let the Shannon capacity of the communication channel be C_1 bits/sec (5). Then the maximum service rate in packets/sec is given by

$$r = \mu C_1. \tag{3.21}$$

Consider a network consisting of i channels (alternative routes). If the arrival rate for channel i is λ_i packets/sec then the mean delay of the ith channel is given by

$$E(T_i) = 1/(\mu C_i - \lambda_i) \quad \text{(sec)} \tag{3.22}$$

where T_i includes the queueing time and the transmission time. The total network delay in seconds is then given by

$$E(T) = \frac{1}{\sum_{i=1}^{n} \lambda_i} \sum_{i=1}^{n} \lambda_i T_i . \tag{3.23}$$

For more details see (4) and (5).

Exercise 3.2

In a certain switching center, messages arrive at random following a Poisson distribution. The mean arrival rate is 180 messages per hour. The line transmission rate is 12 characters per second. The message length probability distribution is given in the following table.

Message length (l_i) (characters)	60	100	150	220	270
Probability (p_i)	0.15	0.20	0.3	0.25	0.1

(1) On the average, how many messages will be waiting or serviced in the switching center? What will be the mean waiting time if the service times are exponentially distributed with mean of 10 seconds?

(2) If the service times are each 10 seconds, find:
- the mean waiting and queueing times, and
- the mean and standard deviation for the queue size.

Solution.

(1) The average message length is

$$E(L_M) = \sum_i p_i l_i$$

$$= (0.15 \times 60 + 0.20 \times 100 + 0.3 \times 150 + 0.25 \times 220 + 0.1 \times 270)$$

$$= 156 \text{ characters.}$$

Then the average service time is

$$E(t_s) = 156/12 = 13 \quad (s).$$

Since the mean arrival rate is

$$E(n) = 180/3600 \quad (\text{messages/s})$$

then from Equation (3.7) the utilization factor is

$$\rho = 0.65.$$

From the *exponential service time* distribution curve in Figure 3.29 for $\rho = 0.65$ we obtain

$$E(t_q)/E(t_s) = 3.0.$$

The average queueing time is, therefore,

$$E(t_q) = 39 \quad (s)$$

and, hence, the mean waiting time from Equation (3.4) is

$$E(t_\omega) = 26 \quad (s).$$

Information on the queue size can be obtained from Figure 3.28. Using the exponential distribution, for $\rho = 0.65$ we obtain the mean queue size (the average number of messages waiting on being served in the switching center) is

$$E(q) = 1.8 \quad \text{(messages)}.$$

Also from Figure 3.28, we note that the standard deviation is

$$\sigma_q = 2.4.$$

(2) For *constant service times* we obtain using the same procedures as above that

$$E(t_q) = 26 \quad (s)$$
$$E(t) = 13 \quad (s)$$
$$E(q) = 1.3 \quad \text{(messages)}$$

and

$$\sigma_q = 1.4.$$

3.12 PROBLEMS

3.1 A data message consisting of x bits is to be sent via a path in a network consisting of k hops and $k-1$ intermediate nodes. The propagation delay is d seconds per hop and the data transmission rate is b bit/s.

(a) Obtain expressions for the delay between the start of the message transmission at the sender to the end of the transmission at the receiver for the following network types:

  (i) a circuit switched network with a circuit set up time of s seconds
  (ii) a message switched network where the node processing is m seconds per node
  (iii) a datagram-style packet switched network where the packet size is p bits and the node processing takes n(secs) per node per packet.

(b) Compare the three cases if there are three intermediate nodes, the message length is 10,000 bytes, and the transmission rate is 48 kbit/s. Also assume

$$s = 2 \text{ (s)}$$
$$m = 0.3 \text{ (s)}$$
$$p = 1024 \text{ bits}$$
$$n = 10 \text{ (ms)}$$
$$d = 0.1 \text{ (ms)}.$$

3.2 A certain circuit switched network is required to provide a total message delay (including set up time) for a 10 kbyte message which is less than that of a virtual circuit style packet switched network. Find the maximum set-up time for the circuit switched network given that both networks interpose four hops between the source and destination. Each hop has a propagation delay of 0.1 msec with a transmission rate of 10 kbit/s.

The virtual circuit network has the following characteristics:

(1) call set up time: 150 msec per end node plus 80 msec per transfer node
(2) packet processing delay: 40 msec for each node in the transmission path
(3) packet size: 128 data bytes (ignore header)

3.3 Numbers in the range 0 to 9999 are to be represented by code characters. How many bits are required if the code to be used is

(1) ASCII
(2) EBCDIC
(3) Binary ?

3.4 Give the character code for the word "Engineer" using

(1) ASCII
(2) EBCDIC

3.5 In a system using a 2400 bit/s modem, the data source is a device generating 11-bit long characters (one start, eight data, two stop bits). How many characters per second can be transmitted?

3.6 (1) Calculate the average message delays for each link in the network shown in Figure 3.30. Assume that each message consists of 150 ASCII characters (1200 bits). All messages are destined for the Zanadu node.

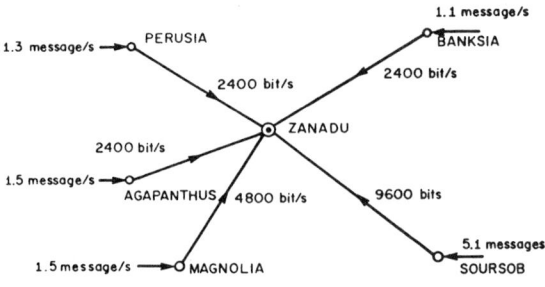

Figure 3.30

(2) Calculate the average message delay for the network as a whole.

(3) Calculate the average message delay for the link Agapanthus-Zanadu in Figure 3.30 assuming that the arrival rate has increased to 2 messages/s. If the problem of an infinite time delay occurs how could it be resolved?

3.7 Calculate the average delay time for each link to the metropolis of Soursob from the destinations listed in Table P.3.7. The distribution of messages originating in each node, the capacity of the links to Soursob and the mean message arrival rates are given in the table.

Table P.3.7
Links to Soursob.

Originating node	Message length in bits	Probability	C capac. (bps)	Mean arrival rate (messages/s)	T
Magnolia	1000	0.4			
	1250	0.2	9600	4	
	2000	0.1			
	1100	0.1			
	1200	0.2			
Banksia	560	0.4			
	660	0.3			
	450	0.2	9600	6	
	880	0.1			
Agapanthus	2000	0.6			
	1500	0.4	4800	2	
Zanadu	1200	1.0	9600	3	
Perusia	1200	1.0	1200	0.5	

3.8 For the M/G/1 queueing system, show that if two service time distributions have equal means, the one with the larger standard deviation will produce a longer waiting time.

3.13 REFERENCES

1. P.H. Enslow Jr., "What is a Distributed Data Processing System?" *Computer*, Vol. 11, pp 13–21, January 1978.

2. B.H. Liebowitz and J.H. Carson, *Distributed Processing, IEEE*, Long Beach California, 1978.
3. W. Chou and H. Frank, "Topological Optimization of Computer Networks," *Proc. IEEE*, Vol. 60, pp. 1385 et seq, November 1972.
4. A.S. Tanenbaum, *Computer Networks,* Prentice-Hall, 1981.
5. H. Kobayashi, *Modeling and Analysis,* Addison-Wesley, 1978.
6. I.B.M., *Analysis of some Queueing Models in Real Time Systems,* I.B.M. Publication No. GF20-007-1.
7. J. Martin, *Systems Analysis for Data Transmission,* Prentice-Hall, 1972.

Chapter 4

ERROR CONTROL IN DIGITAL NETWORKS

4.1 INTRODUCTION

In this chapter we examine several techniques of ensuring that the signals sent through digital transmission systems and networks are delivered to the user without errors. This is particularly important in data networks where high reliability is essential.

As we have seen in earlier chapters, data networks rely on a variety of transmission media for interconnection between terminals, nodes, and computers. Transmission systems such as telephone cables or terrestrial and satellite radio links inevitably suffer occasional transmission errors. In order to maximize system reliability the digital network designer must provide for appropriate protocols and coding procedures.

Several approaches to error control are possible:

(1) forward error detection—the inclusion of parity check bits generated from a code, chosen so as to detect (but not correct) any transmission error patterns
(2) automatic-repeat-request (ARQ)—the utilization of forward error detection and retransmission of data received in error to ensure that data delivered to the user is error-free,
(3) forward error correction (FEC)—the use of parity check bits generated from codes chosen for their ability to locate and correct transmission errors, and
(4) hybrid ARQ systems—the combination of FEC procedures with error detection and retransmission.

It is common to judge the merits of an error control system for a link in a network by the following two criteria:

(1) *Reliability*—the probability that the data will be error-free when delivered to the user

(2) *Throughput efficiency* (or simply *"throughput"*)—the ratio of the average number of information bits successfully accepted by the receiver per unit time to the total number of bits that could be transmitted per unit time.

In any practical error control system design, there may need to be a compromise between these two criteria. For example, in an ARQ system, the reliability can usually be increased by the use of greater numbers of parity check bits for error detection. However, this will reduce the throughput, especially when the channel is quiet. In addition to the above two criteria, it is necessary to consider cost. Many powerful error correcting codes have been designed to correct large numbers of single errors or error bursts. However, the complexity and cost of the required decoder may mean that the codes are impractical for some applications.

This chapter initially reviews the error control options which are available for digital transmission or data storage systems. Then the principles of error-control codes for error detection are summarized and some implementation schemes are discussed. Particular attention is given to error detection procedures used in packet-switched data networks and in computer memories. Some recent results concerning the selection and performance of error detection codes are then presented.

ARQ schemes are the most popular method for error control on most public data networks. This is because they can be more readily designed to guarantee very high reliability than FEC schemes. Various ARQ systems will be examined in this chapter. Recently, several new hybrid ARQ schemes have been developed that provide better throughput performance on high-speed digital links with significant transmission delays, such as long terrestrial or satellite links. The protocols and performance of these schemes will be summarized.

Error detection or error correction techniques rely on the use of codes. Error control codes were the subject of a great deal of theoretical study in the 1960s and early 1970s. Many classes of codes were developed along with various procedures for their implementation in error correction applications. Despite this theoretical development, relatively few practical applications were in evidence. The complexity of the required encoders and decoders, particularly the decoders, was a major deterrent. System designers usually found it more economic to seek other means for ensuring high reliability. Now the situation has been changed considerably by the development of microcircuits and microcomputers. Coding schemes for error detection and error correction are being incorporated in many digital systems, particularly in digital data networks, computer storage hardware, and in digital recording systems.

Error control techniques using detection and correction codes are being applied not only in digital networks and transmission systems, but also are becoming commonplace in memory storage systems. On magnetic tapes or disks, material or surface defects or dust can give rise to errors. Often the resultant errors occur

Errors and Erasures 165

in bursts. Appropriate error detection and correction procedures are incorporated in current computer designs to ensure high reliability of data storage and transfer.

Even in microprocessor systems currently available, provision is made for error control, usually by a combination of error correction and detection procedures. Errors can occur in random-access memory (RAM) systems as a result of noise, crosstalk, chip failure, or because of radiation from traces of radioactive elements in the encapsulation. This may cause a storage cell to change its state or the sense amplifier value to be incorrect. Error control techniques have been developed for RAM applications which guarantee that errors will either be corrected or, with high probability, be detected.

4.2 ERRORS AND ERASURES

There is a simple pencil-and-paper game that illustrates how codes can be designed for error detection, correction of erasures (that is, errors in known locations), and error correction. The game was suggested by McEliece (1). Consider the three interlocking circles shown in Figure 4.1(a). There are seven different segments within the diagram. We will treat them as though they are the seven binary bit locations in a 7-bit code word. Four of the segments shown already have binary symbols, 0 or 1 assigned to them in Figure 4.1(a). We will treat these as the information bits. The values assigned are quite arbitrary.

Next, we consider the other three remaining bits that occupy the remaining three empty segments. These will represent redundant parity bits to be used for detection or correction. Let the coding rule for the assignment of these three parity bits be determined as follows. Each circle contains four segments. One of them contains a parity bit. The parity bit must be chosen so that there is an even number of 1's in the circle. Figure 4.1(b) shows how the parity bits must be assigned to meet this requirement for the previously chosen data bits. The seven bits constitute a code word. The order of the bits is not important except that the decoder must know how they are ordered. We call this a (7,4) code in which each code word contains seven bits, four of them being data bits.

We now examine the error correction decoding operation that might be performed after this code word has been passed through a noisy channel or after it has been read in and out of a storage medium. Let the received code word be that shown in Figure 4.1(c). The decoder checks each circle in turn to determine whether each still contains an even number of 1's; the top circle does but the two bottom ones do not. The bit shown with an asterisk is, therefore, assumed to be in error since it is the only segment common to both the bottom circles but not to the top one. The decoder can proceed to reverse that bit and error correction is complete. It is simple to verify by creating errors in any one of the

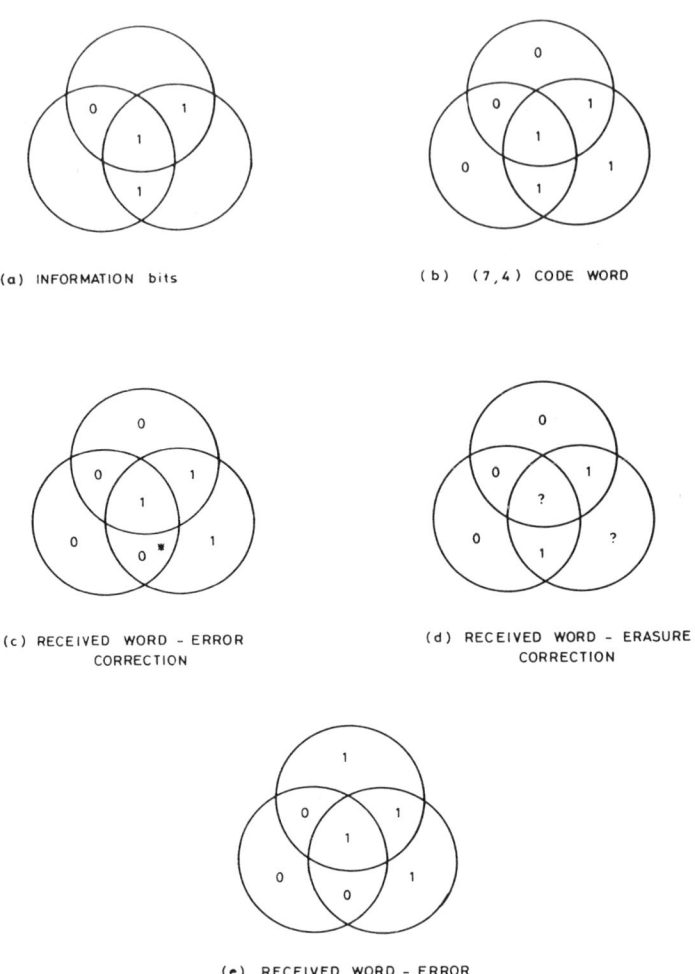

Figure 4.1 A coding game (from McEliece, (1)).

seven locations that the decoder has no difficulty in locating and correcting a single error.

However, the code is not capable of correcting double error patterns. In such cases, the decoder will make an error. To see this, try changing any two of the bits in the code word and then check for even parity in each circle. A decoder error results in each case.

We now consider the problems of decoding for correction of erasures. Figure 4.1(d) shows a code word in which the contents of the two bit locations are

Errors and Erasures

unknown. These are known as erasures. The locations of the erasures are assumed known but the contents are not. By examining Figure 4.1(d) and checking each circle in turn for correct parity, it is apparent that the erasures can be corrected without ambiguity. In this case, they should both be 1's. This code is capable of correcting up to two erasures in any received code word.

Next, consider the method of detection. Figure 4.1(e) represents a possible received word. Note that it does not satisfy the parity formation rules in each circle. Therefore, it is apparent that errors must have occurred to change some original code word into this received word. The number or location of the errors that have occurred cannot be determined.

It follows that error detection involves the receiver in simply checking whether the received word is a code word. If it is not, then errors are detected. It is not difficult to check that this (7,4) code is capable of ensuring that any single or double error pattern will be detected. Not all patterns of three or more errors are detectable.

We can summarize the above ideas as follows. An (n,k) block code is one in which k-bit information blocks are encoded into n-bit code words before transmission. The $n-k$ additional parity bits can be used for error detection, error correction, erasure correction, or a combination of these.

The structure of a code is determined by the way the data bits are combined to give the parity bits. The following exercise illustrates this.

Exercise 4.1

For the (7,4) code described above, determine the parity-check equations. These describe how the parity bits are formed. List all the possible code words in the code.

Solution. Each code word may be represented in the form

$$\bar{v} = (v_0, v_1, v_2, v_3, v_4, v_5, v_6)$$

where v_0, v_1, and v_2 are the three parity bits and v_3, v_4, v_5, and v_6 are the four information bits. From the description given of the encoding procedure, each parity bit is a function of three of the data bits as follows

$$v_0 = v_3 + v_4 + v_5$$
$$v_1 = v_4 + v_5 + v_6$$
$$v_2 = v_3 + v_5 + v_6 \quad (4.1)$$

These are modulo-2 sums. The modulo-2 addition rules are

$$0 + 0 = 0 \quad 0 + 1 = 1 \quad 1 + 0 = 1 \quad 1 + 1 = 0.$$

The equations (4.1) are the parity-check equations for the code. The number of different code words in the code is

$$2^k = 2^4 = 16$$

For each of the 16 4-bit information blocks, the associated parity bits can now be found using Equation (4.1). The resultant code words in the (7,4) code are listed in Table 4.1.

Table 4.1
A (7,4) block code

Information bits	Code word
0000	0000000
0001	0110001
0010	1110010
0011	1000011
0100	1100100
0101	1010101
0110	0010110
0111	0100111
1000	1011000
1001	1101001
1010	0101010
1011	0011011
1100	0111100
1101	0001101
1110	1001110
1111	1111111

4.3 ERROR DETECTION USING BLOCK CODES

Error detection refers to the use of one or more parity bits for detection of transmission errors in a block of data. In this section, fundamental ideas concerning the parity bit generation and error detection procedures will be reviewed. Methods for determining the code performance are outlined. Some bounds on the probability of error of a detection system are presented.

4.3.1 Single Bit Parity Detection

The simplest form of error detection scheme is to transmit data in blocks each containing a single parity bit. This simple scheme is discussed here as a means

Error Detection Using Block Codes

of introducing some elementary ideas and notation. We assume the data to be in binary form. Let the input data be arranged in blocks of k binary bits. Let $\bar{d}$ represent a k-bit data block. Each data block is then encoded using an (n,k) linear block code C_o for which a code word $(v_o, \bar{d})$ consists of $n = k+1$ bits. The bit v_o represents the single parity bit.

Since there are k data bits, there will be 2^k different possible data blocks. That is, there will be 2^k different possible code words in the code C_o, each of length n. We assume that a *systematic code* is used, that is one for which a code word consists of the original data bits plus the parity. If we write the data as:

$$\bar{d} = (d_0, d_1, d_2, \ldots, d_{k-1})$$

where $d_0, d_1, d_2, \ldots$ are the binary data bits (either 0 or 1), then the transmitted block (code word) is

$$\bar{v} = (v_0, \bar{d})$$
$$= (v_0, d_0, d_1, d_2, \ldots, d_{k-1})$$

where v_0 is the single parity bit. The bit d_{k-1} is taken to be transmitted first. The encoding process to generate $\bar{v}$ can be simply represented by the matrix multiplication

$$\bar{v} = (d_0, d_1, \ldots, d_{k-1}) \cdot \begin{bmatrix} p_{00} & 1 & 0 & 0 & \cdots & & & 0 \\ p_{10} & 0 & 1 & 0 & 0 & \cdots & & 0 \\ p_{20} & 0 & 0 & 1 & 0 & 0 & \cdots & 0 \\ \vdots & & & & & & & \\ p_{k-1,0} & 0 & 0 & 0 & \cdots & & & 0 & 1 \end{bmatrix}. \quad (4.2)$$

That is,

$$\bar{v} = \bar{d}\, G \quad (4.3)$$

where the $k \times n$ matrix G is called the *generator matrix* of the code C_0. The elements $p_{00}, p_{10}, \ldots$ in G are each either 0 or 1.

For any systematic block code, the generator matrix G consists of a parity submatrix P with k rows and $n-k$ columns and a $k \times k$ identity submatrix I. The choice of G completely describes the code. For this case, the single parity check bit v_0 is given by

$$v_0 = d_0 p_{00} + d_1 p_{10} + \ldots + d_{k-1} p_{k-1,0} \quad (4.4)$$

where addition is modulo-2, that is v_0 is either a 0 or a 1. For an "even-parity-bit" checking system, the parity bit v_0 is chosen so that the total number of 1's in each transmitted code word (including the parity bit) is even. Examination of

Equation (4.4) shows that this would be achieved by using a code C_0 with generator matrix such that

$$p_{00} = p_{10} = p_{20} = \ldots = p_{k-1,0} = 1.$$

At the receiver, the error detection system determines whether one or more transmission errors have occurred by counting the number of received 1's in each block and determining whether the result is odd or even. That is, the receiver checks whether the received block or word satisfies the encoding Equation (4.4). For an even-parity-bit checking system, an odd number of 1's indicates that the received block or word is not one of the valid words in the code set C_0.

The error detection scheme is by no means 100 percent reliable. If, for example, errors occur in an even number of positions in the block, the system will fail to detect the presence of errors. In such cases, the error pattern has changed the transmitted code word into another code word in the code set.

In any error detection scheme, including those using more than one parity check bit, we can therefore, see that the error detection circuits must perform as follows:

(1) When a block is received, check whether it is a valid code word in the set of 2^k possible code words.
(2) If it is not a valid code word, then errors have occurred and are detected.
(3) If it is a valid code word, then either no errors have occurred or the error pattern is undetectable by the code.

Clearly, the choice of good codes for use in error detection must be such that the most likely error patterns do not change one code word into another valid code word. Before we examine this further, we consider first how the reliability of the single parity detection system can be evaluated.

4.3.2 Weight Distribution of a Code

The total number of 1's in a code word is known as its *Hamming Weight* (or simply "weight"). For example, if a code word is

$$\bar{v} = (1\ 0\ 0\ 1\ 1\ 0\ 0\ 1)$$

then its weight is 4.

The *weight distribution* of all the code words in a code is useful in determining how well the code will perform in detecting errors. If for the 2^k possible code words in a code set,

$$A_0 \quad \text{have weight 0,}$$

$$A_1 \quad \text{have weight 1,}$$

..., and

A_n have weight n,

then the numbers

$$A_0, A_1, \ldots, A_n$$

are called the "weight distribution" or the "weight spectrum" of the code.

Exercise 4.2

Consider an (8,7) single-parity check code for which the single parity check bit is given as the modulo-2 sum of the seven data bits. That is,

$$v_0 = d_0 + d_1 + \ldots + d_{k-1}.$$

Find the weight distribution of the code.

Solution. Each code word contains seven data bits and one parity bit. There is one code word for which the seven data bits are all 0, and therefore, the parity bit is 0. This code word has weight 0. Hence,

$$A_0 = 1.$$

There are seven code words for which six of the data bits are 0 and the remaining bit is a 1. In these code words, the parity bit will be a 1. As a result, these code words have weight 2. These are not the only code words of weight 2. We must also consider those code words for which two of the data bits are 1's and the other six bits are 0. There will be

$$\binom{7}{2} = \frac{7!}{2!5!} = 21$$

such cases. In these cases the parity bit will also be zero so the code words have weight 2. In total we obtain

$$A_2 = 28.$$

By extending this method to considering all possible 7-bit data sequences, it is easy to show that the code weight distribution is

$$A_0 = 1, A_2 = 28, A_4 = 70, A_6 = 28, A_8 = 1. \qquad (4.5)$$

4.3.3 Error Detection Reliability of the Single-Parity Code

We can now describe the error patterns that are undetectable by the single-parity code. If a block is received with one of the errors these error patterns, the receiver will not be able to detect and will incorrectly assume the block is error-free.

For example, if a transmission error occurs in the first and last position of the 8-bit block, we represent the error pattern as

$$\bar{e} = (1\ 0\ 0\ 0\ 0\ 0\ 0\ 1)$$

where a 1 represents an error and a 0 represents a correct transmission.

When the vector $\bar{e}$ is added modulo-2 to the transmitted vector the result is the received block or vector $\bar{r}$.

To illustrate, if the transmitted block is $\bar{v} = (1\ 0\ 1\ 0\ 1\ 0\ 1\ 0)$ and the error pattern is $\bar{e} = (1\ 0\ 0\ 0\ 0\ 0\ 0\ 1)$ then the received block is

$$\begin{aligned}\bar{r} &= \bar{v} + \bar{e} \\ &= (1\ 0\ 1\ 0\ 1\ 0\ 1\ 0) + (1\ 0\ 0\ 0\ 0\ 0\ 0\ 1) \\ &= (1+1\quad 0+0\quad 1+0\quad 0+0\quad 1+0\quad 0+0\quad 1+0\quad 0+1) \\ &= (0\ 0\ 1\ 0\ 1\ 0\ 1\ 1).\end{aligned}$$

The weight of the error pattern $\bar{e}$ in the example is 2. If the channel errors are assumed to be randomly distributed with bit-error rate p, then the probability that the above error pattern of two errors will occur is

$$P_2 = p^2(1-p)^6.$$

If we can sum the probabilities of all the undetectable error patterns, then we obtain the probability $P_u(E)$ that the code fails to detect an error.

It is an important property of all linear block codes that the only error patterns that can change one code word into another valid code word must themselves be code words in the code. This is because a "closure rule" applies to linear codes. That is, the modulo-2 sum of any two codes must also be a code word in the code. For details refer to Lin and Costello (2). Therefore, the only undetectable error patterns are those which are the same as code words in the code. The list of all possible code words is also a list of all possible undetectable error patterns.

For the single-parity check code, it follows that the code weight distribution $A_2, A_4, \ldots, A_8$ must also be the weight distribution of the undetectable error patterns. Note that A_0 is the weight of the all-zero error pattern so that is omitted since an all-zero error pattern represents no errors. It follows that we can determine the probability of failing to detect an error $P_u(E)$ for any parity detection linear (n, k) block code by computing

$$P_u(E) = \sum_{i=1}^{n} A_i p^i (1-p)^{n-i}. \tag{4.6}$$

Exercise 4.3

Compute $P_u(E)$ for the $(8,7)$ single-parity check code for a random error channel with bit error rates of 10^{-4} and 10^{-3}, respectively.

Solution. For the single-parity (8,7) code, we obtain from Equation (4.5) and (4.6)

$$P_u(E) = 28p^2(1-p)^6 + 70p^4(1-p)^4 + 28p^6(1-p)^2 + p^8. \quad (4.7)$$

If the channel bit-error rate is $p = 10^{-4}$, we obtain from Equation (4.7) that

$$P_u(E) = 2.8 \times 10^{-7}.$$

If $p = 10^{-3}$ then

$$P_u(E) = 2.8 \times 10^{-5}.$$

In the latter case this means that there is a probability of 2.8×10^{-5} that a block will be received containing an error pattern such that the error detection code will fail to detect that errors have occurred.

If the block length is doubled (that is, a (16,15) code is used), then for $p = 10^{-3}$, we find that

$$P_u(E) = 1.2 \times 10^{-4}.$$

As the block length is increased, so the performance of the single-parity check system deteriorates. Next we consider the use of more powerful error detection codes to provide higher reliability.

4.3.4 Linear Block Codes for Error Detection

In many applications, more effective error detection procedures are required than those obtained from single-parity check systems. For example, in data transmission networks using packet-switching, extremely high reliability is demanded so that users can be assured that their received data is error free.

In packet switched networks, as discussed in Chapter 3, packets may contain up to approximately 1024 bits. Before transmission, each packet is encoded using a powerful error detection code which generates 16 parity check bits known as the frame checking sequence (FCS). The choice of a code to be used for error detection must ensure high reliability, that is low $P_u(E)$. Typically, it is demanded that $P_u(E)$ should be less than 10^{-10} on any one link.

Error detection schemes that guarantee low probability of undetected error are also essential in data storage systems such as those using random-access-memory (RAM) arrays and magnetic disks. In general, it is necessary to encode the data to be protected in blocks, using an (n,k) block code which generates $n-k$ parity bits for each block. These parity bits are formed as linear sums of combinations of data bits. The choice of the code determines how these sums are formed and how well the error detection will perform.

The CCITT Recommendation X.25 recommends the use of a certain (32767,32751) block code for error detection in packet-switched data networks.

This code generates $n-k = 16$ parity bits which could be used to check up to 32751 data bits. In practice, the number of data bits encoded is considerably less.

It is currently of considerable research interest to find ways of determining how to compute the probability of undetected error $P_u(E)$ for different error detection codes. This is an essential step in the search for good codes for error detection systems. We observed that $P_u(E)$ could be calculated using Equation (4.6) if the code weight distribution $A_1, A_2, \ldots, A_n$ is known. Unfortunately, it is not known how to calculate this distribution for all but a few selected codes. For short codes, a computer can be used to generate each possible code word and its associated weight, and hence, to find the weight distribution for the whole set of code words.

At first thought, it might seem a simple task to write a computer program to compute the weight distribution for a block code. If there are k data bits, there are 2^k different code words. If k is large, the task of generating each of 2^k code words and counting the number of 1's in each becomes impossible. This is certainly true for the 2^{32751} code words in the CCITT X.25 code!

4.3.5 Minimum Distance of a Code

For any block code, an important measure of its performance is the *minimum distance* $d_{\min}$. In order to define this, it is necessary to use the concept of Hamming distance. The *Hamming distance* (or simply "distance") between any two n-bit code words in a code set is equal to the total number of positions in which the two words differ. For example, the two words $\bar{v}_1 = (1\ 0\ 0\ 1\ 1\ 0\ 1\ 0)$ and $\bar{v}_2 = (1\ 0\ 0\ 1\ 1\ 1\ 0\ 1)$ differ in the last three positions so the distance between them is three.

Now consider that for a given code, each of the 2^k code words is compared against every code word in the code. The minimum distance between any pair of code words is called the minimum distance of the code.

Exercise 4.4

Find the minimum distance of the code listed in Table 4.1.

Solution. By examining the distances between all the pairs of code words in Table 4.1, it is apparent that the minimum distance of that code is

$$d_{\min} = 3.$$

If say, a code is such that its minimum distance $d_{\min} = 3$, then when a code word is transmitted through a noisy channel, there must be transmission errors in at least three positions to change that code word into another code in the code set. For example, if the code word transmitted is

$$\bar{v} = (1\ 0\ 0\ 1\ 1\ 0\ 1\ 0)$$

then an error pattern represented by

$$\bar{e} = (0\ 0\ 0\ 0\ 0\ 1\ 0\ 1)$$

of weight 2 would change $\bar{v}$ into the received block

$$\bar{r} = (1\ 0\ 0\ 1\ 1\ 1\ 1\ 1).$$

This cannot be a code word because it differs by less than d_{min} from a code word. A check against Table 4.1 confirms this.

We observed previously that an error detection system can detect only error patterns which do not cause a transmitted code word to be altered into another code word. Therefore, if a certain code has minimum distance of d_{min}, the code can detect all error patterns containing t_d errors providing

$$t_d \leq (d_{min} - 1). \tag{4.8}$$

Actually, it can also detect quite a large number of other error patterns (but not all patterns) containing d_{min} or more errors. To illustrate, it is easy to see that the single-parity check $(k+1,k)$ code discussed earlier has $d_{min} = 2$. Therefore, this code can detect all of the error patterns containing single errors per block. However, this does not give a very accurate indication of the probability of undetected error of the code since it can also detect all error patterns with an odd number of errors.

4.4 CYCLIC CODES FOR ERROR DETECTION

For any (n,k) block code, Equation (4.3) indicates how parity bits are generated. First, the code generator matrix G must be specified. Then the parity bits for any given message can be generated by hardware or software capable of carrying out matrix multiplication.

Among all the possible block codes, there is an important subclass known as cyclic codes. An (n,k) linear block code is called a cyclic code if every cyclic shift of a code vector results in another vector in the code.

By restricting the choice of codes to the subset of cyclic codes, the coding and error detection implementation can be considerably simplified and variable message lengths can be encoded readily. For these reasons, cyclic codes are popular for error detection. We will briefly review here those properties of cyclic codes sufficient to understanding the encoding and decoding processes. Further details can be found in Lin and Costello (2).

4.4.1 Polynomial Representation

In discussing cyclic codes, it is convenient to associate a polynomial with each code word. For instance, if a code word is given by

$$\bar{v} = (1\ 1\ 0\ 0\ 1\ 0\ 1)$$

then it is represented in the form of a polynomial

$$v(X) = X^0 + X^1 + X^4 + X^6 \qquad (4.9)$$

corresponding to 1's in the first, second, fifth, and seventh positions. The symbol X has no significance except that its power represents position in the code word.

4.4.2 Generator Polynomial

Each cyclic code is characterized by a *generator polynomial* $g(X)$ which is used to generate the code words. The generator polynomial for cyclic codes plays a similar role to that of the generator matrix G used for block codes (see for example Equation (4.3)). Tables of generator polynomials for various cyclic codes are given in coding theory texts such as Lin and Costello (2).

The generator polynomial for the code recommended by CCITT (Rec. X.25) for packet-switched data networks is given by

$$g(X) = 1 + X^5 + X^{12} + X^{16}. \qquad (4.10)$$

For a given cyclic code, the code generator polynomial $g(X)$ divides into every code word polynomial. That is, if conventional polynomial division is used, the remainder $r(X)$ consists of $n-k$ 0's. To illustrate, let the polynomial $v(X)$ in Equation (4.9) represent a code word generated by the generator polynomial

$$g(X) = 1 + X + X^3. \qquad (4.11)$$

The following shows how the division of $v(X)$ by the generator polynomial $g(X)$ proceeds. Note that in these polynomial operations, addition and subtraction are done modulo-2. That is

$$0 + 0 = 1 + 1 = 1 - 1 = 0$$

and also

$$X - X = X + X = 0$$

The steps in the polynomial division are as follows

$$
\begin{array}{r}
X^3 + 1 \\
X^3+0+X+1 \overline{\smash{\big)} X^6+0+X^4+0+0+X+1} \\
\underline{X^6+0+X^4+X^3} \\
X^3+0+X+1 \\
\underline{X^3+0+X+1} \\
\text{Remainder } r(X) = \underline{0}
\end{array}
$$

Cyclic Codes For Error Detection

The remainder is zero as postulated, that is, $g(X)$ divides into $v(X)$ without remainder. This guarantees that $v(X)$ must be a polynomial in the code generated by $g(X)$.

We will next examine the encoding and decoding processes necessary to carry out error detection using cyclic codes.

4.4.3 Generation of Parity (Encoding)

Consider that k message bits are to be encoded using an (n,k) cyclic code to generate $n-k$ parity bits. The k bits may include data, address, and any other control bits which must be protected from error. Let the resultant code word to be transmitted be of the format shown in Figure 4.2(a). The code word consists of the k original message bits and $n-k$ additional parity bits. A code which generates code words of this form is known as a *systematic code*.

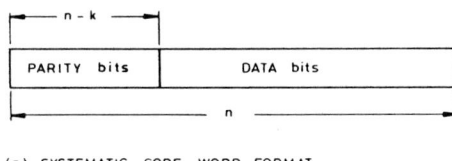

(a) SYSTEMATIC CODE WORD FORMAT

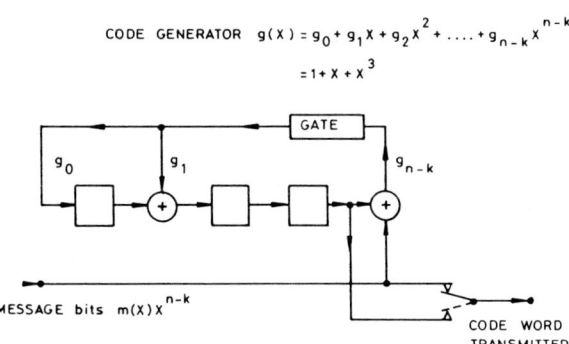

(b) ENCODER SCHEMATIC

Figure 4.2 Encoder for a cyclic code.

To illustrate the encoding procedure consider a (7,4) cyclic code. Let $m(X)$ represent in polynomial form a 4-bit message block to be encoded. In particular, a 4-bit messge (0 1 0 1) is represented by

$$m(X) = X + X^3 \qquad (4.12)$$

The encoder generates the required $n-k=3$ parity bits associated with $m(X)$ by first multiplying $m(X)$ by X^{n-k}, then dividing the result by the code generator $g(X)$. The remainder after this polynomial division becomes the parity check polynomial $p(X)$ representing the 3 required parity bits. That is, the parity check polynomial $p(X)$ is formed using

$$p(X) = \text{rem}\left[\frac{X^{n-k}\, m(X)}{g(x)}\right] \qquad (4.13)$$

where rem[.] denotes the remainder. For example, let $m(X)$ be given by Equation (4.12) and $g(X)$ by Equation (4.11).

Then the parity polynomial is

$$p(X) = \text{rem}\left[\frac{X^3\,(X + X^3)}{1 + X + X^3}\right]$$

and the polynomial division proceeds as follows:

```
                    X³ +          1
     X³ + X + 1  ) X⁶ + X⁴
                  X⁶ + X⁴ +  X³
                             X³
                             X³ + X + 1
      Remainder =                 X + 1
```

Therefore $p(X) = X + 1$ which represents the parity bits (1 1 0). The resultant code word can now be represented in polynomial form by

$$v(X) = p(X) + X^{n-k}\, m(X) \qquad (4.14)$$

where the term $p(X)$ represents the $n-k$ parity bits and the term $X^{n-k}m(X)$ represents the k message bits. As illustrated in Figure 4.2(a), these are the high-order bits in $v(X)$. Multiplication of $m(X)$ by X^{n-k} is equivalent to placing the message bits in the right-most positions. For our example, we obtain from Equations (4.11), (4.12), (4.13), and (4.14)

$$v(X) = 1 + X + X^4 + X^6$$

so that the transmitted code word is (1 1 0 0 1 0 1). The first three bits are parity and the last four bits are the original message.

Cyclic Codes For Error Detection

In practice, the operations denoted by Equation (4.13) and (4.14) can be performed with relatively simple hardware. Figure 4.2(b) shows an encoder circuit using linear feedback shift registers for polynomial division. The message bits are shifted in first with the gate closed. Also the switch is as shown so the message bits are shifted directly out to the channel. After the last of the k shifts, the shift register contains the remainder $p(x)$. That is, the circuit has carried out long division. Now the gate is opened, the switch changed over and the shift register contents shifted out to the channel. It is easy to verify that the circuit generates the required parity bits.

It is perhaps instructive to compare the operation of the circuit in Figure 4.2(b) with our "long-hand" method of performing long division. When we divide polynomials by hand, we reduce the degree of the dividend (input) one step at a time by subtracting off copies of the divisor $g(X)$ until the degree of the dividend becomes less than that of the divisor. This final dividend we call the remainder. The feedback shift register does the same thing. When the register is shifted, if a 1 is the output on the feedback path, we use it to add a copy of the divisor to the dividend. Since the operation is binary modulo-2, the addition is equivalent to binary subtraction.

4.4.4 Encoder for Rec. X.25 Frame Check Sequence

The method of calculating the 16-bit frame check sequence (FCS) for data network protocols recommended in CCITT Rec. X.25 (and used in SDLC, HDLC, and ADCCP standards) differs somewhat from that described above. Figure 4.3 shows the encoder circuit.

The generator polynomial was given in Equation (4.10). The encoder operates as follows:

(1) Before encoding begins, the 16 flip flops are set to the all 1's state.
(2) The message bits are entered with gates A and C open and B closed.

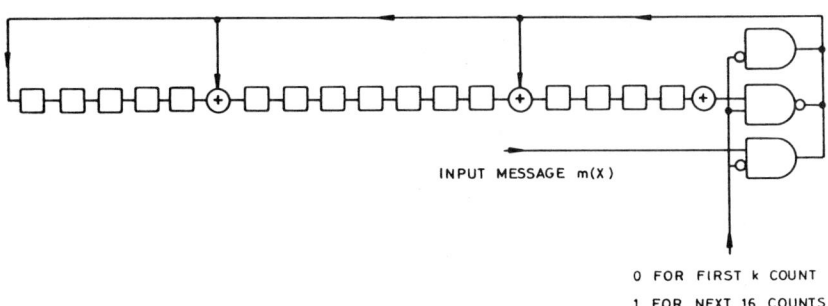

Figure 4.3 Encoder for CCITT Rec. X.25 code.

(3) After the message bits are entered, gates A and C close and B opens.
(4) The 16 check bits are then read out in complemented form.

The transmitted code word $v(x)$ can be related to the message $m(x)$, as follows. Since the encoder contains 16 1's initially, it is actually encoding the polynomial

$$m(X) + X^{k-16}(1 + X + X^2 + \ldots + X^{15})$$

The encoder premultiplies by X^{16} and then calculates the remainder after dividing by $g(X)$. Then the register contains

$$p(X) = \text{rem}\left[\frac{X^{16}m(X) + X^k(1 + X + X^2 + \ldots + X^{15})}{g(X)}\right]. \quad (4.15)$$

This 16-bit polynomial is then complemented, which is equivalent to forming

$$p(X) + (1 + X + X^2 + \ldots + X^{15}).$$

When appended to $X^{16}m(X)$ for transmission, we obtain the code word

$$v(X) = X^{16}m(X) + p(X) + 1 + X + X^2 + \ldots + X^{15}. \quad (4.16)$$

The above encoding operation (and the associated decoding) can be performed using a simple commercially available integrated circuit such as the Motorola MC6854.

In general, the best method of implementation depends on the required bit rate. For low speeds ≤100 kbit/s, a microprocessor may be appropriate. For medium speeds, say 100 kbit/s to 100 Mbit/s, bit serial feedback shift registers can be used. For higher speeds, byte-wide serial shift registers are used.

For byte-serial encoder implementation, it is usual to have the number of parity bits $n-k$ an integral number of bytes. Then an $n-k$ stage encoder is used, each stage consisting of a byte of say b bits. The encoder encodes one byte at a time as illustrated in Figure 4.4. This scheme effectively encodes b times as fast as a conventional encoder. The subcircuits $M_0, M_1, \ldots$ are required to multiply their output bytes by a fixed $b \times b$ matrix. For $b = 8$ bits per byte, this might be achieved using an 8×256 read-only-memory (ROM).

4.4.5 Decoding for Error Detection

In order to determine whether a received n-bit block contains errors, the decoder must determine whether the block is a valid code word. If it is not a code word, then one or more errors must have occurred. If it is a code word, then the decoder must assume that it is the same code word as the one originally sent.

Consider that a cyclic code is to be used. To check whether a recovered block is a valid code word, the decoder simply has to determine whether it is a multiple

Cyclic Codes For Error Detection

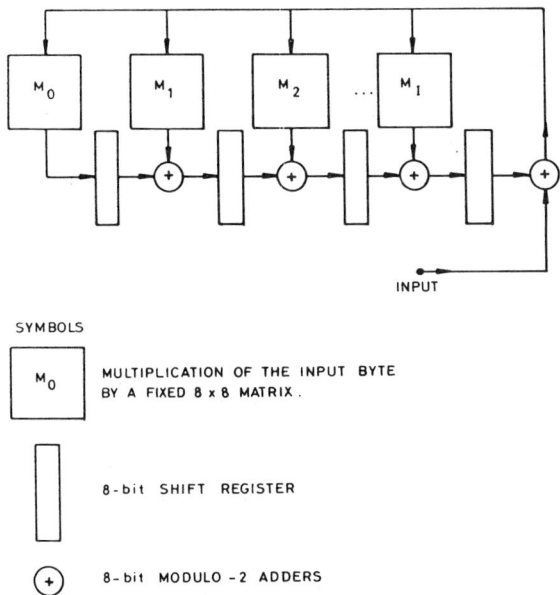

Figure 4.4 High-speed byte-by-byte encoder.

of $g(X)$, the genrator polynomial for the code. All code words are multiples of $g(X)$.

Error detection, therefore, consists of dividing the received polynomial $r(X)$ by the generator polynomial $g(X)$ and checking the remainder $s(X)$. The $n-k$ bits represented by $s(X)$ are known as the *syndrome*. Hence, we can write the $n-k$ syndrome bits in polynomial form as

$$s(X) = \text{rem}\left[\frac{r(X)}{g(X)}\right]. \tag{4.17}$$

If $s(X)$ is not zero, the block contains errors. Otherwise, it is assumed error-free.

The decoding (error detection) process is, therefore, very similar to encoding. Figure 4.5 shows an error detection circuit for the (7,4) code generated by $g(X)$ in Equation (4.11). The received bits are shifted in to the feedback shift register circuit with feedback elements determined by the nonzero terms in $g(X)$. After the n bits have been shifted in, the shift register contains the syndrome. If all $n-k$ syndrome bits are zero, the received block contains no detectable errors.

We can think of $r(X)$ as having two parts, data and parity. Thus

$$r(X) = r_p(X) + X^{n-k}r_m(X) \tag{4.18}$$

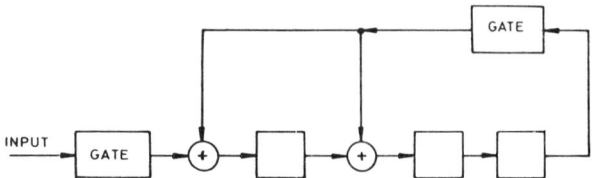

Figure 4.5 Error detection circuit for the (7,4) code generated by $g(X) = 1+X+X^3$.

where $r_p(X)$ are the received parity bits and $r_m(X)$ are the received message bits. This suggests an alternative error detection circuit. At the receiver, let the received message bits $r_m(X)$ be encoded using an encoder identical to that at the transmitter.

The syndrome $s(X)$ can easily be shown to be the sum of $r_p(X)$ and the result of encoding $r_m(X)$. That is, if encoding $r_m(X)$ produces the same parity bits as $r_p(X)$, then $s(X)$ will be zero. Then, $r(X)$ must be a valid code word and is assumed error-free.

4.4.6 Error Detection for the CCITT Rec. X.25 Code

As with its encoder, the error detection procedures used in the Rec. X.25 paritycheck system differ a little from the conventional decoder described above. Figure 4.6 shows a decoder suitable for the X.25 code. It is very similar to the encoder circuit of Figure 4.3, wihout the control gating.

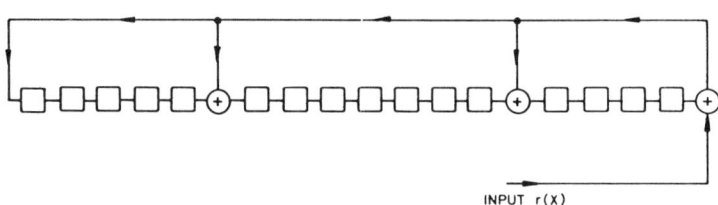

Figure 4.6 Error detection circuit for CCITT Rec. X.25 code.

Error detection proceeds as follows:

(1) The 16 flip-flops are initially set to the 1's state.
(2) The received polynomial $r(X)$ is entered.
(3) If $r(X)$ is a valid code word, the register will contain the bit pattern

$$1\ 1\ 1\ 1,0\ 0\ 0\ 0,1\ 0\ 1\ 1,1\ 0\ 0\ 0$$

which is

$$F\quad 0\quad B\quad 8$$

in hexadecimal notation. Otherwise, it contains some other pattern.

To see this, consider the case where no errors occur and so the received block $r(x)$ is the same as the trasmitted block $v(X)$. For the first k shifts of the decoder, the k received message bits are shifted in. The circuit is identical to the encoder so the register then contains $p(X)$ as defined by Equation (4.15). The remainder of the received block which is waiting in the channel is

$$p(X) + 1 + X + X^2 + \ldots + X^{15}.$$

The next 16 shifts cause this polynomial to be added to the register contents, the sum to be multiplied by X^{16} and the remainder after division by $g(X)$ calculated. Hence, the final register contents are:

$$\begin{aligned} c(X) &= \text{rem} \left[\frac{X^{16}[p(X)+p(X)+1+X+X^2+\ldots+X^{15}]}{g(X)} \right] \\ &= \text{rem} \left[\frac{X^{16}+X^{17}+\ldots+X^{31}}{1+X^5+X^{12}+X^{16}} \right] \\ &= 1+X+X^2+X^3+X^8+X^{10}+X^{11}+X^{12} \end{aligned}$$

which is equivalent to F0B8 (hexadecimal). Note that when no errors occur on the channel the register contents after decoding are not zero as in the conventional syndrome decoder illustrated in Figure 4.5. The use of a nonzero syndrome for an error-free block avoids the possibility of false "no error" indications caused by circuit failures which result in the all 0's word after decoding.

4.4.7 Variable Block Lengths—Shortened Cyclic Codes

In many data transmission schemes in which a cyclic code is used for error detection, the number of message bits may not be constant. This is often the case in data networks in which data is transmitted in variable length blocks.

Consider an encoder based on a generator polynomial $g(X)$ which generates an (n,k) cyclic code. Let the number of message bits to be encoded be $k' \le k$. The encoder produces a fixed number $n-k$ of parity bits. The resultant code word is of length

$$n' = k' + (n - k).$$

This word is said to be one of the code words in a *shortened cyclic* (n',k') code produced by $g(X)$.

It is important to know how the error detection and correction performance of this (n',k') shortened code compares with the original (n,k) code. The code's minimum distance $d_{\min}$ is an important measure of the code's performance, as was discussed in Section 4.3.5. In particular, if the code is to be used to correct

t_c errors and simultaneously to detect $t_d \geq t_c$ errors, then the minimum distance of the code must satisfy

$$d_{\min} \geq t_c + t_d + 1. \tag{4.19}$$

The minimum distance of a shortened cyclic code is at least as great as that of the original code. (Note that the shortened code is no longer cyclic.) If the shortened code is to be used for error detection only, then all error patterns of weight $d_{\min} - 1$ or less will be detected as for the original (n,k) code.

It might appear that the probability $P_u(E)$ that an error pattern will go undetected will be at least as good for the shortened code as for the original code. Unfortunately, that is not necessarily so. This will be discussed in the next section.

Error detection decoding of a code word can only be performed if the length of the code word is known at the decoder. In schemes such as in CCITT Rec. X.25 where variable message lengths are permitted, it is necessary to incorporate special flag sequences at the beginning and end of a code word. The receiver uses these to locate the beginning of each code word and also to determine the number of bits n' to be entered into the syndrome computation circuit before checking the resultant syndrome pattern.

4.4.8 Probability of Undetected Error

For any (n,k) binary code, there are 2^k possible code words which can be transmitted and 2^n possible n-bit blocks which could be received. Out of the set of 2^n blocks, a fraction

$$2^k/2^n = 2^{-(n-k)}$$

are code words.

Likewise there are 2^n possible error patterns. As we have seen, an (n,k) linear block code is capable of detecting all error patterns except the 2^k patterns which have the same form as code words. Hence, the ratio of undetectable error patterns to the total number of possible error patterns is $2^{-(n-k)}$. It can be shown by averaging over all possible (n,k) linear block codes that there exists at least one (n,k) block code for which the probability of undetected error on a binary-symmetric channel with bit-error rate p is

$$P_u(E) \leq 2^{-(n-k)} [1 - (1-p)^n]. \tag{4.20}$$

This result shows only that such a code exists. It does not tell us how to design a code which satisfies the bound. For any code which satisfies Equation (4.20), it is apparent that $P_u(E)$ is strongly dependent on the number $(n-k)$ of parity bits. The bracketed term in Equation (4.20) is always less than 1 and for most practical values of p, this term is only weakly dependent on the block length n. See Lin and Costello (2) for details.

Recently Kasami, Klove, and Lin (14) have derived another useful upper bound on $P_u(E)$ and demonstrated that certain classes of codes do indeed satisfy the bound. They proved that (n,k) codes exist for which

$$P_u(E) \leq 2^{-(n-k)} [1 - 2(1-p)^n + (1-2p)^n]. \tag{4.21}$$

The classes of codes which satisfy this performance bound include the following well-known codes:

(1) distance-4 Hamming codes of lengths $n = 2^m - 1$ and $n = 2^m$
(2) distance-6 BCH codes of lengths $n = 2^m - 1$ and $n = 2^m$, and
(3) distance-8 BCH codes of lengths $n = 2^m - 1$ and $n = 2^m$ with m odd and $m \leq 5$.

In each case, m is a positive integer. The code recommended by CCITT for error detection in packet-switched networks is a distance-4 shortened-Hamming code. All of the above codes are based on "cyclic" code structures.

The probability of undetected error for shortened cyclic codes is a function of block length n' as well as the channel bit error rate p and the number of parity bits $(n'-k')$. Figure 4.7 shows how the CCITT X.25 code performance varies as a function of n' and p.

In general, a useful performance measure to indicate whether a code is a "good" code for error detection is the following upper bound. We say an (n',k') shortened block code with m parity bits is "good" for error detection in a particular application if it satisfies the bound

$$P_u(E) \leq 2^{-m}[1 - 2(1-p)^{n'} + (1-2p)^{n'}] \tag{4.22}$$

for all channel error rates p and for all message lengths n' such that $n_m < n' < n$ where n_m is the minimum required code word block length for that application. See Miller (14) for details.

4.5 FORWARD ERROR CORRECTION

4.5.1 Types of Codes

So far we have emphasized the use of codes for detection of transmission errors. Sometimes it is desirable that a code be used not only to detect the presence of errors, but also to locate and correct them. This is called forward error correction.

Coding systems used for forward error correction can be classified into two classes. They are block codes, and convolutional codes, respectively.

We have seen that in an (n,k) systematic *block coding* scheme, for each block of k information bits, $n-k$ parity bits are generated algebraically to obtain an

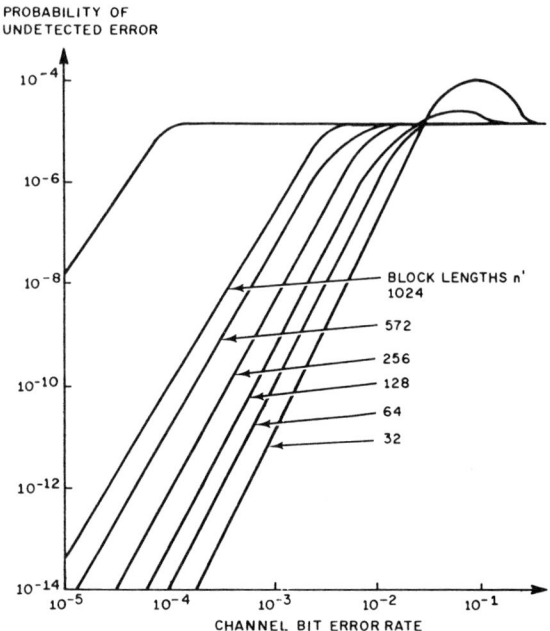

Figure 4.7 Error detection performance of the CCITT Rec. X.25 code.

overall code word of n bits. The most important codes for error correction are cyclic codes. These include the following:

(1) Bose-Chaudhuri-Hocquenghem (BCH) codes: These are the best constructive block codes for correction of random error patterns.
(2) Reed-Solomon codes: These are more complex codes using m bits for each of the n symbols in each code word. The codes are very popular for correction of combinations of random errors and errors occurring in short bursts. Their development has been encouraged for military communications.
(3) Golay code: This is a very special (23,12) cyclic code capable of correcting three errors in any received code word. The code has been widely used as a (24,12) code by adding an extra parity bit as a parity check over the other 23 bits. It can be decoded very simply.

In (n,k) *convolutional codes*, n code bits are generated by the encoder for every k message bits as for block codes. However, convolutional codes are characterized by the fact that the n code bits associated with a given message block are a function not only of that message block but also depend on preceding

Forward Error Correction 187

blocks of message bits. That is, these codes are said to have memory. The longer the memory, the greater the potential error correcting capability of the code.

The most popular convolutional codes are half rate (2,1) codes. The best (2,1) codes for given memory lengths have been found by computer search. See for example, Lin and Costello (2). A (2,1) code with memory K symbols can be generated using a K-stage shift register. For each input bit, two output bits are obtained by addition modulo-2 of two selected sets of the contents of the shift register.

Figure 4.8(a) shows a memory 2 encoder. Information bits are shifted in at the left. For each information bit, the outputs of the modulo-2 adders provide two channel bits.

Methods of decoding convolutional codes can be described in terms of their associated tree or trellis diagrams. The convolutional encoder of Figure 4.8(a) can be described by the code tree of Figure 4.8(b). Each branch of the tree represents a single input bit. An input 0 corresponds to the upper branch from any node and an input 1 corresponds to the lower branch. The associated output sequence is labelled on the branch. Any sequence of input bits traces out a particular path through the tree. For example, an input sequence of 10110 traces out the 11 01 00 10 10 output sequence.

A code tree contains repeated (redundant) information which can be eliminated by merging nodes at a given time level. A node corresponds to a given encoder state. (The state of the encoder at any time is prescribed by the contents of the shift register.) The redrawing of a tree with merged nodes is called a trellis. Figure 4.8(c) represents the code trellis for the encoder of Figure 4.8(a).

The error correction performance of block and convolutional codes is a function of the minimum distance of the codes. The minimum distance d_{min} of a block code was defined in Section 4.3.5. For convolutional codes, the most important distance parameter is known as the minimum free distance d_{free} of the code. It is the minimum Hamming distance between all pairs of possible code sequences, which differ in their first branch of the code tree. For many applications, the best (n,k) codes for given values of n and k are those that have largest minimum distance. (An exception to this occurs in certain cases when coding is combined with redundant signal sets as described in Chapter 1.)

4.5.2 Soft Decision Decoding

In communication systems with forward error correction (FEC), the encoded sequence is transmitted over the channel. At the receiving end, the demodulator (in the case of a radio system) or the regenerator (in the case of a line or fiber system) estimates which of the possible symbols was transmitted. The estimate in any symbol interval is based on the observation of the received signal. For example, consider a binary PSK transmission system in which a correlator type

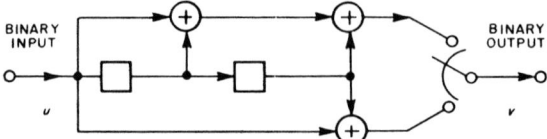

(a) A (2,1) ENCODER WITH MEMORY 2.

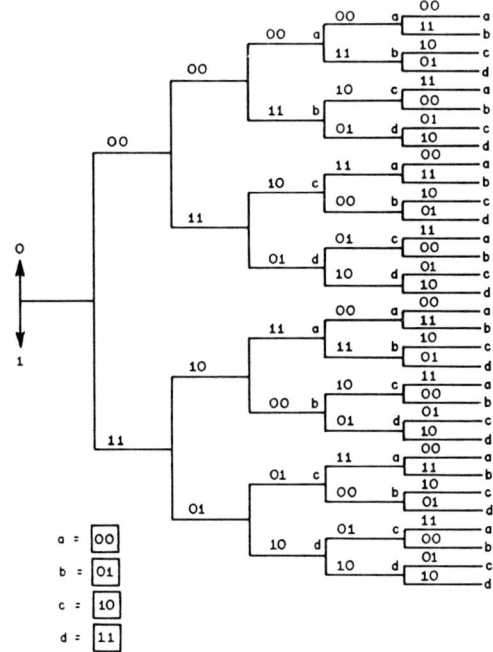

(b) TREE FOR THE ENCODER IN (a).

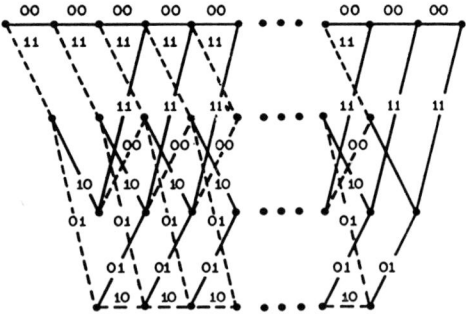

(c) TRELLIS FOR THE ENCODER IN (a).

Figure 4.8 Convolutional coding.

of receiver is used. The symbol estimate is based on the sampled value z of the correlator output at the end of each symbol interval. As discussed in Chapter 1, z will be a Gaussian random variable with positive mean (if a 1 is sent) and a negative mean (if a 0 is sent).

The analogue sample voltage z can be sent to the FEC decoder. In practice, it is normally quantized. If binary (1 bit) quantization is used, we say that a *hard decision* has been made on the correlator output as to which symbol was actually sent. An FEC decoder can overcome the effect of transmission noise more effectively if *soft decisions* are used. A soft decision demodulator is one which simply provides the decoder with the quantized value of the analogue voltage estimate z. Typically, 3-bit quantization is used. The decoder can use this quantized signal value to obtain information about whether z is above or below a decision threshold. More importantly, it has available a "confidence indicator" in that the value specifies how far the demodulator output was from the decision threshold. This can lead to much more effective error control. Soft decision decoding translates to an additional gain of 2 dB over hard decision decoding in most cases. Soft decision decoders are in general much more complex than hard decision decoders.

Note that the channel with erasures discussed in Section 4.2 represented a very simple form of soft decision system. That example demonstrated the increased effectiveness of a soft decision decoder over a hard decision type.

4.5.3 Decoding Techniques

In systems using block codes, the main methods of decoding are

(1) *table look-up decoding* (using error patterns stored in read-only-memory), and
(2) *algebraic decoding* (using techniques based on the algebraic structure of the code).

The first of these techniques can be implemented with minimum complexity. However, it can only be used with codes for which the number of parity bits $(n-k)$ is not larger than about 12–15. Algebraic decoding techniques are more powerful. Unfortunately, if large numbers of errors are to be corrected in each code word, the decoder can become very complex. Unfortunately, space does not permit a description of these techniques here. See for example, Lin and Costello (2).

For convolutional codes, the decoding problem can be thought of as finding the path through the code tree or trellis which provides the nearest match of the received sequence. Decoding algorithms for doing this are as follows:

(1) The *Viterbi decoding* algorithm is the most important decoding technique as it provides optimum performance in a maximum likelihood sense. Given

the received sequence, a "metric" is computed for every possible path through the trellis. Each metric is a measure of the distance between the received sequence and that prescribed by the path through the trellis. The algorithm then discards a number of incoming paths at each node that exactly balances the number of new outgoing paths that are created. It maintains a relatively samll list of paths that eventually converge on the maximum likelihood choice. The complexity of the decoder increases exponentially with code memory K so that in practice, Viterbi decoding is usually restricted to codes with memory 6 or less.

(2) *Sequential decoding* is a slightly suboptimum technique which has the advantage that decoder complexity is relatively independent of code memory length. It can, therefore, be used with very powerful codes when very low bit error rates are required. Given the received sequence, the sequential decoder searches for the most probable code word by searching in a local area of the entire code tree. The search is performed in a sequential manner, operating on a single path, but the decoder can back up and change previous decisions. A major problem is that the number of decoder computations per data bit is a random variable so that the decoder performance is limited by the possibility of buffer overflow.

Further details on these decoding techniques and their performance can be obtained in Lin and Costello (2).

4.6 AUTOMATIC-REPEAT-REQUEST (ARQ) SYSTEMS

When errors are detected in data transmission, then they may be corrected by using either forward error correction (FEC) or an ARQ system.

In FEC systems a block or convolutional code is used to encode each k data bits into an n-bit code word for transmission. The (n,k) code must be chosen so as to provide sufficient redundancy to ensure that error correction at the receiver results in acceptable error rates after decoding. Ideally, this should be achieved by a code for which the rate $R = k/n$ is as large as possible.

An alternative approach to error control is to use an ARQ scheme. The data to be transmitted is divided into blocks and encoded using a high-rate systematic (n,k) block code to produce a code word consisting of the original data plus parity-check bits designed for detection of transmission errors. At the receiver, a decoder checks for the presence of errors in each received code word. Then an acknowledgement (ACK) or a negative-acknowledgement (NAK) indication is sent back to the transmitter to indicate that the received block did not or did contain detectable errors, respectively. On receipt of a NAK, the transmitter retransmits the block received in error. If necessary, this is repeated a sufficient number of times until the block is successfully received without detectable errors.

ARQ schemes have been by far the most popular choice for error-control on links in digital data networks. They provide a relatively simple and highly reliable means of eliminating the effects of transmission errors. In an ARQ system, high reliability can be achieved if the error detection code is properly chosen. Only a relatively small number of parity check bits are needed to ensure that there is a very small probability of undetected error.

In contrast with ARQ systems, it is difficult to achieve high reliability with FEC systems. It is necessary to ensure that, even when the transmission channels are noisy, all errors will be corrected. This may require the use of a powerful code with relatively low rate, that is a high percentage of redundant bits. Most practical channels are designed to be relatively error-free for most of the time. The code rate of an FEC system is fixed so that when the channel is "quiet," the redundant bits represent an unnecessary reduction in throughput efficiency. For an FEC system, the throughput is fixed at a fraction given by the code rate, so that the throughput is

$$\eta_{FEC} = k/n.$$

Typically in FEC systems, the codes used have a rate of approximately 0.5.

When channel error rates are low, the throughput of an ARQ system will be much higher than for an FEC system because the rate of the ARQ error detection code will be higher. Also, the complexity and costs of implementing ARQ systems with a given reliability are often lower than for an FEC system. In view of these factors, it is not surprising that ARQ schemes have become so popular.

We next survey a number of different ARQ schemes. They represent alternatives to the design of retransmission protocols, particularly the modes in which the transmitter stores, orders, and retransmits blocks that have been received in error.

These different schemes have arisen primarily in an attempt to combat the problem that when channel error-rates increase, the throughput of an ARQ system may deteriorate very rapidly. This is because of the time "wasted" in retransmitting blocks containing errors. This problem becomes particularly severe if there are significant round-trip delays between transmission of a block and receipt of its error status information (ACK/NAK) back at the transmitter. Delays are inevitable when satellite or long terrestrial circuits are being used.

Another approach to error-control is through the use of *hybrid ARQ* schemes, which incorporate forward-error correction with retransmission. Hybrid schemes offer the potential for better performance if appropriate ARQ and FEC schemes can be properly combined. Either block or convolutional codes may be used for forward error correction. In a later section we will also discuss different classes of hybrid ARQ schemes and their performance.

4.6.1 ARQ Procedures

Stop and Wait ARQ

The *stop and wait* scheme represents the simplest ARQ procedure and was implemented in early error-control systems. For example, the IBM Binary Synchronous Communication (BSC) procedure discussed in Chapter 3 is of the stop and wait type.

Figure 4.9 illustrates a typical transmit and receive sequence for a stop and wait system. The transmitter sends a code vector (a block containing information plus parity) to the receiver and waits for an acknowledgement. If an ACK is received, the transmitter proceeds to send the next block waiting to be transmitted. If a NAK is received, the transmitter resends the faulty block and again waits for an ACK or NAK. Retransmissions continue until the transmitter receives an ACK.

This scheme is simple but is inherently inefficient unless transmission delays are negligible. The fraction of idle time the transmitter spends waiting for an ACK for each transmitted block can be large and as a result the throughput efficiency will be very low. One possible remedy is to make the block length n extremely long. However, the use of a very long block length does not really provide a solution since the probability that a block contains errors increases with the block length. Hence, using a long block length reduces the idle time but increases the frequency of retransmissions for each block. Moreover, a long block length may be impractical in many applications because of restrictions imposed by the data format.

Go-back-N ARQ

By the 1970s, ARQ systems were in extensive use in packet-switched and other data networks. Higher data rates and utilization of satellite circuits with long round-trip delays established the need for continuous transmission strategies to replace the stop and wait procedures. International standards organizations such as ISO and CCITT began making efforts towards protocol standardization. This resulted in the HDLC (high-level data link control) protocol, and the CCITT X.25 recommendations for node-to-node operation, as discussed in the previous chapter. These envisaged the use of a Go-back-N ARQ system on full duplex links. This is sometimes referred to as "sliding-window" ARQ.

The Go-back-N procedure is illustrated by the block schematic in Figure 4.10(a). In Figure 4.10(b), a typical block transmission sequence is shown.

Each block is numbered and contains the data and parity bits for error detection. In cases where the number of data bits is variable, it is necessary to make provision to indicate the size of the block to the receiver. This is usually done by inserting at the beginning and end of each block, a "flag" sequence consisting of a unique bit pattern (for example 01111110).

Automatic-Repeat Request (ARQ) Systems

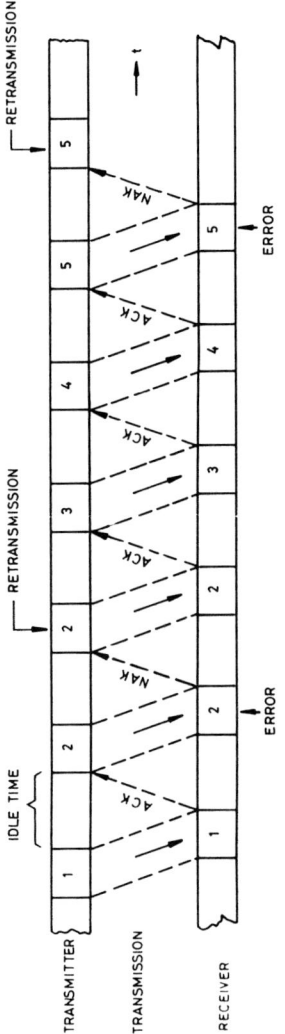

Figure 4.9 Stop-and-wait ARQ: Transmitter and receiver block sequence.

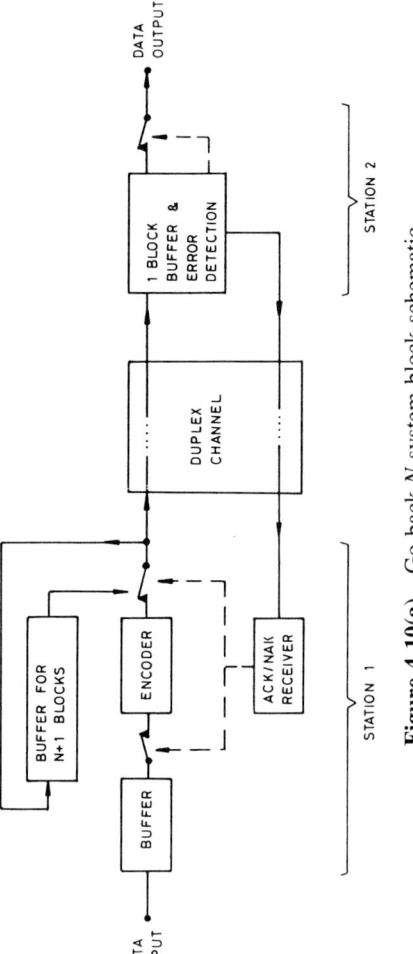

Figure 4.10(a) Go-back-*N* system block schematic.

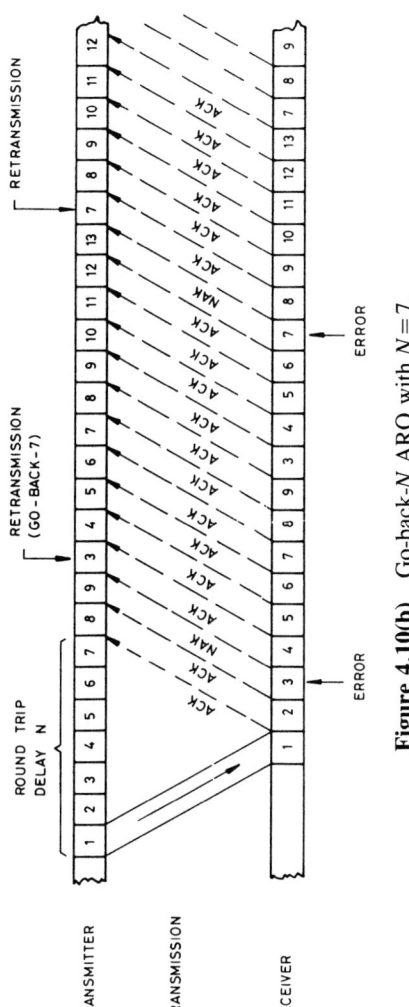

Figure 4.10(b) Go-back-N ARQ with $N=7$.

The transmitter continuously transmits new blocks in order and then stores them, pending receipt of an ACK/NAK indication for each. On receipt of a block, the receiver uses the parity check bits to determine whether one or more errors are detectable in the block. It then sends an ACK or a NAK to the transmitter. In full duplex systems, the ACK/NAK signal may form part of the control byte in a block being used to transmit data in the opposite direction over the same link.

If block i say, is found to contain transmission errors, the receiver discards all N-subsequent blocks whether or not they are received error-free. It sends a NAK and waits until block i is received again.

When the transmitter receives a NAK indicating that block i was received in error, it stops transmitting new blocks. Then it goes back to block i and proceeds to retransmit that block and all $N - 1$ blocks subsequently transmitted during the previous round-trip delay period.

If the first retransmission attempt of block i is successful, the transmitter, on receipt of the ACK indication for block i, proceeds to transmit new blocks in sequence providing it continues to receive ACK indications. Otherwise, it repeats the Go-back-N retransmission procedures until the faulty block is received without detectable errors.

In implementation of the Go-back-N procedure, a number of additional practical details must be considered. Blocks must be numbered in sequence up to some maximum value. Then the numbering cycle is repeated. For example, if 7 bits in the block are to be used for the sequence number, then blocks could be numbered modulo-128. The sequence number length must be large enough to ensure that it is always greater than the round-trip delay (N). Also provision is usually required for handling the situation when expected blocks fail to arrive (perhaps because they were incorrectly routed to some other destination in the network) or when the transmitter fails to receive expected ACK/NAK indications.

CCITT Recommendation X.25 (1) provides appropriate control procedures for such situations. We will not examine them further here.

4.6.2 Throughput of Go-back-N ARQ

The main drawback of the Go-back-N ARQ procedure results from the receiver discarding all $N - 1$ blocks received after a block arrives with detectable errors. This procedure avoids storage and overflow problems but it wastes transmission intervals. This waste of transmissions can result in severe deterioration of throughput if large round-trip delays are involved.

For example, consider a satellite circuit with round-trip delay of approximately 700 milliseconds. If blocks are of a size $n = 1000$ bits and the bit rate is 1 Mbit/s then the round-trip delay is $N = 700$ blocks. Therefore, whenever a block is initially received with errors, the receiver discards the following 699 blocks

Automatic-Repeat Request (ARQ) Systems

and the transmitter sends them again whether or not they were originally received error-free. If errors occur often enough, the system throughput may fall off very rapidly.

To examine this problem, consider the following evaluation of throughput for this ARQ system. To simplify the analysis, we assume equal length blocks of length n bits of which $n-k$ are overhead bits. Also consider a random error channel with bit error rate p. We further assume no errors in the return ACK/NAK channel.

Let the random variable L be defined as the number of transmission attempts of a particular block for it to be received without detectable errors. Then the probability $P_L(l)$ that the required number of attempts is l is given in terms of the block-error probability P_B as

$$P_L(l) = (1-P_B)P_B^{l-1}. \tag{4.23}$$

This follows from the fact that, of the l transmission attempts, one was successful and $l-1$ were unsuccessful. The block-error probability can be obtained using

$$P_B = 1 - (1-p)^n. \tag{4.24}$$

For a block to be successfully accepted by the receiver, the average number of block transmissions (including the original transmission and any subsequent retransmissions) is obtained using (4.23) as

$$T_{GBN} = (1-P_B) + (1+N)(1-P_B)P_B + (1+2N)(1-P_B)P_B^2 + \ldots$$

$$= \frac{1 + P_B(N-1)}{1 - P_B}.$$

Then the throughput efficiency of the Go-back-N system is

$$\eta_{GBN} = \frac{1}{T_{GBN}} \cdot \frac{k}{n} \tag{4.25}$$

so that

$$\eta_{GBN} = \frac{1 - P_B}{1 + P_B(N-1)} \cdot \frac{k}{n}. \tag{4.26}$$

Figure 4.11 shows how throughput varies with bit error rate p for various values of block length n. Throughput values are computed for two values of round-trip delay, namely a delay of 30 milliseconds such as might be the case for a long terrestrial circuit and namely a delay of 700 milliseconds which is typical for a satellite circuit. In both cases, the bit-rate is 64 kbit/s. Each block is assumed to contain 32 bits of overhead such as parity and control bits. The

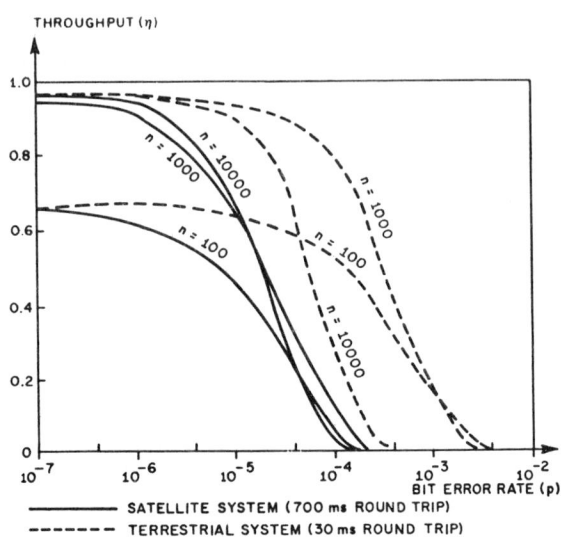

Figure 4.11 Throughput versus channel BER for Go-back-N ARQ for various block lengths (n). Bit rate = 64 kbit/s.

remaining $n-32$ bits are data bits. Studies have also been made of the expected delays and queue lengths (4). Figure 4.12 shows values for the two cases described above.

4.6.3 Other ARQ Procedures

Go-back-N variations

The throughput of the conventional Go-back-N ARQ scheme deteriorates rapidly with bit-error rate if there is a significant delay in the channel. A number of variations on the basic Go-back-N procedure have been suggested.

Sastry (5) suggested a scheme such that when the transmitter receives a NAK for a given block, then it repeatedly retransmits only that block until it receives an ACK. When bit error rates are very high, this procedure can result in higher throughput. Figure 4.13 shows how the throughput of this, and other, Go-back-N types of schemes vary with bit error rate. For most values of bit error rate, Sastry's scheme is inferior to the conventional Go-back-N ARQ.

Morris (3) proposed the use of a buffer at the receiver to store the good blocks following a block received with errors. The transmitter operates in the conventional Go-back-N mode so that the $N-1$ blocks following an erroneous block will each be transmitted twice. As a result, the throughput for this scheme is

Automatic-Repeat Request (ARQ) Systems

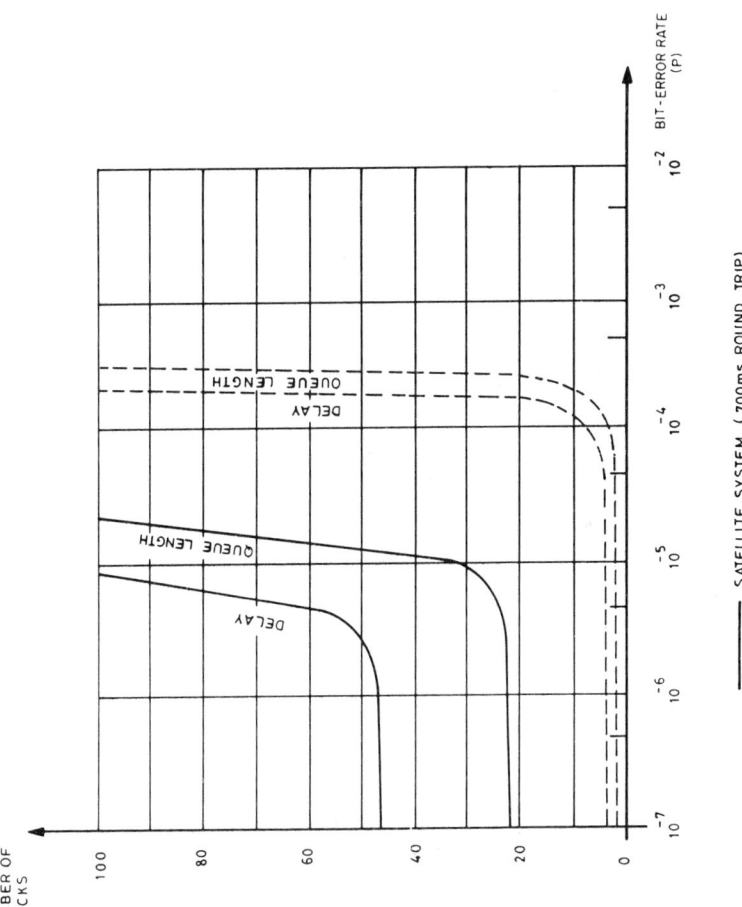

Figure 4.12 Queue length and wait times for Go-back-N ARQ. Bit rate = 64 kbit/s.

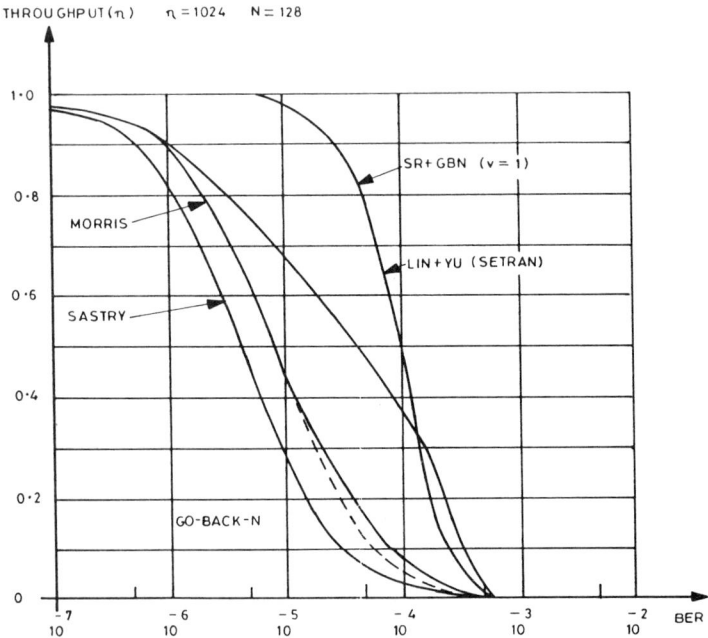

Figure 4.13 Comparison of Go-back-N Schemes.

marginally better than for the Go-back-N when the channel bit error rate is such that the throughput is less than 50 percent. This is illustrated in Figure 4.13.

Lin and Yu (7) proposed a scheme which they called SETRAN ARQ in which the receiver also stores up to $N - 1$ good blocks following an error. However, the scheme has a different retransmission strategy. When the transmitter first receives a NAK, it backs up to the NAKed block and retransmits that block and any subsequently NAKed blocks. The time slots following receipt of ACKs are used to repeat the first NAKed block until an ACK is received for that block. As shown in Figure 4.13, this scheme does provide higher throughput efficiency than that for other Go-back-N types of schemes but at the expense of increased complexity. Another scheme denoted SR + GBN is also shown in Figure 4.13. This will be discussed later.

So far as we know, none of these variations on the conventional Go-back-N retransmission strategy have been implemented in any practical systems.

All of these strategies have the common feature that, after the transmitter first retransmits a NAKed block, the next $N - 1$ blocks sent are retransmissions of blocks previously sent. We next consider schemes in which greater throughput efficiency may be achieved by using the time interval following a retransmission to transmit new blocks.

Selective-repeat ARQ

In a selective-repeat ARQ scheme, blocks are also transmitted continuously. However, the transmitter only resends those blocks that are negatively acknowledged. As illustrated in Figure 4.14, after resending a NAKed block, the transmitter continues transmitting new blocks waiting in the transmitter buffer. In any ARQ system, error-free blocks must be delivered in correct order so a buffer must be provided at the receiver to store good blocks until they can be delivered in correct order.

When the first negatively acknowledged block is retransmitted and successfully received, the receiver then releases all blocks previously received error-free. These blocks are delivered in consecutive order from the receiver store until the next erroneously received block is encountered. The throughput efficiency of this scheme is optimal in that it achieves the Shannon capacity (8) for an erasure channel providing that the receiver buffer size is unlimited. For a block to be successfully accepted by the receiver, the average number of retransmissions (including the original transmission) is given in terms of the block-error rate P_B as

$$T_{SR} = 1\,(1-P_B) + 2\,(1-P_B)P_b + 3\,(1-P_B)P_B^2 + \ldots$$
$$= 1/(1-P_B). \tag{4.27}$$

Then the throughput of the selective-repeat ARQ is

$$\eta_{SR} = \frac{k}{n}(1-P_B). \tag{4.28}$$

In this case, the throughput efficiency does not depend on the round-trip delay so that this retransmission strategy appears to offer significant benefits for satellite and long terrestrial channels. However, the problem of receiver buffer provision must be considered. High reliability requirements demand that sufficient receiver buffer must be provided to ensure that no blocks will be lost because of buffer overflow. Theoretically, infinite buffer is needed to achieve this for this ARQ scheme. Despite its superior throughput performance over other ARQ schemes, practical implementations must avoid overflow despite having a finite amount of receiver store. We next consider some schemes which attempt to achieve this.

Mixed-mode ARQ

Selective-repeat types of ARQ schemes using a finite range of block sequence numbers and a finite receiver buffer have been suggested by several authors (8–12). One such class of ARQ schemes is known as *mixed-mode ARQ*. The mixed-mode ARQ schemes require that the transmitter respond to NAK indications by retransmitting the faulty blocks in one of two possible modes.

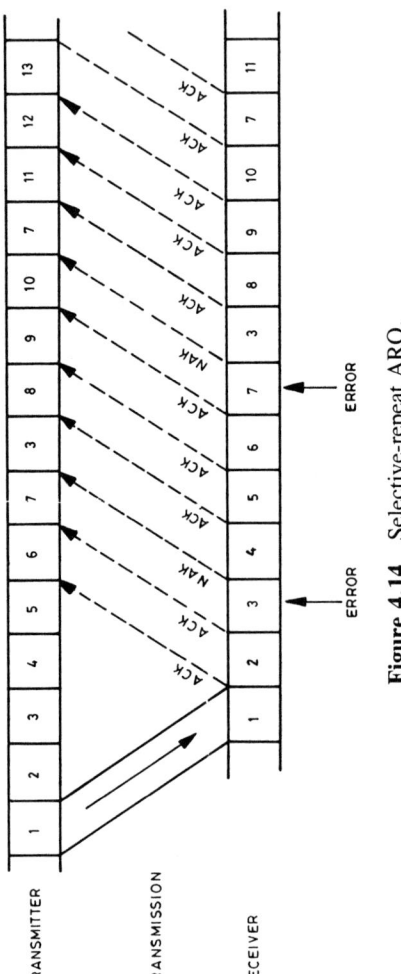

Figure 4.14 Selective-repeat ARQ.

One mixed-mode scheme is the *selective-repeat plus Go-back-N* (SR + GBN) ARQ. When the transmitter first receives a NAK for a given block (say block i), it retransmits that block and then continues transmitting other new blocks as in the conventional selective-repeat (SR) mode. If another NAK is received for block i, indicating its second transmission attempt was unsuccessful, the transmitter switches to the Go-back-N (GBN) retransmission mode. That is, it sends no more new blocks but backs up to block i and resends that block and the $N-1$ succeeding blocks that were transmitted after the previous transmission attempt of block i. The transmitter stays in the GBN mode until the block i is positively acknowledged. At the receiver, when the second transmission attempt of block i is detected in error, all the subsequent $N-1$ received blocks are discarded regardless of whether they were received error-free or not.

This scheme can achieve superior throughput performance compared with any of the Go-back-N schemes. This is because of the benefits gained from the use of the SR mode for the first retransmission attempt. The use of the secondary mode (GBN) guarantees that buffer overflow cannot occur at the receiver so long as storage is provided for N blocks. The scheme can have even higher throughput if more receiver store is provided and the transmitter designed to permit more than one retransmission attempt for a given NAKed block in the SR mode before switching to the GBN mode. Figure 4.15(a) illustrates a typical sequence and Figure 4.15(b) shows how the throughput of the SR + GBN scheme varies with bit-error rate. Results are shown for different values of v where v is the number of SR retransmission attempts allowed for any repeatedly faulty block before the transmitter switches to the GBN mode.

Also shown in Figure 4.15(b) are throughput values for a related mixed-mode scheme called the *selective-repeat plus stutter* (SR + ST) ARQ scheme. This is the same as the SR + GBN scheme except that, instead of using the GBN mode after v retransmission attempts of a given block, the transmitter switches to a ST (stutter) mode. In this mode it repeatedly retransmits that block only until it receives an ACK.

The throughput characteristics of Figure 4.15(b) show that for best performance, at least the first retransmission of a block following an error should be in the selective-repeat mode. The choice of the secondary mode does not have a significant bearing on throughput. It should be such that it prevents receiver buffer overflow.

Weldon (8) later suggested a variation of these mixed-mode protocols. When the transmitter is in the SR mode following the receipt of a NAK, it retransmits the NAKed block n' times. Then it proceeds to transmit other blocks waiting in the transmitter buffer in sequence from where it left off. The number n' can be chosen to provide maximum throughput for given error rate and delay. Typically, $n' = 3$ provides good results. If all n' retransmissions of a block are received

Error Control in Digital Networks

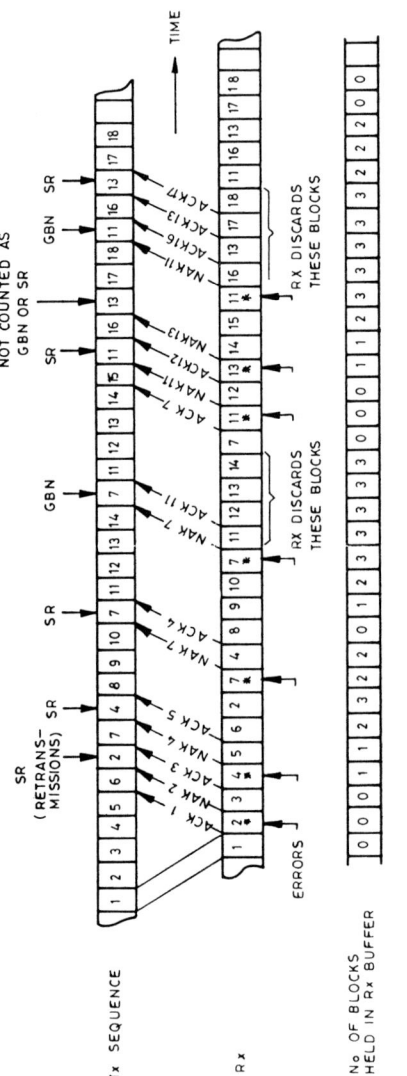

Figure 4.15(a) SR+GBN Scheme for $v = 1$.

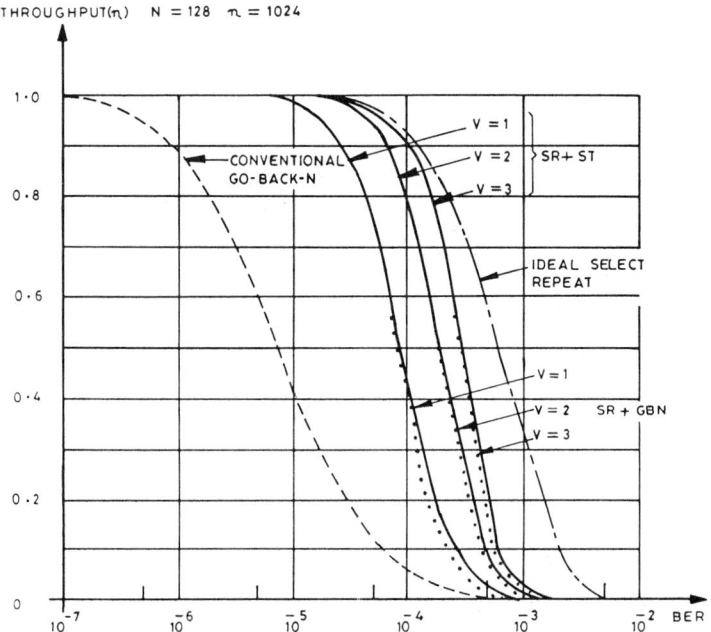

Figure 4.15(b) Throughput performance for various values of v (SR mode retransmission attempts).

with detectable errors, then the transmitter reverts to the GBN mode as in the other mixed-mode schemes.

In all of the mixed-mode types of schemes, provisions must be made for the occasion when the received block number is unknown because of errors. To resolve this case, it is necessary to synchronize the transmitter and receiver, by keeping a continuous check on round-trip delay.

The mixed-mode ARQ schemes are more complex than the Go-back-N schemes. However they do offer considerable improvements in throughput performance for channels with significant errors and large round-trip delays.

4.7 HYBRID ARQ SCHEMES

In hybrid ARQ schemes, forward-error correction is combined with a block retransmission protocol. This is done in an attempt to retain high throughput even when the channel becomes noisy.

In some hybrid ARQ systems known as *Type I hybrid* schemes, the message block to be transmitted is first encoded with a block code suitable for correcting some error patterns and detecting most of the remaining patterns. At the receiver, if an error pattern is detected, then a decoder attempts error correction. If the decoder fails to correct all transmission errors, then a retransmission is requested. In this class of schemes, each transmitted block must include a significant number of redundant parity bits for error correction. As a result, the throughput will be lower than for normal ARQ systems when the channel is quiet. Details on Type I hybrid ARQ schemes are available in Lin, Costello, and Miller (15).

More recently, a new class of hybrid ARQ schemes has been proposed in which the number of parity bits in any transmitted block is made to vary adaptively with the channel bit-error rate. These schemes are referred to as *Type II hybrid* schemes or *parity retransmission ARQ*. When the transmitter receives a request for a retransmission of a given message, it sends a block of parity bits for error correction rather than retransmitting the original message block. Either block codes or convolutional codes may be used to generate the parity bits. Parity retransmission ARQ can provide much higher throughput than conventional ARQ schemes when error rates are high. In the following, one such parity retransmission ARQ scheme is described.

4.7.1 Parity Retransmission ARQ Strategy

The parity retransmission ARQ scheme provides that, when errors occur in a given data block, alternate repetitions of data and error correction parity are sent until that block is recovered without detectable errors. One procedure is illustrated in Figure 4.16. It can be summarized as follows:

(1) The initial transmission of any block of data consists of the n bits $(u, P_o(u))$ containing the original k data bits u. These are followed by $n-k$ parity check bits $P_o(u)$ generated by a high rate block code C_0 used for forward *error detection* only. The block $(u, P_0(u))$ is referred to as a "data block transmission."

(2) If the initial transmission of a block is received with detectable errors, then the receiver stores the received message bits $\hat{u}$, and sends a negative acknowledge (NAK) back to the transmitter which then initiates a retransmission.

(3) The transmitter does not just retransmit the block $(u, P_0(u))$. Instead it sends an n-bit block denoted $P_1(u)$ and referred to as a "parity block transmission." $P_1(u)$ represents n parity bits generated from the original data block $(u, P_0(u))$ by a half-rate *error correction* code C_1. A half-rate code generates n parity bits for each n message bits.

(4) On receipt of the "parity block" $\hat{P}_1(u,$ the receiver divides $\hat{P}_1(u)$ by the generator polynomial $g(X)$ for the code C_1. If the remainder is zero, $\hat{P}_1(u)$ contains no detectable errors. In this case, the receiver recovers the wanted

Hybrid ARQ Schemes

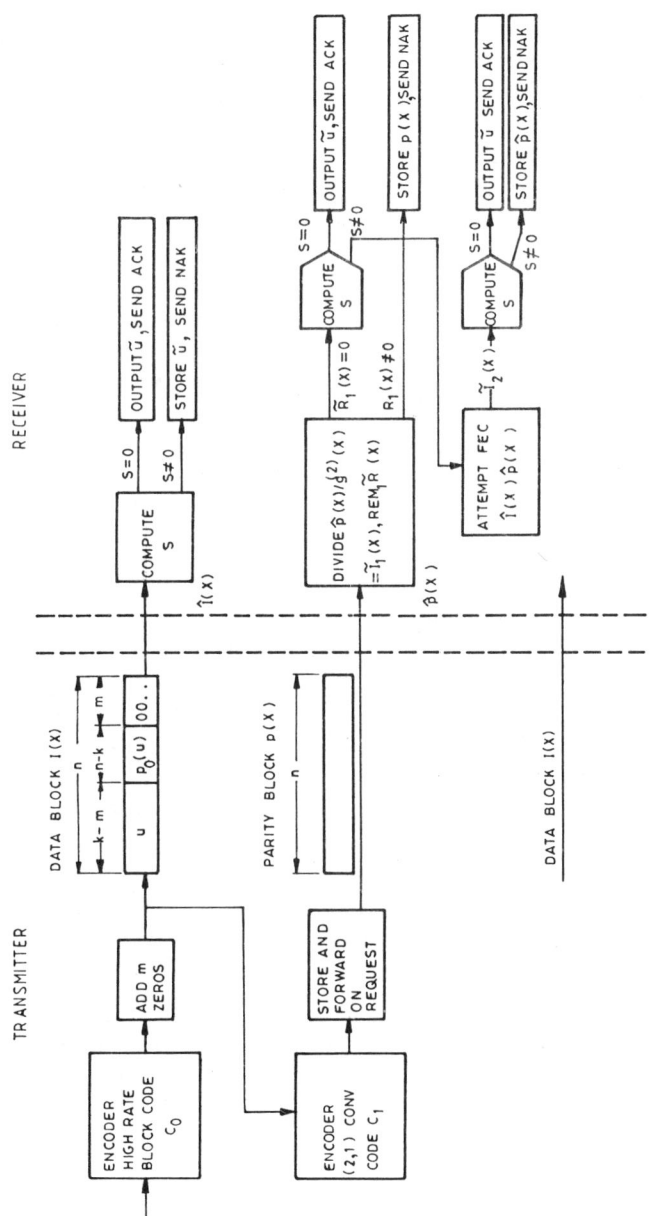

Figure 4.16 Parity retransmission hybrid ARQ.

data $\bar{u}$ from the parity block $\hat{P}_1(u)$ by inversion. In order to make this possible, the code C_1 must be "invertible," that is, the k information bits can be determined from knowledge of only the k parity check bits in $\hat{P}_1(u)$.

(5) If errors are detected in the received "parity block" $\hat{P}_1(u)$, the receiver attempts to correct the errors using a decoder based on code C_1. That is, the n data bits $(u,P_0(u))$ initially received in the "data block transmission" and the n-bit "parity block" $\hat{P}_1(u)$ are decoded. If the errors in the $2n$ bits form a pattern that is correctable by the decoder, then the decoded data bits $\bar{u}$ are delivered to the user. If not, the receiver stores the parity block $\hat{P}_1(u)$ and sends a NAK back to the transmitter to request a retransmission.

(6) For the third and subsequent attempts at transmission of a given packet the transmitter sends alternatively the data block transmission $(u,P_0(u))$ or the parity block transmission $(P_1(u))$ until the receiver sends back a positive acknowledgement (ACK) for that block. This will be done if the receiver detects no errors in a $\hat{u}$ or a $\hat{P}_1(u)$ or if a $\hat{u},\hat{P}_1(u)$ pair contains errors that are correctable by the decoder.

4.7.2 Retransmission Protocols

The above description of a parity retransmission ARQ scheme defines the format of the blocks to be transmitted and the procedures required at the receiver to attempt to recover the data without errors. However, it does not prescribe the retransmission protocols to be used by the transmitter.

The *retransmission protocol* describes the order in which the transmitter sends new blocks and retransmits NAKed blocks. Possible retransmission protocols include Go-back-N, Selective-Repeat and Mixed-mode ARQ schemes. These schemes were discussed in Section 4.5. They are all forms of basic ARQ, that is, ARQ schemes where forward-error detection is performed but not error correction.

For the Type-II hybrid scheme to be effective, it is necessary to use it in conjunction with a retransmission protocol of the selective-repeat type. This includes the mixed-mode ARQ schemes.

4.7.3 Choice of Error Correction Code

The parity retransmission strategy is based on the use of a half-rate code for forward error correction. Either a block code or a convolutional code can be used. Half-rate convolutional codes may be used with simpler decoders than long half-rate block codes, and hence, may offer better performance for a given complexity in hybrid ARQ applications.

For this application, systematic (2,1) convolutional codes can be used with relatively simple sliding-block decoding based on a large-scale read-only-memory (ROM). See (13) for details.

4.7.4 Throughput Analysis

Expressions for the throughput of the parity retransmission strategy in conjunction with various ARQ retransmission protocols have been derived.

Figure 4.17 shows typical calculated values of throughput efficiency versus bit error rate for selected hybrid ARQ schemes in comparison with the conventional Go-back-N protocol. It is apparent that convolutional codes with relatively simple sliding-block decoding can ensure that high throughput is maintained on digital channels using a hybrid ARQ system despite significant bit-error rates and transmission delays. See Lin (15) for details.

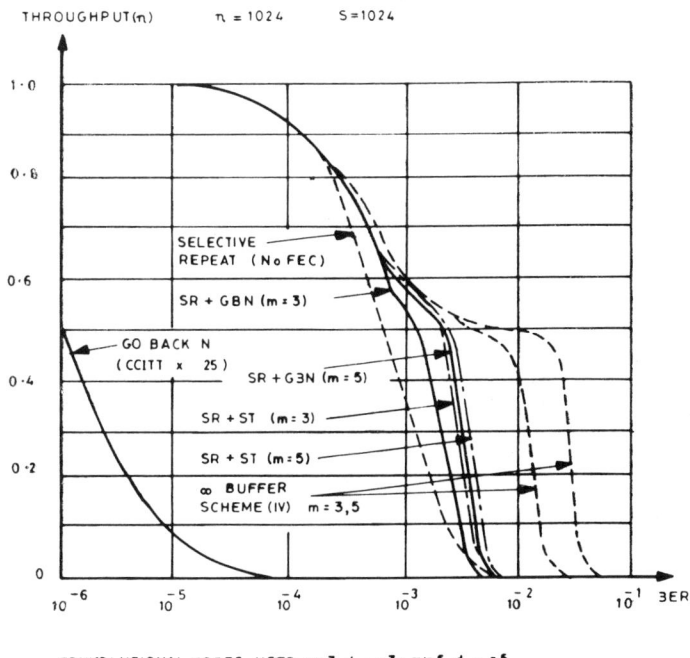

Figure 4.17 Throughput of Type II hybrid ARQ with SR + ST and SR + GBN protocols.

PROBLEMS

4.1 Consider the (7,4) code illustrated in Figure 4.1.

(a) By reversing the locations of the information and parity bits respectively, show how the diagram can be used to describe a (7,3) code.

(b) For your (7,3) code, list all possible information sequences and the associated code words.
(c) Write the parity-check equations for the (7,3) code.
(d) Explain how many errors the (7,3) code can correct.
(e) How many erasures are correctable?
(f) Compare the error detection capabilities of the (7,3) and (7,4) codes.

4.2 Four-bit information blocks are encoded by adding a simple parity bit so that each code word contains an even number of 1's.

(a) Find the weight distribution of the (5,4) code.
(b) If the code words are transmitted through a channel with random errors occuring with probability $p = 10^{-4}$, find the probability of a received code word containing one or more errors.
(c) For $p = 10^{-4}$, find the probability that an undetectable error pattern occurs.

4.3 A (6,3) code is generated by the generator matrix

$$G = \begin{bmatrix} 0 & 1 & 1 & 1 & 0 & 0 \\ 1 & 0 & 1 & 0 & 1 & 0 \\ 1 & 1 & 0 & 0 & 0 & 1 \end{bmatrix}$$

(a) List the code words. Is the code systematic?
(b) Find the minimum distance for the code.

4.4 A (15,7) cyclic code has a generator polynomial

$$g(X) = 1 + X^4 + X^6 + X^7 + X^8 .$$

(a) Find the code word in systematic form for the message $\overline{m} = (1\ 0\ 0\ 0\ 0\ 0\ 1)$.
(b) Draw block diagrams of an encoder and error detection circuit for this code.
(c) Is $v(X) = X^4 + X^6 + X^{10}$ a code polynomial?

4.5 Consider the (15,7) cyclic code generated by $g(X)$ in Problem 4.4.
(a) Find the weight distribution for the code.
(b) For a channel error rate of $p = 10^{-2}$, find the probability of occurrence of an error pattern which is undetectable by the code.

4.6 Consider the CCITT error detection code with generator polynomial $g(X)$ given by Equation (4.10). This code has minimum distance of 4.

(a) Show that $g(X)$ can be factored into two polynomials one of which is $(1 + X)$.

(b) Show that this is equivalent to an overall parity check bit.

4.7 Consider a stop and wait ARQ scheme in which the transmitter signalling rate is r_b bit/s. Let D be the round trip delay defined as the time from the end of transmission of one block to the beginning of transmission of the next.

The transmitted blocks are code words of length n bits, of which $n-k$ are overhead bits. The channel block error rate P_B is given in terms of n and the channel bit error rate p by Equation (4.24).

(a) Show that the throughput efficiency of the stop and wait ARQ scheme is

$$\eta_{SW} = \frac{1}{1+Dr_b/n} \cdot \frac{k}{n}$$

(b) Compare the throughput efficiency of the stop and wait and Go-back-N schemes assuming the following operating parameters:

$n = 1024, k = 960, p = 10^{-5}, D = 500$ ms, $r_b = 2$ Mbit/s.

4.8 For a Go-back-N ARQ scheme calculate the number of blocks N in the round-trip delay if the round trip delay is $D = 0.5$ seconds and the transmitted bit rate is $r_b = 1$ Mbit/s. Repeat for $D = 30$ ms.

4.9 (a) For the Go-back-N scheme with throughput given by Equation (4.26), suppose that $p = 10^{-4}$ and $k/n = 0.9$. Find the block length which maximizes the throughput.

(b) Repeat for the stop and wait ARQ scheme.

4.10 Consider the SR+GBN mixed-mode ARQ scheme with $v=1$ selective repeat retransmission attempts.

(a) Show that the throughput efficiency is given in terms of block error rate P_B and round-trip delay N(blocks) by

$$\eta = \frac{1-P_B}{1+(N+1)P_B^2} \cdot \frac{k}{n}.$$

(b) Compare the performance of this mixed-mode scheme with the Go-back-N scheme for the situation where $N = 100$. Assume $k = 960$, $n = 1024$, and the channel bit-error rate is $p = 10^{-4}$.

(c) For the SR+GBN scheme, what is the receiver buffer size required to ensure buffer overflow is avoided?

4.11 Consider the SR+ST mixed-mode ARQ scheme.

(a) Show that the throughput efficiency is given by

$$\eta = \frac{1 - P_B}{1 + (N+1)P_B^2(1-P_B)} \cdot \frac{k}{n}$$

(b) What is the receiver buffer size required to avoid overflow?

4.12 Consider a hybrid ARQ scheme in which a Parity Retransmission strategy is used in conjunction with a selective-repeat protocol. Assume that the forward error correcting code is sufficiently powerful that no more than two transmission attempts are required for any block to be recovered without transmission errors. If k/n is fixed, and the channel bit error rate increases, explain why the minimum value of throughput efficiency is expected to be

$$\eta_{min} = 0.5k/n.$$

4.13 In a certain ARQ system, the process of error detection is performed by a block code capable of detecting all error patterns of up to 10 errors per block. Error patterns of more than 10 errors are not detectable. Given that the block size including parity bits is 1000 bits, and the channel is a random error channel with bit error rate $p = 10^{-3}$, compute the reliability to be expected of the ARQ system. That is, find the probability that error-free blocks will be delivered to the user.

4.9 REFERENCES

1. CCITT: Recommendation X.25 "Interface between data terminal equipment (DTE) and data circuit-terminating equipment (DCE) for terminals operating in the packet mode on public data networks," VIIth Plenary Assembly, Doc. No. 7, Geneva 1980.
2. S. Lin and D. J. Costello Jr., *Error Control Coding: Fundamentals and Applications*, Prentice Hall, 1983.
3. J. M. Morris, "Optimal Block-lengths of ARQ Error Control Schemes," *IEEE Trans. Comm.*, Vol. COM-27, pp. 488–493, February 1979.
4. D. Towsley and J. K. Wolf, "On the statistical analysis of queue length and waiting times for statistical multiplexers with ARQ retransmission schemes," *IEEE Trans. Comm.*, Vol. COM-27, pp. 693–702, April 1979.

References

5. A. R. K. Sastry, "Improving ARQ performance on satellite channels under high error rate conditions," *IEEE Trans. Comm.*, Vol. COM-23, pp. 436–439, April 1975.
6. J. M. Morris, "On another Go-back-N ARQ technique for high error rate conditions," *IEEE Trans. Comm.*, Vol. COM-26, pp. 187–189, January 1978.
7. S. Lin and P. S. Yu, "An efficient error control scheme for satellite communications," *IEEE Trans. Comm.*, Vol. COM-28, pp. 396–401, March 1980.
8. E. J. Weldon, Jr., "An Improved Selective-Repeat ARQ Strategy," *IEEE Trans. Comm.*, Vol. COM-30, No. 3, March 1982.
9. J. J. Metzner, "Improvements in block-retransmission schemes," *IEEE Trans. Comm.*, Vol. COM-27, No. 2, pp. 525–532, Feb. 1979.
10. P. S. Yu and S. Lin, "An efficient selective-repeat ARQ scheme for satellite channels," *IEEE Trans. Comm.*, Vol. COM-29, No. 3, pp. 353–363, March 1981.
11. M. C. Easton, "Design Choices for Selective-repeat retransmission protocols," *IEEE Trans. Comm.*, Vol. COM-29, No. 7, pp. 944–953, July 1981.
12. M. J. Miller and S. Lin. "The analysis of some selective-repeat ARQ schemes with finite receiver buffer," *IEEE Trans. Comm.*, Vol. COM-29, No. 9, pp. 1307–1315, September 1981.
13. M. J. Miller, "Automatic-Repeat-Request systems," Ph.D. Dissertation, University of Hawaii, 1982.
14. M. J. Miller, M. G. Wheal, G. J. Stevens and S. Lin, "The reliability of error detection schemes in data transmission," *J. of Elec. and Electronics Engineering, Australia*, Vol. 6, No. 2, pp. 123–131, June 1986.
15. S. Lin, D. J. Costello Jr., and M. J. Miller, "Automatic-repeat-request systems," *IEEE Communications Magazine*, Vol. 22, No. 12, pp. 5–17, Dec. 1984.

Chapter 5

INTEGRATED SERVICES DIGITAL NETWORKS (ISDN)

5.1 INTRODUCTION

In this section, a basic overview of the Integrated Services Digital Network (ISDN) is presented in the context of the existing status of the telephone networks. Before such an integrated network is established, the interim thinking for the implementing the ISDN is to utilize the existing digital networks to their economic limit. The more established telephone networks appear to be viable contenders in carrying the ISDN data traffic. Hence, the concepts fundamental to the ongoing evolution of the ISDN, the growth of the telephone network to its present form, and the methodologies (modes) of data transmission for ISDN are presented in this chapter.

5.1.1 Fundamental Concepts of ISDN

The proposed Integrated Services Digital Network is a centralized network for providing expanded services to customers. It promises to bring home the impact of the digital revolution by using new techniques in digital processing, control, and transmission, as well as recent developments in networking and digital switching. A very desirable feature of the network is the extent of customer control of the service functions performed by the network. Examples of the new services are videotex, televideo facility, and encrypted voice capability.

The need for fast and accurate information is steadily increasing. This calls for inexpensive and dependable data networks for easy access to information. Such networks are also a basic necessity for the implementation of the futuristic services. Most countries have some form of information network; the most common being the telephone network. The features and the status of such networks become essential in evaluating the ISDN capability of that particular environment.

Introduction

The ISDN concept has steadily evolved from the exploratory studies of the various communication scientists at AT&T-Bell Laboratories in New Jersey. Here the ideas of computing and communicating were initially merged together. Some of the very early foundations of ISDN can be traced to the late 1960s when researchers were developing special purpose computers to control the telephone networks. At this stage it became clear that some of the computing functions could be made available to the customer and to telephone subscribers. Thus, a new era of exploration of newer economically-viable digital services started.

The fields of computing and communications may appear to have developed somewhat independently in the 1960s through mid-1970s. Over the last decade, these two disciplines have merged dramatically and the design of computers effectively overlaps the concepts of controlling electronic circuits and systems. This is indeed the goal of telephone scientists who actively seek to change and modify the network using the signals as they are received from the subscribers. Similar network modifications occur when control signals are dispatched from electronic switching systems at the central offices. This approach of solving the combined problems of the computing, communicating and actively controlling the networks in an optimal and an interdisciplinary way has offered substantial opportunities in expanding the possibilities of bringing the results of innovation and research to the homes of customers, where most of the telephone lines terminate. As the solution of these problems becomes more feasible, the new services promised by ISDN become more realistic.

The impact of the technological changes is felt from three directions. First, from the field of communications, the major innovations are the electronic switching systems, the analogue and digital carrier systems, the digital networks, digital microwave systems, fiber optic links, satellite communication channels, automatic error detection, and correction systems. Each of these innovations has resulted in the need to unify the control of these systems, channels, links, and networks.

Second, from the field of computing, the innovations and the algorithmic problem-solving approach have been equally effective. Adaptive and intelligent operating systems, software engineering, user friendly systems, expert systems, artificial intelligence, multiple and distributed processor systems, and the algorithms of their effective utilization offer newer and expanded use of the computer.

Third, from the field of control, the recent innovations exert a two-fold impact. In a conventional computing environment, control generally implies the distribution of control signals and/or interpretation of the microcode. In the communication network environment, control generally means the switching of appropriate channels that carry voice and/or digital information. In context of the proposed ISDN, both the implications become necessary since the modern switching systems are massive computers which control the telephone network. The three distinct areas of computation, communication, and control and the recent innovations in these three areas thus blend harmoniously in the implementation of the proposed digital network.

ISDN viability is founded in the successes of three evolutionary stages of the telephone network development: (a) the analogue transmission and switching networks, (b) the hybrid networks using both analogue and digital techniques for the telephone voice messages, and (c) the currently evolving totally digital network. These are discussed in Sec. 5.1.2. It is the ultimate objective of the network proponents that the two major network functions of transmission and switching be performed in an entirely digital mode. The novel techniques of voice and picture encoding facilitate the transmission aspects of the customer information. The trend toward being totally digital can already be traced to the program control concepts in the recent electronic switching systems in the central office to perform the switching functions. When these two functions can be unified by digital techniques of implementation, the communication of customer information (via high-speed digital pipelines for long-distance communication and via digital subscriber lines in subscriber loop environment), appears as a distinct and preferable alternative. The flexibility, control, and accuracy of the digital system recurs in the ISDN.

As the algorithms and techniques to solve the conceptual and implementational obstacles are finalized, two issues linger on. The first issue deals with unified format(s) for transmission and for switching. Considerable effort is being expended by the international committees in resolving this issue. The second issue deals with the security of vast amounts of private, semiprivate, sensitive, and/or confidential information. Security of information and implementation of effective, yet practical methods to prevent the abuse of the digital network are areas of challenge for the communication scientist.

5.1.2 ISDN and Telephone Networks Features

The Analogue Telephone Network

Through the mid-1960s, crossbar systems and twisted pairs were extensively used to fabricate the core of the telephone network. The human voice was the primary medium of communication. Services responsive to voice or to frequencies contained in the allocated telephonic channel bandwidth (typically 200–3200 Hz) were well managed during this first stage of network development. The implementation of the proposed ISDN services on an analogue network would be monumental if not impossible. The economics and the extent of effort would rule out any serious consideration.

Introduction   217

The Hybrid Telephone Network

Electronic Switching Systems (ESS) have been gradually introduced in the network since May 30, 1965. On this date, the first full-scale commercially feasible switching system was introduced in Succasauna, New Jersey. However, in many ways, these early machines were designed to take advantage of the speed, reliability, costs, and compactness of the electronic devices, but these machines were still functionally comparable to the earlier electromechanical switching systems. Novel features at the central offices were added, but the commercial availability of radically new services to the customers was not explored during the 1960s and early 1970s. The introduction of stored program control of the ESS switching capability contributed to the limited version of the customer calling services (such as telephone number abbreviation and call forwarding). Additional customer services included with the introduction of No. 101 ESS in late-1963 were the Touch-Tone telephones, call transfer, and conference calling.

During this same time, the principle of digital carrier systems in the loop environment was introduced, even though the analogue carrier systems (1) has been prevalent and evolved in the various classes of systems (the open wire, the paired and coaxial cable, the terrestrial microwave radio, and/or the satellite) for over sixty years. In the analogue carrier systems, the modulation technique generally chosen is the single side band amplitude modulation or frequency modulation techniques, even though double side band transmitted carrier modulation has been used occasionally for the cabled pairs. In the United States the open wire pair facility was the first (as far back as 1925) to experience four voice circuits per channel. In general, with the single side band amplitude modulated analogue carrier systems, the amplitude of a carrier signal is modulated and the signal is then clipped to generate the single side band effect. In the frequency modulated analogue carrier systems, the voice frequencies are shifted up to a band of frequencies centered around a particular frequency. This frequency chosen to be beyond the voice frequency range is called the carrier frequency. The carrier systems had the effect of enhancing the amount of the information carried on the channels, such as the wire pairs, cables, and microwave links. However, these carrier systems did not stimulate the ISDN concept, even though they were useful in carrying the added traffic and data created by the newer services that ISDN brings with it.

The Evolving Digital Network

The telephone networks in most technologically advanced countries are experiencing a major thrust toward a total digital network concept. This concept penetrates both the transmission and switching aspect of the telephone business. The transmission aspect overlaps the distributed processing and networking concepts in the computer field. The switching aspect overlaps the control and the central processing functions in the computer field. Whereas the field of computing has become almost all digital, we are still converging toward an entirely digital telephone network. The impact of the all digital computing industry has already manifested itself in the telephone network as the novel services to the network customers.

There are two major types of services that can be offered within the framework of evolving digital telephone networks. The first service is similar to the one provided earlier with the analogue networks where voice or voice frequency initiated services continue to be offered. The second service, which truly encompasses the purpose of ISDN, offers end-to-end digital connectivity between customers using the same existing transmission medium and the switching facilities. In a traditional sense, where telephone services are provided on a local, national and international basis, ISDN services also may be offered on a local, national, and an international basis. It appears that there will be considerable unification in the direction in which ISDN trends will evolve in different countries.

ISDN incorporates three major features which tend to make it attractive in this stage of the technology. First, the economies derived by digital systems and their sliding costs tend to create an economic opportunity to offer new services at a very marginal cost. Second, the integration of many types of services offers economies of scale, thus tending to reduce the costs further. Third, the novelty in the services offered will encourage and stimulate the additional utilization of the telephone plant thus creating expansion and revenues over and above what the plain old telephone could create. Hence, the common carriers around the world have an active interest in the introduction, development, and growth of ISDN.

The impact of the growth of ISDN will be felt in at least four directions: the new hardware devices, the new software to control and communicate, and the new standards to interface the networks, and the protocols to establish communication. The domestic common carriers around the world are actively investigating these four major areas of impact and in the next section we present an overview and status of the ISDN activity in the more important national telephone network organizations.

5.1.3 An Overview and Impact of ISDN

The possibility of offering extended digital services to the public telephone subscriber has attracted the attention of the telephone organizations around the world. The trend towards making the public communications networks digital and thus enhancing the nature and the extent of the possible new services has prompted the network administrators and the business community in general to

Introduction

investigate a systematic and early introduction of the ISDN. The digital revolution has provided ample tools techniques and the integrated circuit chips to establish the technical viability of the proposition. Two major issues still linger on. The first issue focuses on the administrative aspects of making and managing the forthcoming transition to the integrated services digital network. The second issue focuses on assuring the compatibility of the ISDN services, facilities, and specialized hardware with those of the existing services, facilities, and specialized hardware. In these areas, most of the problems need committee action, collaboration, and cooperation of the communications organizations around the nation and around the globe. Some of the problems arise because of the inadequate research effort in isolating and resolving questions dealing with the administrative and the compatibility issues posed by the ISDN. It is our belief that these nontechnical issues will be resolved because of the large potential for economic gains in store for the network owners. Such gains will be realized by offering these novel services to the public at an affordable price.

On a global basis, the ISDNs emerge from a very strong trend of the telephone networks to become entirely digital. The ISDN concepts have been evolving steadily during the 1980s and are becoming more and more established as the national and international effort is documented. There have been two distinct phases in the rather steady evolution of telephone networks. In the first phase, these networks were all analogue as far as switching and transmission are concerned. Analogue speech signals are transmitted over wire pairs and switched by analogue devices. The second phase incorporated digital transmission and switching techniques in the telephone networks. Digitized voice and information are transmitted over a transmission medium (wire pairs, cables, microwave links, and satellites) and are switched by digital switches similar to those used in modern computers. An integrated version of an extended facility providing end-to-end digital connectivity and transparency for both voice and data is now emerging. The object still remains that the available and evolving technology be used in providing extended voice and data services by utilizing a common switching and transmission network. The concepts that have been instrumental in the design, development, and implementation of the ISDN are discussed here.

The type of new services that ISDN can bring forward to customers has considerable social and economic impact. The new services that ISDN can generate, and the enhancement in the use of the existing networks has been an attractive feature accelerating its growth. At the same time, the new technology that is necessary to implement the new services and to modify the existing networks to be integrated is available. This technology has been a sizable beneficiary of the technologies evolving in manufacture of the electronic switching systems for the telephone plant and of the fourth and fifth generation computer systems for data processing. The possibilities of offering novel voice and digital services by using the current and evolving technology, together with the potential of sizable economic gains, has catalyzed to rapid growth of the algorithms and techniques for ISDN. The establishment of ISDN standards, including protocol and interfaces; the design and development of the new hardware devices and special purpose integrated circuit chips; and finally, the new software and systems for the implementation, all constitute the wide international interest in the ISDN.

The new services ISDN proposes to bring to the customers vary from nation to nation. However, we present here the broad categories of new services ISDN is capable of offering in countries with sophisticated telephone networks. The proposed time span for these services spans the next five to seven years: (a) digitized and encrypted voice facility, (b) televideo facility with appropriate bandwidth compression for communication, (c) facsimile, graphics for business applications, and (d) enhanced customer services such as videotex, electronic mail, database access, searches, word processing, telemetering, inexpensive computing via high speed terminals, mass data transfers, etc. The broad impact of the proposed ISDN can vary significantly from country to country. However, it is clear that when this network comes into its own, the impact will be far more dramatic than the impact of the telephone and telephone network. One way to visualize the ISDN is to view the public communications networks being integrated into one gigantic mostly digital and computerized communications facility.

5.1.4 Data Rates for ISDN

The transmission of binary data over the telephone lines is not a new concept. At very low rates, such as 300 to 1200 bit/s, the plain old telephone service (POTS) can be readily used. At slightly more enhanced rates (2.4, 4.8, and 9.6 kbit/s), conditioned telephone lines can handle the data traffic. But these rates simply do not suffice for the type of proposed ISDN services. Much higher data rates are now imminent with the progress in customized semiconductor chips and inexpensive specialized components, which are now becoming available. Hence, the ISDN rates proposed are much higher than these subrates available in most networks.

The unit for a digital channel offered to the customers has been chosen as 64 kbit/s. The digital channel unit at this rate is generally abbreviated as B for the ISDN terminology. Similarly, a low speed channel of 16 kbit/s, abbreviated as D, is chosen to offer the digital capability to the subscribers. At this lower speed, the steady telemetering and line monitoring functions (such as security system, line monitoring, call waiting, etc.) are also accomplished. This channel is always open to the subscriber and fulfills the equivalent line scan functions in the POTS environment.

Data rates of 64, 72, 80, 96, 144, 160, and even 192 kbit/s have been considered from time to time. It is also necessary to realize the need for an

Introduction

overhead of 8 to 16 kbit/s for the additional bits for network functions, management, and error recovery, etc. These bits are over and above the rates the subscriber uses, and the line rate (defined as the actual rate on the telephone line) can typically become 160 kbit/s for a (2B + D) subscriber rate. The (2B + D) offers the subscriber two high-speed channels at 64 kbit/s and one low-speed channel at 16 kbit/s. The user has the option of using the channels in any format or fashion he desires. Over the last few years most of the ISDN designers are favoring the (2B + D) rate at a line rate of 160 kbit/s.

5.1.5 The Modes of ISDN Data Transmission

A higher rate of data transmission is a necessity for ISDN services. Some early ISDN designers are envisioning ISDNs as being digital in nature and aiming to accomplish a range of added services with ten to a hundred fold increase in bandwidth over the voice frequency bandwidth. But the bandwidth requirements depend upon the code chosen to encode the transmitted data and the mode of transmission. Standard codes for data transmission have been discussed in Chapters 2 and 3 of Vol. I. However, the mode of data transmission can also play an important role. In the ISDN environment, there are five basic modes to transmit the data; the first four are well known in the literature.

First, the time division multiplexing depends upon isolating the data from the transmitter and the receiver by partitioning the duration into finite blocks of time during which data is either transmitted or received. This technique, also called time compression multiplexing (TCM), while receiving the data, compresses the transmitted data for the duration, and transmits the data for a shorter duration but at a much higher rate during transmission. The bursting of data becomes necessary because the subscriber loop essentially carries unidirectional data at any given instant of time. This feature also has led to the name as the burst mode of transmission for the TCM mode.

Second, the hybrid mode of ISDN transmission forces data in both directions simultaneously on the two-wire subscriber loops. A hybrid device is used to isolate the transmit and received data. It can be seen that the demands on the hybrid can become severe if other supplementary techniques of isolation are not used. To supplement the functions of the hybrid, echo cancelers are used. The echo canceler, in principle, has to be adaptive in order to cope with the changing loop and environmental conditions. For this reason, the hybrid mode using these Adaptive Echo Cancelers is generally abbreviated as the AEC mode of transmission. The sole purpose of these echo cancelers is to suppress the near end echo of the transmitted signal from contaminating the received signal. It can be seen that the received signal has already suffered severe attenuation during its transmission through the channel, whereas the near end transmitted signal is quite high. This fact puts an added burden on the hybrid and the echo canceler, thus calling for delicately balancing out the echo of the transmitted signal from the received signal.

Third, the frequency division multiplexing (FDM) mode separates the received and transmitted data in two bands of frequency and retains the directional isolation by spectral filtering. This method for the transmission of information and retention of the desired isolation is well documented in literature.

Fourth, the four-wire mode retains the isolation between the transmitted and the received data by physically separating out the channels that carry the data. One pair of wires is used for each direction and simultaneous data transmission can occur.

Finally, the hybrid balance method, which is the least well-known mode, depends upon an adaptive balancing network which is so delicately balanced that even a rudimentary hybrid will effectively isolate the received signal from the transmitted signal. This technique is perhaps the hardest to implement inexpensively. Our early simulation studies on this technique proved it to be totally inadequate even for the simplest loop configurations. Hence, no detailed discussion of this technique will be presented.

The four wire mode for the transmission of ISDN data is overly wasteful of cable and conductor resources already invested in the network. Since the loop plant cannot be easily supplemented, the use of four wires for ISDN data transmission is virtually ruled out from most discussions. Frequency division multiplexing is wasteful of the bandwidth available on that channel and also calls for very effective filtering. Modern filters are made from semiconductor devices and the use of such materials to build effective filters seems to defeat the purpose when the very same devices and technology can be used to build the components and devices for the two remaining modes of ISDN data transmission. Hence, the two most viable contenders for transmission are the TCM and the hybrid duplex modes. These modes of transmission are discussed further in Chapter 6 of this volume. However, it can be seen that the status of ISDN (Section 5.2), the influence of the loop environments (Section 5.3), the major limitations for the data transmission (Section 5.4), and the evolving trends in ISDN (Section 5.5) are all influenced considerably by the mode of ISDN data transmission selected.

5.2 THE GLOBAL STATUS OF ISDN

The efforts to establish the ISDN in most of the technologically advanced nations are well along the way. A considerable amount of effort has been expended in the research and development of the Digital Subscriber Lines (2), digital switching techniques incorporating the additional needs of ISDN (3), study of Circuit

The Global Status of ISDN

Switched Digital Capability (CSDC) (4), and its implementation (5). The expertise and learning in these areas of investigation and existing services appears crucial in the direction and development of the ISDN.

Functionally, the ISDN promises to fulfill a set of objectives: (a) a range of data rates, (b) a range of call setup, call clear-down, and holding times, (c) customer options on costs and services, (d) low error rates and a range of secure voice and data transmission, (e) additional economies for the burst and continuous data communication, and (f) a choice and range of services and service grades. The digital facilities used in the implementation of the ISDN would offer most of these aforementioned functions very inexpensively. Further, the proposed topology of the ISDN incorporating the very efficient data carry mediums, such as optical fibers, microwave links, etc. (6, 7) with large scale switching facilities, such as 5 ESS systems (8), enhances the economies foreseen by the implementation of the ISDN.

Initially, the ISDN topology in the United States will use the telephone subscriber as the ISDN customer. In addition, it will use both the circuit switched networks and packet switched networks for data streams. Further, with the evolution of other digital networks and facilities (9), the proposed ISDN in the United States will interconnect with International ISDN's local area networks, private networks, value added networks, international record carriers, etc. The medium of digital communication between the nodal points in the network is the digital pipe. Digital pipes have multiple bandwidths and can consist of any number of digital data communication links. The familiar central offices assume the role of service centers that communicate with the customers, the other networks, vendors, police, firehouses, hospitals, and data base ware-houses. The switching takes place by the user supplied information at the circuit switched networks or by program control for packet networks (10). The customer now has access to this service center via a digital path on multiple bandwidth digital pipe.

On a network basis, data from the user arrives at the local switch over the subscriber loop. Here, it may be switched, or communicated to a tandem switch over interoffice trunk lines, or metro facilities. At the tandem switch or the hub offices, the data is either switched or communicated elsewhere in the network, depending upon the connection desired or the destination code. The average length of the subscriber loop is about 3.7 km. The average length of a typical ISDN customer is about 2.36 km and consists of a two-wire pair between the local switch and the customer terminal. Between the local and the tandem switch high-speed digital wire carriers (T1 at 1.544 megabits per second, T1C or T1D at 3.152 megabits per second or T2 or T1G at 6.312 megabits per second) are used. In addition, large capacity exchange digital coaxial lines and radio systems can also be used as short-haul digital carriers. For the communication between the tandem switches, a host of long-haul carriers are available. The coaxial analogue L-carrier system (1) has experienced steady growth in its channel capacity and bandwidth. The previously designed L4 cable system can be recast to handle 13.29 megabits per second. In addition, terrestrial radio with frequency division multiplexing and frequency modulation systems (TD systems at 4-GHz; 3700–4200 MHz range and the TH systems at 6-GHz; 5925–6425 MHz range) also offer wide bandwidths for enhanced data transmission needs of the ISDN. Finally, satellite and undersea cable systems already provide additional high-speed data paths. The advent of fiber optic data transmission systems also has a profound effect upon the direction and impact of the proposed ISDN.

The local switch can be any one of a series of vendor-supplied small exchange facilities. Typical examples of the switch are 1A or 5 Electronic Switching Systems (ESS) or D-channel bank by Western Electric, 5 EAX by General Telephone and Electric, etc. The tandem switch or the hub office can be 4 ESS by Western Electric or 3 EAX by GTE, etc.

Signaling plays a dominant role in the exchange of information between the switches. Common Carrier Interoffice Switching (CCIS) introduced in 1976 is in the process of modification to (the International Telegraph and Telephone Consultations Committee) the CCITT signaling system No. 7. By 1990 it is expected that about 95 percent of the trunk signaling between the central offices will be compatible with the ISDN requirements. Stored Program Control (SPC), which initiated with the advent of electronic switching, will also become prevalent in most of the ISDN implementation.

This section provides an insight into the directions and the accomplishments of various countries. We start out with United States and Canada in Section 5.2.1 because of the large investment the carriers have made in the expensive analogue switching systems. Here the telephone plant environment is severe because of the size and the rapid growth that these networks have experienced. The transition to totally digital network can only be gradual and evolutionary in this type of a national network.

The status of ISDN in Japan and Europe is discussed in Section 5.2.2. In these countries, both the switching and loop environments are quite different. These differences tend to favor the use of different hardware, software, and associated algorithmic solutions to the same problems. The protocol tends to be standardized because of the ANSI (the American National Standards Institute), the CCITT (the International Telegraph and Telephone Consultative Committee), and the ISO (the International Organization for Standardization) standards.

5.2.1 ISDN in United States and Canada

In the United States and Canada, the ISDN facility attempts to provide a multitude of services. Typical of these already identified are: digitized voice transmission and encryption facility, facsimile and graphics capabilities, video and computer

The Global Status of ISDN 225

communications, and finally other customer services such as telemetry, videotex, data base services, electronic mail, word processing, centralized computational facilities with high-speed links, etc. Major facilities of this nature have to be evolutionary, in context of the two major digital systems already installed: digital switching and digital transmission via the digital subscriber lines, fiber optic systems, and satellites and other digital data services such as TWX, Telex, Telenet, Tymnet, Dataphone Digital services now offered.

The evolution towards the ISDN has already started and is gaining steady momentum. The Public Switched Digital Capability (PSDC) will permit receiving and transmission of voice and data alternatively over the same loop. Time Compression Multiplexing (TCM) and Digital Subscriber Lines (DSL) are being used for data transmission at 56 kilobits per second (2). Initially, the call setup will be done by a touchtone telephone at the subscriber in the voice mode. Additional digits and the number of the called party will facilitate the selection of the digital path and the appropriate signaling. Stored program control, CCIS and switch activated network terminal equipment (NTE) will be used to establish digital connectivity between the parties. Efficient use of the existing 1ESS switching system is thus retained. The PSDC now offered is expected to provide a few of the overall services promised by the ISDN. Typically, it is aimed towards offering secure communication by digitizing the voice and scrambling the data, facsimile, and bulk data transfer applications. It can also provide access to packet switched networks.

Local Area Data Transport Capability (LADT), though not strictly a ISDN service, needs a special mention here in order to indicate the trends in the customer services. In the first mode of operation, this service receives customer data at 1200 bits per second in a special (the International Organization for Standardization Open Systems Interconnection (ISO-OSI) reference mode, level 2, link layer), format and transmits packetized data at 56 kilobits per second to a data base. Local multiplexing is also possible at the concentrators. In the second mode, the data over voice concept is used to send up to 8 kilobits per second data over analogue voice conversations through the existing analogue switch at the central office. Special equipment (known as Data Subscriber Loop Carrier; a pair gain system) is necessary to combine the transmitted voice and data and to separate the incoming voice and data.

The Advanced Communication System (ACS) facility offers the access to packet networks now abundant in the United States with local computers, terminals, and the voice data equipment. Data storage, management, and processing capabilities are featured at the switches which house the packet network port or have a 56 kilobits per second link with the network. Added public networking facilities are also available at these switches.

Other forerunners for the ISDN type of networks have been considered seriously in some geographical areas in the United States. For example, Metropolitan

Area Networks using the fiber optic technology are being investigated for high-density data traffic between the customer and the central hub through an interface unit. The interface unit can be switched between the customers and provides high-speed digital access from a hub or a node in the network to the customer. Next, the Public Packet Switching Network, using separate switches yet sharing the existing digital or analogue networks, offers new digital capability to the customers. The Bell Operating Companies are expected to take full advantage of the Federal Communications Commission approval of the asynchronous to X.25 protocol waiver and offer more and more capabilities to the customer. Digital connectivity is also promised by the Integrated Network Access between the local distribution and the interoffice facilities. The Integrated Special Services Network also provides limited service capabilities before it is eventually incorporated in the full fledged ISDN channel switching and remote access capabilities.

Among the many network architectures studied, an attempt to provide transmission capability on digital pipe and variable bandwidth assignment, common channel signaling and flexible routing and control at switching nodes, centralized intelligence such as routing and control data bases with embedded artificial intelligence concepts, and finally newer switching configurations and architectures are being considered. The divestiture of the Bell System[1] has fragmented the steady progress towards an integrated national effort towards ISDN. However, the T1D1 committees are attempting to provide a leadership position in the standardization of interfaces, codes, components, protocols, etc.

The CSDC (Circuit Switched Digital Capability) offered by a few of the Bell Operating Companies (BOC) in the United States, is designed to provide full duplex synchronous transmission at 56 kbit/s using the TCM mode. The system is designed for the loop population whose loss is up to 48 dB at 72 kHz. This frequency is based upon the TCM line (burst) rate of 144 kbit/s bursting the 200 bits of information every 3 ms for 1.3888 ms. This system uses an optimized analogue equalizer (2) capable of compensating for loop loss up to 48 dB at 72 kHz. However, since the central office environment can become noisy and crosstalk limitations can become severe, a 40 dB loss limit at 72 kHz is more realistic. The peak penetration of this capability is computed to be about 93 percent of all nonloaded loops in the 1973 loop survey. The CSDC with the TCM mode of data transmission should be considered as an intermediate step toward the ISDN. As the newer adaptive echo cancelers (devices for handling the near end echos at the transmitter) and decision feed back equalizers (see

---

[1] The Bell System underwent a major reorganization including the divestiture of the various operating companies during the early 1980s. The final phase of the divestiture was completed by January 1, 1984.

Chapter 6) become refined and inexpensive, the continuous duplex system at the (2B + D) rate of 144 kbit/s is gaining steady acceptance for the future.

In addition, Pacific Northwest Bell is active in offering digital service for the ISDN type of services in the metropolitan areas of the United States west coast. These trials will use Northern Telecom's DMS-100 switching systems. Mountain Bell in Central United States has also made public statement about providing ISDN access during 1986. Illinois Bell using the American Telephone and Telegraph's (AT&T's) 5 Electronic Switching System. It is expected that the ISDN network will be deployed during 1986.

Field trials and ISDN realizations by various operating companies in the United States are reported in (11). Six (BellSouth, Ameritech, Pacific Telesis, Southwestern, Nynex, and Bell Atlantic) of the seven regional Bell Operating Companies have reported the extent of the ISDN implementation in their companies' networks. Other organizations, such as GTE and AT&T, are also actively participating in making the ISDN-type services available to the network customers. Nynex's efforts in building high capacity Metro Net at 100 Mbit/s is also a clear commitment to providing high quality metropolitan area networks in addition to the 144 kbit/s ISDN facilities. In addition, corporate networks at T1 (1.544 mbit/s) are also being contemplated for major clients in the Nynex serving areas. Bell Communications Research, serving as a focus of the research activity for the seven regional Bell Operating Companies, is actively defining the architecture and establishing the protocols for ISDN. The nature and type of services to be performed by the evolving intelligent networks such as the ISDN is being actively investigated in light of their architecture and protocol distribution.

The carrier serving area concept introduced by AT&T offers an elegant solution to the incorporation of remote terminals in office buildings, shopping centers, and other large population centers that have central offices (and thus provide high speed data links). Feeder distribution interface is provided to serve a multiplicity of subscribers in a distribution area in the remote population centers. Fiber optics and cable pair groups are both used to share the data traffic according to the volume encountered in the different segments of the network. More recently, New York Telephone's concept for establishing digital centers, call the Service Area Multiplexing Sites (SAMS) a step towards the placement of high-capacity digital nodes at strategic metropolitan locations.

Bell Canada is attempting to evolve an integrated digital network (IDN) as a precursor to the ISDN. There is already a high penetration (50%) of digital transmission in the local facilities and expected to grow higher (about 65%) by 1990. Intertoll digital facilities should also grow rapidly (from 17% in 1980 to about 65% by 1990). Fiber optic technology is expected to enter the local network involving DS-1, DS-2, and DS-3 based system. The CCITT #7 (Common Channel Signalling, CCS-7) recommendation is adopted for the IDN for the wide range of new services. These signalling links are designed to operate at 64 kbit/

s. In Canada (as in the United States) the early ISDN efforts are at 64 and 80 kbit/s in the TCM mode. Facilities center concept has been considered to perform storage, retrieval, and processing of the ISDN data. In the more recent efforts in Canada, a variety of rates from 16 (B/4) to 1536 (24∗B) kbit/s are being considered for the wideband channel structure. The plan calls for establishing a tariffed wideband digital MEGAROUTE service between business customers. The service will be initially intracity, nonswitched, and end-to-end digital connectivity. The established T1-carrier provides an excellent option in the loop plant to extend the digital service to the end user. Megaroute over the T1-carrier with 26 AWG with 0.7 km repeater spacing is proposed. The proposed methodology promises to use the fiber optic and twisted wire pairs effectively as the network evolves. Devices and IC's still have to be designed, and systems have to be tested. The very wideband channel capacities of 6144 and 43008 kbit/s, also contemplated in the Canadian network, is consistent with the new DS-2 rate and with the proposed Bell Communications Research's SYNTRAN rate at the 672∗B or at the rate of 43.008 Mbit/s.

The more recent efforts in the United States have been at 144 kbit/s (line rate of 160 kbit/s) in the hybrid duplex mode of transmission. There is considerable optimism that the adaptive echo canceler systems with decision feed back equalizer and band efficient codes will be eventually used for ISDN systems. Very detailed simulation studies are warranted before the introduction of the architectures, systems, and components because once they enter the very complex communication networks in the United States they become an integral part of the network possibly for many decades. Even the replacement of channel units can be a monumental task. The New York Telephone Company's serving area multiplex sites (SAMS) and Bell Canada's megaroute for high-speed high-density traffic are some of the cornerstones in the evolution of the ISDN in America. High capacity fiber optic links are seen as the digital paths among central offices and between switching networks and remote switching modules. Two types of ISDN subscriber loops are proposed: twisted wire pairs at the (2B + D) rate and broadband optical fiber loops for high resolution video services.

5.2.2 ISDN in Other Countries

The International Telegraph and Telephone Consultative Committee (CCITT) plays an integrating role in unifying the ISDN efforts. Guidelines and features for the new services are being drawn up. The areas of recommendation cover the service capability, network architecture, and the user-network interfaces. During the next few years, the components, systems, networks, and interfaces for ISDN are expected to experience some standardization and consolidation.

The effort of most of the European countries is to implement the digital subscriber lines at 144 kbit/s, even though some trials are proposed and implemented at lower rates. The effective usage of the bandwidth by using newer line codes (WAL 1, WAL 2, 4B3T, 2B1Q, etc.) is being emphasized. The CMOS implementation for the integrated circuits to realize the circuit functions has been favored.

In Europe and Japan, the efforts to realize ISDN concepts have required considerable efforts. A clear direction for the rates, modes, and interfaces is not established. The CCITT recommendations are expected to be adopted when they become available. In Britain, a trial system at 80 kbit/s for digital facsimile and a low scan television is expected to be developed by British Telecom. Line cards with 25 mm CMOS IC's are expected to be used. The low demand for these IC's is a definite economic obstacle. The problems faced in the effective utilization of the fiber optics in the ISDN have also been addressed in Europe and Japan.

In France, a multiservice network, in Biarritz (12), is proposed and some early experiences with the network are reported. Two categories of services were initially proposed. The first one consisted of telephone and videophone and uses the switched services subsystem. The second one, Hi-Fi and TV programs, uses the distributed services subsystem. The central office side has a multiservice connection unit and the customer side has a network termination unit. Standard protocol (LAPB × 25) is proposed to establish the connectivity between the customer and the network. Interesting novel features (such as local picture control, videotex message insertion, hands-free dialing, high quality sound) are proposed and realized in the field trials. The possibility of offering end-to-end digital connectivity at 64 kbit/s using INTEL 8031 microprocessor is also explored in the French telephone network as a precursor the full fledged ISDN services. More recently, the digital service has become available. CCITT recommendations have been followed through during this trial period. This step is similar to the Circuit Switched Digital Capability adopted in the United States in 1982 as a precursor the entire ISDN capabilities to be developed by the late 1980s.

Two optical fiber cables link the subscriber and the network node. Each fiber carries unidirectional data to and from the user having two types of services: (a) telephone and videophone as switched services and (b) Hi-Fi and TV as distributed program services. The service is being offered with both telephone and TV features and is reported to have a limited subset of ISDN type of services (13).

In addition, the French National Center for Telecommunications is proposing to test an asynchronous time division switching network as an integral part of the PRELUDE project. The techniques used for the multi-bit-rate data transmission are derived from packet switching and from the telephone network time division circuit switching. Fixed packet length of 128 bits with 120 bits for information and 8 bits for label is adopted. Maximum terminal bit rates are used for transmission over coaxial cables. For picture transmission, this rate is about 50–100 Mbit/s. Fairly elaborate multiplexing-demultiplexing and frame allocation techniques are used. Though the early experiments have confirmed the feasibility of the proposed techniques, the IC components need to be designed and an overall system test needs to be performed.

In Sweden, the ISDN system trial is well underway since 1982 and offers telephony, videotex, teletex, and facsimile. The customer has both an alphanumeric display and a keyboard. The ERICSSON group, responsible for the system, has trials in Europe, especially in Italy, and has adopted the three layered protocol according to the OSI reference model and the new CCITT draft recommendation I.320. The channel interface structure handles data at the (B + D) or the (2B + D) rates. Loop measurements for three subscriber transmission systems (TCM, FDM, and Adaptive Echo Canceler (AEC)) were carried out at 80 kbit/s. The AEC system, with decision feed back equalizer, is reported to have the capability of handling the data traffic over 95–98 percent of the subscriber loops with bridged taps. However, the average length of the subscriber loops and the number and distribution of the bridged taps in the other (especially the American) networks offer far more severe conditions than those in the European countries. In the recent trials, the switching system includes the AXE 10 packet switching application; the AXB 30, which is a data switch for circuit switched application; and an MD 100, which is a digital PABX for voice and data. This trial offers some insight into the possible architectures, protocols, and the terminal configurations for the evolving ISDN in the European network, using the existing telephone subscriber loops.

In the United Kingdom, Switched Star integrated services network was proposed for providing a pilot service in early 1985, followed by a full service during 1985 and 1986. It is proposed that this network use coaxial cables for the early systems of an average length of about 300 m. The use of optical fibers is also explored in the network and these primary links are very promising. The network promises to offer distribution TV (basic, subscription and pay per view) services, interactive videotex with both alphanumeric and photographic facility, and individual, immediate access to video libraries. British Telecom appears to be more inclined towards the switched star architecture over the tree and branching networks.

In addition, ISDN systems trials have been performed in Sweden, and some of the problems associated with subscriber access to ISDN with subchanneling capability have been investigated in Italy. From the recent contributions from the various administrations in Europe, a unified effort in the systematic implementations of the ISDN is lacking. The major reason for this is the vast number of component, system, and interface problems still to be resolved. It is still early in the development of the global ISDN and a unified approach will take time. The possible economic rewards of the national and international ISDN have prompted much early development and design. The user responses to these newer services will become crucial to the eventual economic viability of ISDN.

In Japan the implementation of the ISDN appears to be more clearly established. The LSI and VLSI components necessary for the end-to-end digital connectivity in the Nippon Telegraph and Telephone Public Corporation have been designed and realized for the Information Network System. Work is reported on architectural aspects, interface structure, user network interface, timing extraction, etc. for 80 kbit/s, TCM system. The system has been in service in the outskirts of Tokyo since late 1984. Considerable effort and documentation exists for the successful development of the LSI effort for line termination circuits and an optimized $\sqrt{f}$ equalizer for ISDN rates of 80 to 200 kbit/s in the TCM mode.

IC design and realizations of the specialized components for the customer access for ISDN are being as actively investigated in Europe and Japan as they are in Canada and United States. Some of the IC chips for different line codes have been available and tested for possible adaptation as standard interfaces for the ISDN implementation. The selection of the (2B + D) rate and the 2B1Q line code by the Bell Operating Companies during 1986 will have to be factored in the design of these special purpose IC's for ISDN.

5.3   ISDN AND SUBSCRIBER LOOP ENVIRONMENTS

The subscriber loop servicing the individual customers has to eventually bear the data at enhanced rates to accommodate the new ISDN services. The loop environment thus becomes the critical channel of communication between the central offices and the eventual user of these ISDN services. The major networks around the World have attempted to study their loop plants in order to characterize the physical and electrical characteristics. Physical characteristics change gradually over the decades because the cables buried once remain in service until they are physically removed or altered. However, the bandwidth over which information has been transmitted has changed dramatically over the decades.

Voice frequency characteristics have been studied and documented for many of the important national subscriber networks over the last two decades. The carrier frequency characteristics have gained prominence since the 1950s and 1960s. Analogue and digital carriers each need specialized characterizations. Hence, the Bell System (as it existed prior to the divestiture in the early 1980s), the Independent companies, and some of the major networks of America, Europe, Japan, and Australia have periodically conducted surveys of their subscriber loop plants in order to facilitate newer innovations such as the digital carriers and enhanced capabilities such as the ISDN services. In this section, we present the important results from such surveys as they pertain to the ISDN rates of data transmission. The study of such characteristics in context to the ISDN rates was undertaken in the United States at Bell Telephone Laboratories utilizing the 1973 Survey data. The 1973 survey characterization for the voice frequency was published in 1978 (14) and the characterization for the ISDN data rates was published in 1982 (15). A more recent survey was initiated in 1981 prior to the divestiture of the Bell System. Results from the survey are now being scrutinized at Bell Communications Research in New Jersey. The most recent study (16) was undertaken by Bell Communications Research during the mid-1980s. The data used for the current ISDN study is the 1983 loop survey data.

Surveys of this nature have also been undertaken in Canada, Italy, France, Germany (to a limited extent), Japan, Australia, and other countries. The results have not been systematically generated and published in the open literature. However, some of the important characteristics are available in the justification of the data rate, the mode, or the approach these countries have adopted for the proposed ISDN services. In this section, we assemble this information for the other national networks. This information is valuable in developing and standardizing the ISDN rate(s), mode(s), and devices for the proposed digital network in different loop environments.

5.3.1 Physical Characteristics of Loop Plants

The United States, Australia, the United Kingdom, and most Western European countries all have telephone networks in which a number of different diameters of the wires are used. This gives rise to gauge changes in the subscriber loop plant. The junction points between wire of different diameters are sources of reflection and constitute nonuniformity in the cable characteristics.

Open circuited cable sections tapped off the main loop between the central office and the subscriber (known as bridged taps) are also abundant in United States, Canada, Japan, Italy, and Australia. In the other Western European countries, the data on bridged taps is not positive. Data for the developing countries is not available even though it may be interpreted that if the resistance rule (Section 5.4.1) was used in designing the loop, multi-gauge sections would be present.

Combined gauge discontinuities and bridged taps are generally present in the United States, Japan, Canada, Italy, and Australia. Hence, the system tailored for the bridged taps should also accommodate discontinuities that result from gauge changes. At this stage in our discussion, we partition the results into two subsections. The first section deals with the physical characteristics of the pre-divestiture Bell System (referred to as the American) loop plant. Detailed surveys have been made and the results have been published for this loop plant. In the second section, the results as they are applicable for ISDN rates for the other countries are presented and compared against the American loop plant statistics.

American Subscriber Network

(1) Loop Statistics

The 1973 Loop Survey presents a cross section of loops that may be expected to carry the data traffic. The loop length is generally limited to 18000 ft or about 5.5 km, because of the presence of loading coils beyond this distance. In this section, an extensive physical characterization of the loops is presented.

Out of all the (1098) loops available in the 1973 Loop Survey, about 24 percent have loading coils, are longer than 18000 ft (5.486 km), or have nonstandard cable sections. Eliminating these loops, about 76 percent (831 loops) of the loop population (1098) surveyed constitutes the truncated data base over which the digital data transmission studies are based. Of the 831, 178 have five or more sections, 119 loops have four sections, 100 loops have three sections, 179 loops have two sections, and finally 165 loops are single cable sections. A histogram relating the population to distribution of the loops by their length and the number of cable section(s) is shown in Figure 5.1. Table 5.1 presents the numerical data together with a tabulation of the bridged taps in each of the 18, one thousand feet bands of loop length.

There are four dominant wire sizes used. The finest diameter wire generally encountered is the #26 American Wire gauge (AWG) and is roughly equivalent of the 0.4039 mm wire used in the European countries. The coarse wire is the #19 AWG with a diameter of 0.9119 mm. The two intermediate wire sizes used in the loop plant are the 24 (0.5105 mm) and 22 (0.6426 mm) AWG.

Table 5.2 presents the average length of loops surveyed in the particular band of a thousand feet or 0.3048 km. The overall composition of an average loop (in the truncated population) which is 7748 ft. or 2.362 km long, with 4500 ft. (1.37 km) of #26 AWG cable with 2408 ft. (.734 km) of #24 AWG cable 797 ft. (.243 km) of #22 AWG cable and finally with 42 ft. (0.13 km) of #19 AWG cable. Figure 5.2 depicts the gauge distribution in each of the 18, one thousand foot (0.30480 km) bands in the truncated data base. All of the loops which can be potentially used for ISDN application add up to about 1219.25 miles (1962.2 km) of all the loops in the truncated population. The average length of the loop gets shortened from 11,412 ft. or 3.48 km for the entire population to 7748 ft (2.36 km) for the truncated population consisting of standard cables, nonloaded under 18000 ft. (5.48 km). Seventy-six percent of the loops are included in the truncated population and the average length of population excluded from the present data base is estimated to be about 23000 ft. (7 km).

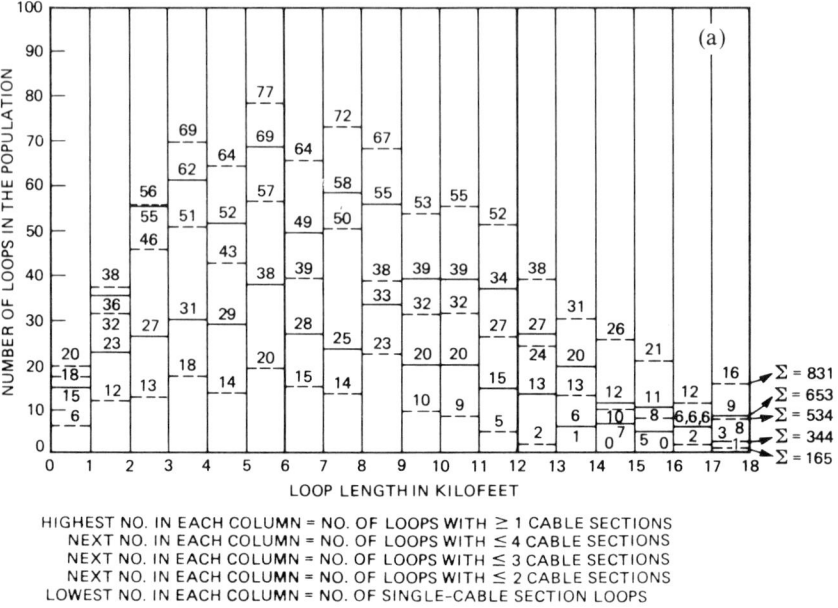

HIGHEST NO. IN EACH COLUMN = NO. OF LOOPS WITH ≥ 1 CABLE SECTIONS
NEXT NO. IN EACH COLUMN = NO. OF LOOPS WITH ≤ 4 CABLE SECTIONS
NEXT NO. IN EACH COLUMN = NO. OF LOOPS WITH ≤ 3 CABLE SECTIONS
NEXT NO. IN EACH COLUMN = NO. OF LOOPS WITH ≤ 2 CABLE SECTIONS
LOWEST NO. IN EACH COLUMN = NO. OF SINGLE-CABLE SECTION LOOPS

Reprinted with permission from the *AT&T Technical Journal*. Copyright 1982 AT&T.

Figure 5.1 Histogram of loop population distribution by number of cable sections in each kit.

(2) Bridged Tap Statistics

The truncated data base consisting of 831 loops has a total of 1365 bridged taps (see Table 5.2) on them. The average loop is 7748.63 ft. (2.362 km) long, has approximately four sections, has 1.64 bridged taps on it. The length of average tap is 922.42 ft. (0.282 km) consisting of 54 percent of #26 AWG and less than one percent of #19 AWG cable composition. The median distance between the central office and the first tap is 6523 ft. (1.99 km) and the median distance between first and second taps is 1071 ft. (0.326 km). Loops between 0 to 1000 ft. (0.3048 km) have the longest average length of bridge tap of 1333.3 ft. (.406 km) with 1.15 bridged taps for each loop. The average composition of each bridged tap is 59.3, 37., 3.6, 0. percents of #26, #24, #22, and #19 AWG cables. The total number of loops in this one thousand feet band is 20, seventeen of these do have bridged taps. The average loop length is 776 ft. (0.237 km). Loops between 17000 to 18000 ft. (5.18 km to 5.486 km) have the highest average number (2.87) of bridged taps per loop. The average loop length is about 17331 ft. (5.282 km) and the average bridged tap is only 626.9 ft. (0.191 km). The loop population in this band consists of 16 loops, 14 of these have taps. The tap

Table 5.1
Bridged-tap distribution by loop lengths (0 to 18 kft)

Loop Length in Feet	No. of Bridged Taps	No. of Loops, No. Bridged Taps	No. of Loops
0–1000	23	3	20
1000–2000	48	6	38
2000–3000	82	4	56
3000–4000	104	13	69
4000–5000	104	12	64
5000–6000	115	16	77
6000–7000	114	5	64
7000–8000	115	7	72
8000–9000	121	5	67
9000–10,000	97	3	53
10,000–11,000	96	8	55
11,000–12,000	87	8	52
12,000–13,000	59	3	38
13,000–14,000	50	5	31
14,000–15,000	45	1	26
15,000–16,000	32	3	21
16,000–17,000	27	0	12
17,000–18,000	48	2	16

Total no. of loops = 831: No bridged taps = 104
With bridged taps = 727

Reprinted with permission from the *AT&T Technical Journal*. Copyright 1982, AT&T.

population consists of 46 taps and the average composition of the bridged tap is 10, 74.0, 16.0, and 0.0 percents of #26, #24, #22 and #19 AWG cables. Loops between these extremes (shortest loops with longest bridged taps and longest loops with shortest bridged taps) lies a population whose average is fairly consistent and summarized in the previous paragraph. Even though the population density is sparse in the extreme bands of length the population density in the intermediate bands is relatively trustworthy.

In Fig. 5.3 the percentage of loops having bridged taps are depicted. Percentage to loops with taps (<1, <2, <3, etc.) is shown in this figure. Figure 5.4 shows the distribution of the gauges in the bridged taps. The similarity between Fig. 5.2 which displays similar data for the loop population and Fig. 5.4 is evident. These distributions indicate the similar patterns of the dominance of the #26 AWG cable being gradually replaced by #24 AWG cables as the length of the loop increases. Numbers 22 and 19 AWG cables do not dominate either the bridge tap or the loop lengths.

Table 5.2
Average loop length and average bridged-tap length distribution

Loop Length in Feet	Average Loop Length (ft)	Average No. of Bridged Taps	Average Bridged-Tap Length (ft)
0–1000	776.00	1.15	1333.35
1000–2000	1559.74	1.26	995.58
2000–3000	2548.87	1.46	782.12
3000–4000	3544.38	1.51	1121.13
4000–5000	4539.55	1.62	939.42
5000–6000	5445.13	1.49	962.14
6000–7000	6481.23	1.78	740.68
7000–8000	7483.33	1.60	957.54
8000–9000	8541.78	1.81	904.93
9000–10,000	9446.68	1.83	848.39
10,000–11,000	10572.91	1.75	843.49
11,000–12,000	11466.83	1.67	1092.56
12,000–13,000	12435.87	1.55	938.66
13,000–14,000	13383.77	1.61	932.66
14,000–15,000	14412.73	1.73	917.27
15,000–16,000	15447.76	1.52	926.06
16,000–17,000	16490.17	2.25	1016.19
17,000–18,000	17331.19	2.87	626.91
Total no. of bridged taps = 1365		Average length of bridged taps = 922.42 ft.	

Reprinted with permission from the *AT&T Technical Journal.* Copyright 1982, AT&T.

In the real physical sense, the loop plant environment is far from an ideal situation where uniform gauge wires connect the central office to the subscriber. Any design of the ISDN data transmission facility has to accommodate the wide disparity of cable compositions and bridged tap configurations. The range and statistics of these variations can vary significantly depending upon the country and the design rules that have been enforced during the evolution of the telephone network. In a sense, the loop plant statistics reflect the rate of growth of the telephone services, the demographic trends, the geography of the country, and finally what type of efforts the network managers have taken to monitor the network growth patterns of that country.

Other Subscriber Networks

Gauge discontinuities and bridged taps are also present in Japan, Canada, Italy, and Australia. The digital subscriber systems (Chapter 6) tailored for the bridged taps should also accommodate discontinuities because of gauge changes. Additional data is available from other countries, and the reader should refer to the

ISDN and Subscriber Loop Environments 237

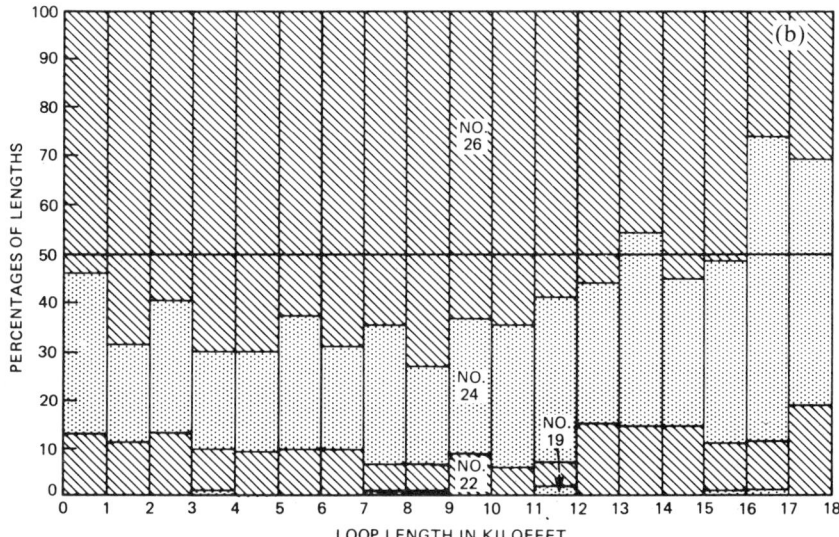

Reprinted with permission from the *AT&T Technical Journal*. Copyright 1982 AT&T.

Figure 5.2 Distribution of gauge numbers versus loop length.

original publications regarding national statistics of loop plants. The information presented here is for comparison and to provide an overview of the other national networks in the context of the implementation of the ISDN.

Particulars of the surveys are summarized in Section 5.3.1. Particulars of the physical characteristics, gauge or size statistics, and bridged tap distribution are presented here. Other relevant ISDN data is presented in some of the references. It is to be realized that the statistics can vary substantially between the urban and suburban loops, between domestic and business loops, between metropolitan and rural loops, etc. Hence, a certain amount of caution is necessary before generalizations can be made. As far as possible, the information presented here will be in the context of the proposed ISDN services.

The predivestiture Bell System operating company loops in 1983 numbered about 95 million with an average loop length of about 2.56 km. The independent companies in the United States have about 30 to 40 million loops and their statistics are believed to be similar to those of the other operating company loops.

With only four dominant wire gauges (#26, #24, #22, and #19 AWG) used in the American loops and 80 percent of the loops have gauge changes with an average of 2.3 gauge changes per loop. The percentage distribution of gauges being 58 percent of the #26 gauge, 31 percent of the #24 gauge, 10 percent of the #22 gauge, and <0.5 percent of the #19 gauge. PIC (Polyethylene Insulated Cables) dominate the loop plant. The paper insulated cables are very

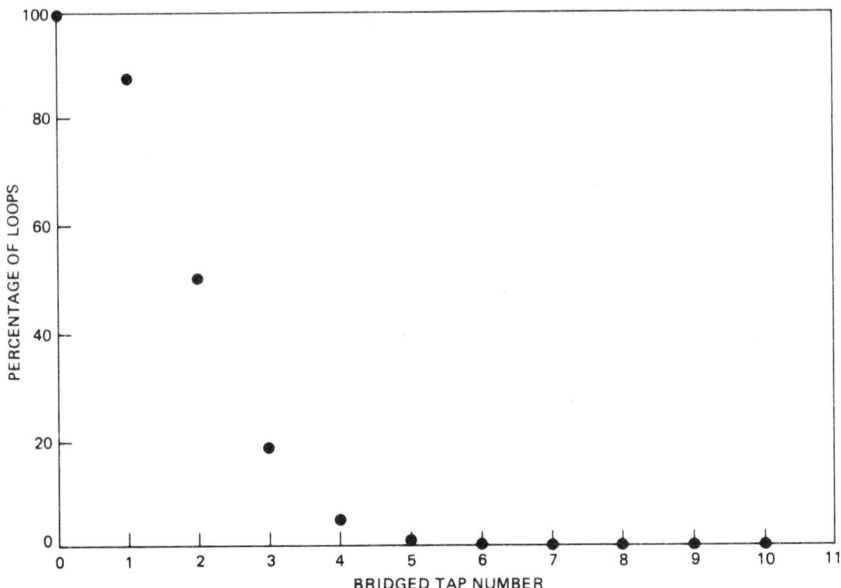

Reprinted with permission from the *AT&T Technical Journal*. Copyright 1982 AT&T.

Figure 5.3 Percentage of loops (Y-axis) having bridged taps equal to or greater than the bridged-tap number (X-axis) plotted.

scarce. Eighty-seven percent of the loops have at least one bridged tap and the average bridged tap length is 281 m. The percentage gauge distribution in the bridged taps is about the same as that in the loops.

The Canadian loops also have the same gauges found in the American loops, but only 55 percent have gauge discontinuities. The two finer gauges (#26 and #24) account for 65 percent of the loop length and 80 percent of the loops have bridged taps and 99 percent of the bridged taps are less than 1.6 km. Both PIC and paper isolated cables may be expected in the loops.

In the European countries, the physical characteristics of the Italian loops are available. Here 0.4 and 0.6 mm PE insulated cables are used. Size discontinuities are present. In the main population of the loops, .4 mm cables account for 80 percent of the length and .6 mm cables account for the remaining 20 percent. In the secondary population .6 mm cables are used. Twenty percent .6 mm cables are used. Twenty percent of all loops have bridged taps with the average tap length in the main population of loops being about 75 m, and in the secondary population being about 38 m.

In the United Kingdom, a wide mixture of sizes and insulations may be expected. The cooper and aluminum cable sizes may be .32, .4, used. In France, the .4 and .6 mm cables; in Sweden, .4, .5, and .6 mm cables exist. Size

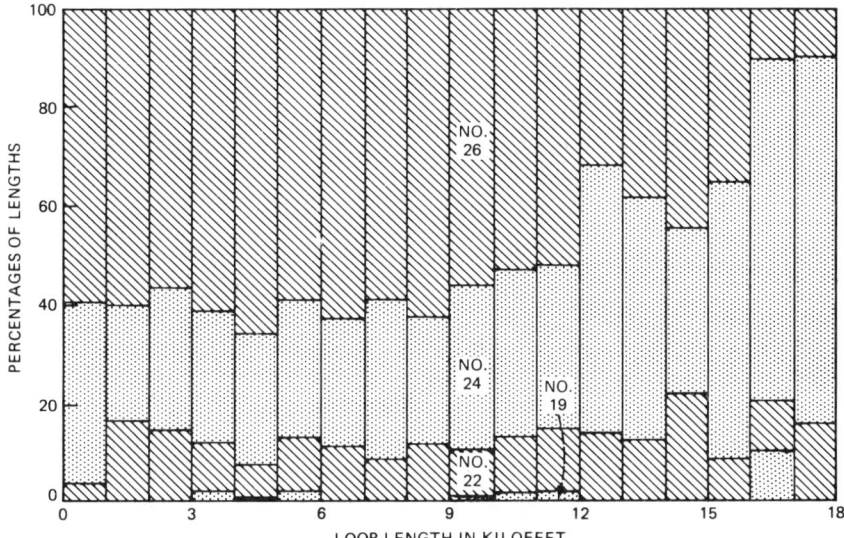

Reprinted with permission from the *AT&T Technical Journal*. Copyright 1982 AT&T.

Figure 5.4 Distribution of percentages of gauge sizes in the bridged taps.

discontinuity also occurs in most of these networks. The bridged tap data for these countries is not readily available.

In Japan, five cable sizes (.33, .4, .5, .65, and .9 mm) exist. Forty-eight percent of the loops have gauge discontinuities and the percentages of the loops have at least one bridged tap and only 14 percent have more than one tap. The average bridged tap length is about 200 m.

The Japanese NTT network had about 43 million loops (early 1980s). Out of these, about 5500 loops were surveyed. An average length of about 2 km is estimated for ISDN services and 90 percent of the loops are less than about 4.1 km. About 90 percent of the Canadian loops are under 4.2 km and the average length of the nonloaded loops is only slightly longer than the Bell System loops. The average length is about 1.7 km with 88 percent of the loops being shorter than 2.7 km. In the United Kingdom, the average length can be expected to be between 1.7 to 2 km. Eighty-eight percent of the French loops are less than 5 km, the national mean of the Swedish loops is about 1.7 km, and 92 percent of the German loops are less than 4 km. Finally, 93 percent of the loops in Norway are less than 4 km.

The Australian subscriber plant uses .32, .4, .51, .64, .9, and 1.2 mm copper cables and .52, .81, 1.15 mm aluminum cables. Ninety-five of the loops have size discontinuities with an average of 3.8 discontinuities per loop. The size distribution in the loops is 0.4 mm PIUT 42 percent, 0.51 mm PIQL 21 percent, 0.64 mm PIQL 15.5 percent, 0.4 mm PEIUT 10 percent, etc. Sixty percent of the loops have bridged taps with 54 percent having at least one tap and 4.7 percent having more than one tap. The 0.4 mm PEIUT cable accounts for 65.5 percent of the bridged tap length.

5.3.2 Electrical Characteristics of the Subscriber Networks

The cable and conductor access exists between the central office and the subscribers in the conventional telephone plant. These metallic paths have electrical properties. Four such parameters uniquely characterize a uniform section of the cable or the conductor. Many such sections may constitute an individual loop. But the characterization of the entire loop may be derived if the make up of the loop defining these sections and the four parameters for each section is known. The four parameters defined as the primary constants for each section are the resistance (R), the inductance (L), the conductance (G), and the capacitance (C) per unit length. These four parameters vary with the type of cable or conductor, its size, its temperature, its type, and its insulation. To a larger extent these parameters also change with the frequency at which the measurements are taken. In this section, we present the combined electrical characteristics of the subscriber loops wherever they are available. Again, the details of the American loops are published and the details of the other national networks are sketchy. For this reason, we divide the presentation of the electrical characteristics in the two following sections.

The American Loop Plant for ISDN

Digital data transmission at 2.4, 4.8, 9.6, and 56 kbit/s is presently available in selected areas in the United States. Enhanced transmission rates envisioned for the ISDN services range from 56 to 400 kbit/s. The set of bit rates (28, 32, 40, 64, 96, 144, 216, and 324 kbit/s) is chosen to span the range of interest. From the 1973 Loop Survey, we present the electrical characteristics of a cross section of loops that may be expected to carry the data traffic. These loops are generally limited to a length of 18000 ft. or about 5.5 km because of the presence of loading coils beyond this distance. In the more recent studies, with the 1983 Survey the focus at an ISDN frequency spectrum is between 40 to 200 kHz. The possibility of using the more band-efficient codes in the hybrid duplex mode and offering the initial services at $(2B+D)$ or a line rate of 160 kbit/s has prompted the spectral investigations for ISDN application in the 40 to 80 kHz region.

An extensive characterization of the loops is presented, first with all bridged taps stripped and next with all bridged taps intact. In the first configuration, the effect of temperature variation is also presented by tabulating losses at 0, 60, and 140° F. Image impedances (at the half bit rates) are also presented. Results are presented in condensed form and selected electrical parameters are also depicted as scatter plots (2).

Loops with Bridged Taps Removed

The results are organized by the possible ISDN data rates of 64, 144, 216 and 324 kbit/s. Correspondingly, there are four half-bit rate frequencies of 32, 72, 108, and 162 kHz for broadly classifying the data presented. There are three temperatures at each of the frequencies under consideration. Next, there are four major categories of results at each temperature. First, the statistical summary of the data is presented as scatter plots. Second, the loop attenuation of the 831 loops is shown in dB vs the physical loop length, and also, against the equivalent length of a #22 AWG cable. Third, the image impedance[2] (i.e., the resistive and the reactive component for each of the loops) is plotted against the physical length of the loop. Two such impedances are computed because of the lack of symmetry of the loop. The resistive and reactive components are separated out. Fourth, the loop image reactance for both sides of each loop is plotted against the loop image resistances.

There are three sets of results for the three temperatures (140° F, 60° F, and 0° F) at which the computation is done for all the loops in the data base. In this section, typical results at 60° F are presented. The electrical characteristics summarizing the average image impedances from Central Office and subscriber side are also presented. In Figures 5.5 and 5.6, the loop attenuations are depicted as scatter plots of the individual attenuation of each loop. Figure 5.6 contains the data when the attenuation is plotted against the converted loop length. The conversion of the actual loop length of its equivalent #22 AWG cable length is done by first determining the equivalency numbers between cable length for a given attenuation and then by accumulating the length of each loop section in terms of its equivalent length. In Figures 5.7 and 5.8, the resistive and the reactive components of the image of the impedance (as seen from the Central Office) are plotted against the loop length. Figures 5.9 and 5.10 hold the corresponding data when the image impedance is determined from the subscriber end. The dominant collection of the points along the horizontal lines at approximately 147 ohms (Figure 5.7) and 94 ohms (Figure 5.8) indicates the dominance of the #26 AWG cable leaving Central Offices towards the subscriber. Similar but less dominant lines do get formed for the #24 AWG cables. In comparison,

---

[2] Defined as $\sqrt{AB/CD}$ for Central Office to subscriber image impedance and as $\sqrt{DB/CA}$ for subscriber to Central Office image impedance. Here *ABC* and *D* are the four elements of the composite *ABCD* matrix for the entire loop.

Figs. 5.9 and 5.10 show relatively low dominance of such horizontal lines as seen from the subscriber side.

At each frequency and each temperature, the high value, the low value, and the average value of the most important parameters are tabulated. Two tables summarizing the results are presented. In Table 5.3, the real and imaginary components of image impedance from the Central Office side and the subscriber side are presented. The three sections under each side indicate the maximum, the minimum and the average values of these components of the image impedance.

Loops with Bridged Taps Intact

Bridged taps have a profound influence on the overall loop plant capability to carry the ISDN data. In the initial stages of the feasibility studies, it was envisioned that it may be necessary to remove the bridged taps for the loop plant the ISDN data traffic. The relative electrical characteristics of the plant with and without the bridged taps are reported here to indicate the degradation in the electrical characteristics of the predivestiture Bell System network with bridged taps. Equally indicative is the fact that the terminating channel units at the

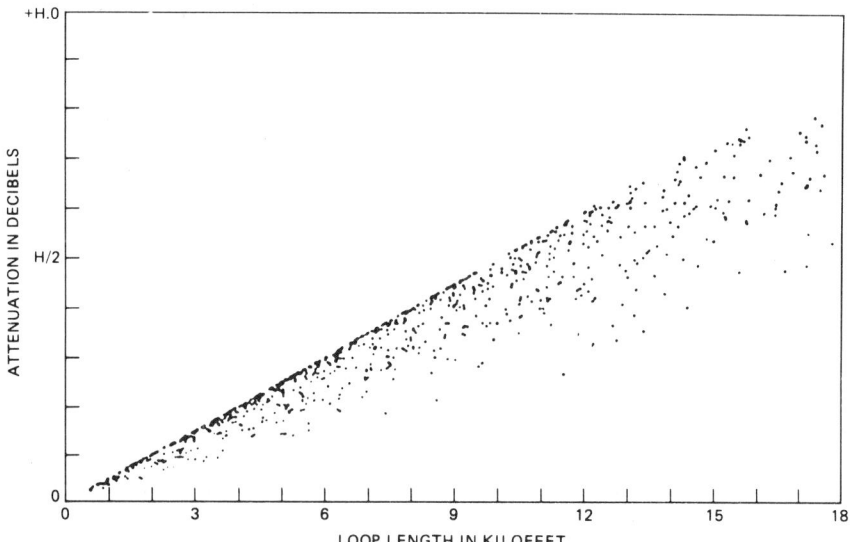

Reprinted with permission from the *AT&T Technical Journal*. Copyright 1982 AT&T.

Figure 5.5 Scatter plot of loop attenuations plotted against the loop length at 32 kHz and 60° F. The *X*-scale is 1 kft per division and *Y*-scale is 5.0 dB/division.

ISDN and Subscriber Loop Environments

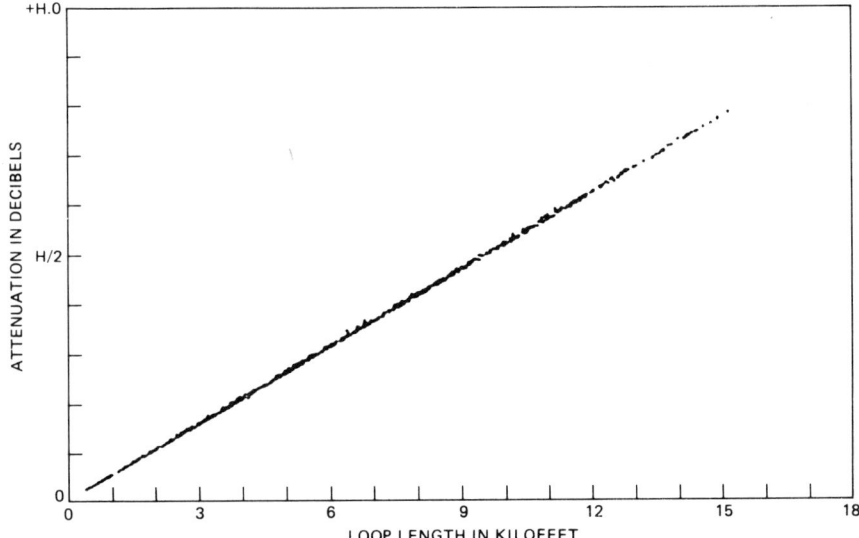

Reprinted with permission from the *AT&T Technical Journal*. Copyright 1982 AT&T.

Figure 5.6 Scatter plot of loop attenuations plotted against the equivalent loop length of No. 22 AWG cable. The frequency is 32 kHz and the temperature is 60° F/. The *X*-scale is 2 kft/division and *Y*-scale is 5.0 dB/division.

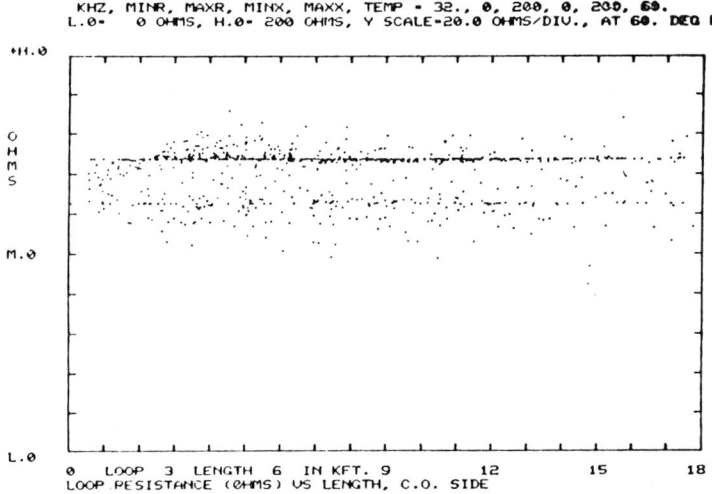

Reprinted with permission from the *AT&T Technical Journal*. Copyright 1982 AT&T.

Figure 5.7 Real (Z) vs length.

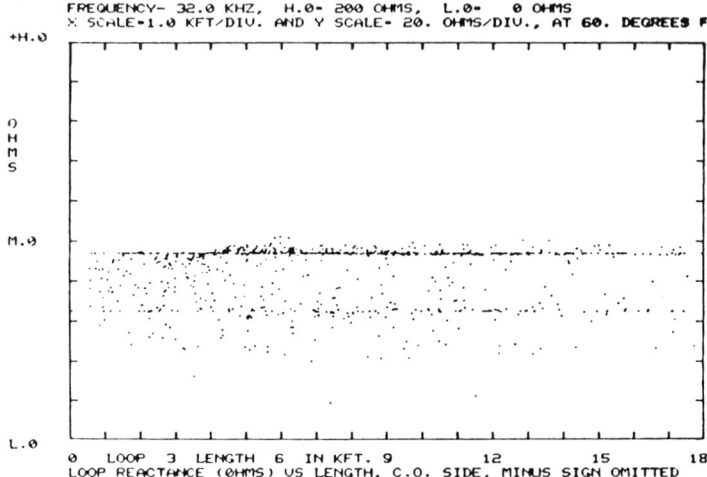

Reprinted with permission from the *AT&T Technical Journal*. Copyright 1982 AT&T.

Figure 5.8 Imaginary part of Central Office Impedance (Z) versus length.

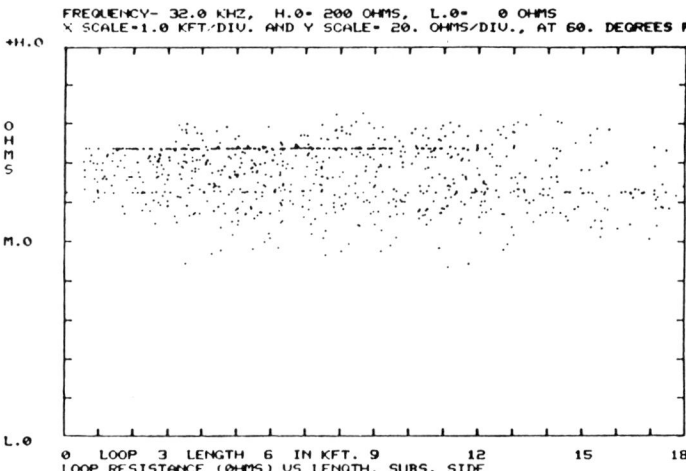

Reprinted with permission from the *AT&T Technical Journal*. Copyright 1982 AT&T.

Figure 5.9 Real (Z).

subscriber and the central office have to be redesigned to handle the loop plant with the bridged taps intact. The extent to which the channel units have to accommodate the degradation of the loop plant characteristics due to bridged taps is also indicated.

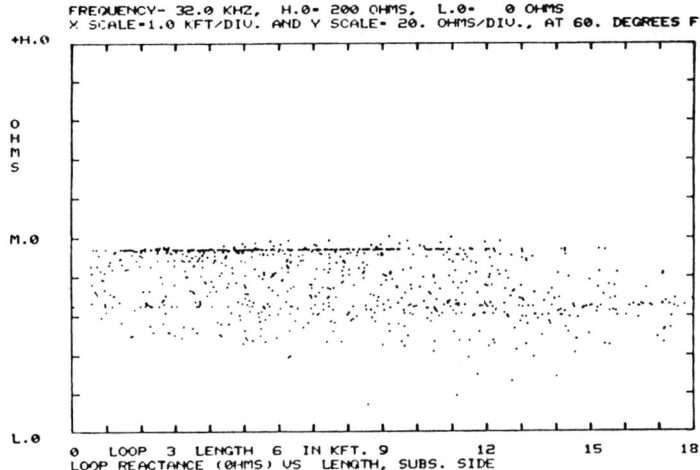

Reprinted with permission from the *AT&T Technical Journal*. Copyright 1982 AT&T.

Figure 5.10 Imaginary part of subscriber side impedance (Z) versus length.

Typical Results Generated (Bridged Taps Intact)

In this section, the electrical characteristics of the loop plant for ISDN data transmission are presented. We also focus upon the effects of bridged taps in the American loop plant. The main object for this comparison is to estimate the relative effects of the bridged taps in other national networks that have no bridged taps or relatively fewer bridged taps. The bridged tap population is the highest in the United States and the effects are most dramatic in this environment.

In Table 5.2 the statistical average of loop lengths is about 7750 ft. (or 2.36 km) and the equivalent length (about 12900 ft. or about 3.93 km) of #22 AWG cable. The equivalent length is based on the average length of #26, #24, #22, and #19 AWG cable percent in an average loop. The loss of such a loop is about 16.75 dB at 60° F and at 32 kHz. However, the average loss computed for the entire loop population based upon the bridged taps being intact is 18.25 dB at 60° F and at 32 kHz. Thus, the average loss of all the loop with bridged taps intact is about 1.75 dB higher than the corresponding loss with the bridged taps stripped at 60° F and at 32 kHz. At this frequency, the equivalent length of #22 AWG with the same average loss increases by about 1400 ft. or 0.427 km.

The effects of the bridged taps on the image impedances are also considerable. The subscriber side impedance changes from (134.36–j 78.15) ohms for loops without bridged taps to (99.8–j 68.73) ohms, whereas the Central Office impedance does not experience such major change because a larger population of the bridged taps is located nearer to the subscriber than to the Central Office. This

Table 5.3
Maximum, minimum, and average values of loop impedances at different frequencies (with bridged taps)

kbit/kHz	Temperature °F	Central Office Side of Loops (Z_{in})						Subscriber Side of Loops (Z_{in})					
		Maximum		Minimum*		Average		Maximum		Minimum*		Average	
		Re	Im	Re	Im	Re	Im	Re	Im	Re	Im	Re	Im
2.4/1.2	140	744	−696	400	−400	649.3	−628.8	720	−544	416	−38	632.0	−634.1
	60	680	−644	276	−376	607.6	−586.9	664	−660	388	−352	593.9	−592.3
	0	632	−616	368	−332	572.7	−552.0	640	−616	352	−344	561.7	−557.0
4.8/2.4	140	544	−500	272	−270	466.4	−443.1	512	−500	264	−265	446.6	−443.8
	60	504	−456	256	−256	473.1	−412.5	480	−456	264	−240	420.4	−414.4
	0	480	−432	240	−240	412.5	−387.0	456	−344	256	−228	398.2	−389.6
9.6/4.8	140	400	−365	190	−180	336.6	−311.2	380	−350	180	−190	318.5	−307.1
	60	370	−325	190	−160	316.7	−288.7	360	−325	165	−160	300.4	−286.1
	0	360	−300	180	−155	299.9	−269.8	340	−290	162	−155	285.2	−268.4
64/32	140	170	−110	97	−22	144.6	−93.1	170	−110	85	−16	183.3	−86.8
	60	170	−101	92	−16	140.4	−83.9	165	−100	85	−14	134.4	−78.2
	0	168	−94	92	−14	136.9	−76.3	160	−96	85	−12	131.3	−70.9
144/72	140	145	−67	90	−17	120.4	−15.3	150	−70	90	−8	117.4	−47.2
	60	145	−60	90	−15	119.4	−45.3	147	−61	86	−7	116.6	−41.7
	0	145	−55	90	−11	118.7	−40.5	145	−37	83	−6	116.1	−37.1
216/108	140	134	−54	85	−6	114.1	−37.4	140	−50	89	−6	112.0	−34.7
	60	134	−50	86	−5	114.0	−32.9	140	−46	90	−4	112.1	−30.5
	0	134	−45	86	−4	114.1	−29.3	136	−43	94	−1	112.2	−27.1
324/162	140	130	−40	86	−6	110.1	−27.4	134	−42	87	−5	108.1	−25.7
	60	132	−34	88	−3	110.6	−24.1	132	−40	86	−3	108.8	−22.7
	0	134	−36	90	−1	111.1	−21.5	134	−39	86	−1	109.3	−20.3

* The low numbers in these columns are not very accurate, since they are not program generated; instead, they are all scaled from the scatter plots.

Reprinted with permission from the *AT&T Technical Journal*. Copyright 1982, AT&T.

Table 5.4
Maximum, minimum, and average values of loop impedances at different frequencies (with bridged taps)

kb/kHz	Temperature °F	Central Office Side of Loops (Zin)						Subscriber Side of Loops (Zin)					
		Maximum		Minimum		Average		Maximum		Minimum		Average	
		Re	Im	Re	Im	Re	Im	Re	Im	Re	Im	Re	Im
2.4/1.2	140	704.8	−639.7	176.8	−513.2	560.0	−499.2	649.9	−640.	178.3	−120.6	497.47	−533.24
	60	652.3	−597.5	160.4	−478.8	520.8	−467.7	609.6	−597.5	161.5	−114.7	471.19	−497.8
	0	606.8	−562.1	147.8	−449.7	488.8	−440.9	575.8	−562.1	148.6	−109.5	448.5	−468.0
4.8/2.4	140	520.4	−453.0	123.4	−364.3	411.8	−343.1	476.3	−449.9	131.8	−78.8	340.5	−340.47
	60	486.0	−418.9	120.3	−337.3	383.6	−321.7	440.7	−419.5	123.9	−73.5	323.9	−346.9
	0	456.8	−393.5	117.3	−315.3	360.2	−303.4	412.1	−394.4	117.3	−69.5	309.6	−326.6
9.6/4.8	140	393.1	−324.8	91.5	−259.8	303.3	−230.7	351.6	−314.1	80.4	−57.2	234.5	−252.5
	60	371.5	−302.1	83.3	−239.0	284.2	−215.8	331.0	−292.6	78.6	−49.2	223.9	−236.5
	0	349.4	−281.4	77.6	−221.83	268.4	−203.2	312.2	−274.3	78.3	−44.5	214.9	−222.8
56/28	140	204.4	−129.7	48.8	−97.2	150.3	−100.7	198.8	−128.9	31.2	−9.5	106.6	−82.4
	60	198.7	−119.5	44.7	−87.3	145.2	−82.4	188.9	−117.6	30.5	−7.0	103.8	−76.0
	0	193.9	−112.4	41.8	−79.9	141.1	−67.1	178.0	−109.2	30.2	−5.4	101.6	−70.6
64/32	140	196.0	−120.1	47.1	−88.9	144.3	−99.1	193.2	−121.0	28.9	−4.3	102.3	−74.7
	60	194.7	−112.2	43.8	−79.6	139.9	−82.0	188.3	−112.2	29.1	−0.9	99.8	−68.7
	0	192.9	−105.5	40.3	−71.9	136.4	−67.0	181.3	−103.2	28.6	+2.6	98.0	−63.7
80/40	140	187.8	−108.9	44.7	−76.1	135.7	−95.2	177.9	−105.0	25.5	+3.1	96.1	−63.2
	60	189.8	−99.4	43.9	−67.7	132.3	−82.2	181.3	−100.5	24.9	+8.9	94.2	−57.8
	0	189.6	−94.8	41.8	−60.8	129.7	−70.2	183.1	−96.6	24.8	+15.3	92.9	−53.4

**Table 5.4
Continued**

kb/kHz	Temperature °F	Central Office Side of Loops (Zin)							Subscriber Side of Loops (Zin)						
		Maximum		Minimum		Average		Maximum		Minimum		Average	
		Re	Im	Re	Im	Re	Im	Re	Im	Re	Im	Re	Im
144/72	140	187.1	−101.7	40.3	−49.3	120.5	−67.5	157.6	−84.3	21.5	+27.8	85.2	−39.7
	60	190.5	−98.4	40.1	−43.2	119.4	−62.4	164.1	−84.5	21.3	+43.7	84.6	−35.9
	0	192.0	−93.0	40.1	−38.3	118.7	−57.6	175.3	−85.7	21.2	+54.6	84.2	−32.8
192/96	140	206.7	−79.7	38.6	−39.5	115.8	−49.6	180.3	−84.1	19.4	+49.7	82.6	−31.5
	60	233	−80.2	83.6	−34.5	115.4	−49.6	186.8	−38.5	20.3	+54.4	82.5	−28.1
	0	253.3	−81.1	38.7	−30.5	115.2	−50.6	200.7	−85.1	18.9	+56.8	82.4	−25.4
216/108	140	212.2	−105.5	37.9	−36.1	114.4	−38.0	180.1	−82.7	16.3	+59.4	81.8	−29.0
	60	198.8	−111.5	38.0	−31.5	114.2	−36.8	194.0	−81.4	16.5	+68.5	82.0	−25.9
	0	201.8	−114.2	38.1	−27.8	114.2	−37.5	203.1	−84.4	16.8	+74.9	82.1	−23.3
324/162	140	216	−119.7	35.1	−26.7	110.2	−34.6	184.8	−100.6	15.2	+74.5	81.5	−22.4
	60	233.4	−133.4	35.4	−23.4	110.7	−27.9	213.6	−105.3	14.0	+81.4	79.2	−20.3
	0	261.5	−145.5	35.6	−20.8	111.2	−19.8	242.0	−108.1	13.3	+88.0	79.8	−18.6

Reprinted with permission from the *AT&T Technical Journal*. Copyright 1982, AT&T.

ISDN and Subscriber Loop Environments

is indicated in the summary of statistics of physical characteristics in Section 5.3.1.1. A summary of these impedance at the seven discrete frequencies (28, 32, 40, 72, 96, 108, 162 kHz) is tabulated in Table 5.4. In Table 5.5, the maximum, and average attenuation of the loop at these frequencies are also presented.

Any design of the bidirectional data transmission for ISDN services has to accommodate the wide disparity of loop characteristics such as its image and input impedances, highly variable frequency dependent losses as they are measured from either end of the loop. The range and statistics of these variations are presented in this section as in the form of tables and scatter plots for the use in the design of the terminal equipment and the channel units for the digital subscriber systems. This will be discussed in greater detail in Chapter 6 of this volume.

The statistics from individual Bell Operating Companies (BOC) in the American telephone network have been accumulated in the 1983 Loop survey. The results from this most recent survey are available. (See Ref. 16). However, based upon the 1973 survey results, a summary of the electrical characteristics is presented here for comparison with the electrical characteristics of the other national networks.

At 32 kHz the average loss of loops for consideration of the ISDN data traffic is about 18.5 dB increases to 23.2 dB, 25.2 dB, and 29.7 dB at 72, 96, and 262 kHz, respectively. The matching impedances at 32, 72, 96, and 162 kHz is about (140−j82), (119−j62), (115−j50), and (79−j20) ohms, respectively, at the central office side of the loops. The corresponding numbers at the subscriber side are (100−j70), (85−j36), (82−j28), and (79−j20) ohms, respectively. The average propagation velocity is about 6.56 μsec/km at 162 kHz. The wave length thus becomes 0.94098 (6.1728 μsec divided by 6.56 μsec/km) km. The quarter wave length[3] is about 327 m and this number is close to 281 m, which happens to be an average bridged tap length. Hence, if the TCM mode of transmission is selected at 144 kbit/s, then the TCM rate of 324 kbit/s will suffer a very high intersymbol interference. This interference is evidenced by the eye opening simulation studies of the loop plant scatter plots, which are discussed in Section 6.3.2. The extent of such intersymbol interference at other frequencies in other networks and other codes can be estimated. Similar details for the other American independent companies are not available.

---

[3] Quarter wave length bridged taps cause the maximum intersymbol interference since the interval for the pulse to travel up and down the open bridged tap is one half the period at half bit rate frequency. A binary +1 and a binary 0/+1 (which is coded as 0/−1 in the bipolar code) would be separated by this interval and thus the 0/−1 pulse "sees" the highest amplitude of the reflected signal at the instant it is being scanned. Unless decision feedback or other sophisticated equalization strategy (Section 6.4.1) is used, the eye diagrams (Section 6.3.2) can become severely closed.

Table 5.5
Maximum and average attenuation of loops

kbit/kHz	Temperature (°F)	Attenuation in dB	
		Maximum	Average
2.4/1.2	140	11.7	4.93
	60	10.7	4.49
	0	9.9	4.13
4.8/2.4	140	15.6	6.92
	60	14.1	6.29
	0	13.3	5.78
9.6/4.8	140	21.5	9.63
	60	19.5	8.73
	0	18.0	8.01
56/28	140	45.5	19.93
	60	41.0	17.75
	0	37.4	16.02
64/32	140	47.6	20.82
	60	42.9	18.50
	0	39.1	16.66
80/40	140	50.6	22.27
	60	45.6	19.72
	0	41.7	17.72
144/72	140	60.0	26.30
	60	53.3	23.22
	0	47.9	20.84
192/96	140	66.2	28.46
	60	60.5	25.21
	0	56.4	22.73
216/108	140	69.1	29.44
	60	61.9	26.13
	0	56.4	23.62
324/162	140	76.0	33.15
	60	69.0	29.68
	0	63.5	27.08

5.3.2.2 Electrical Characteristics of the Other Networks

In Japan, the average loop loss at 100 kHz is 17 dB and about 90 percent of the loops have under 12 dB loss. The resistive component of the impedance is about 110 ohms. The propagation duration is about 5.6 μsec/km at 100 kHz. The Canadian loops have an approximate average attenuation of about 22 dB at 80 kHz. The resistive component of the matching impedance lies between 110 to 120 ohms. The Australian loops suffer an average of 17 dB at 72 kHz. The Italian network has an average loss of 19.5 dB at 128 kHz with a resistive component of the matching impedance at about 120 ohms and the average duration per km at 128 kHz is about 5 μsec. The maximum loss for the surveyed loops in France is about 40 dB at 88 kHz. The German network has a loss in the range of 8.6 to 23.2 dB for the surveyed loops with a 6 μsec/km duration at kHz. Loops from Norway have an average loss of about 20 dB at 128 kHz with a resistive component of about 135 ohms and the duration is 4.6 μsec/km. Results of the other networks are not readily available. However, some initial interference can be drawn from the data presented here.

In this section, we have summarized the data from different networks in the spectrum of interest for the ISDN capabilities. The American loop plant have the most severe conditions with 2.3 gauge changes and 1.64 bridged taps per loop. Japanese networks offer fewer loops with discontinuities and fewer bridged taps. The average length is shorter and the average attenuation is also about 8 dB lower at about 100 kHz. Canadian loops have a fractional proportion of paper-insulated cable and the percentage is unavailable, hence the impact of loop discontinuities is not clearly evident. The properties of loops that do have one or more bridged taps is about 80 percent as compared to 87 percent in United States. There is considerable penetration of 22 AWG and 19 AWG cables in the network (35% vs 10.7%) as compared to the loops in United States. The average bridged tap length is not available, and hence, the combined severity of higher proportion of 22 and 19 AWG cables with bridged taps cannot be easily compared against the severity in the American loop plant. Average loop attenuations are only slightly lower (22 dB at 80 kHz vs. 23.2 dB at 72 kHz).

Australian loops have a greater variety of cables and more gauge discontinuities but fewer bridged taps. The average length of the taps (or multiple) is also shorter. The effect of cable discontinuities between the dominant gauges (.4mm PIUT, .51mm PIQL, .64mm PIQL, .4mm PEIUT, and .64mm PIUT) depends on the differences between their primary characteristics (R, L, G, & C) and the effect of these differences in the spectral band of ISDN frequencies is not evident, even though the necessary simulations can be easily performed. The average attenuation at 72 kHz (Nyquist frequency for 144 kbit/s) is also about 8 dB less than that for the loops from the American operating companies. The impact of this loop environment as far as the image impedance (for hybrid echo cancellation techniques, if it is considered in this environment) matching should be studied in detail. In a general sense, the loop environment appears considerably less hostile than the environment in the American operating companies in spite of gauge discontinuities. The extent of the gain for the introduction of the ISDN services from the more friendly loop environment conditions can be computed.

Italy also offers a more friendly loop network with only two gauge discontinuities, shorter loops, fewer loops with bridged taps (20 percent vs. 87 percent), shorter bridged tap (38–75m vs 281m) than the American telephone network. The average loop attenuation (19.35 dB at 128 kHz vs 26.13 dB at 108 kHz) is also considerably less than for the loops from the American environment.

England offers a larger variety of gauge sizes but few or no bridged taps. The average attenuation is also about 8 dB (17.5 dB at 100 kHz vs 25.2 dB at 96 kHz) lower than the American loops. Norway has an average loop loss about 7–8 dB (20 dB at 128 kHz vs 26.13 dB at 108 kHz) lower. The loop lengths for both the United Kingdom and Norway are also about 0.75 km shorter.

The data from the other countries is either too sparse or unavailable for any meaningful comparisons. The rapid expansion of the telephone network has contributed to the harsh loop environment in the American telephone network. The most severe deterrent appears to be the presence of the bridged taps, and their average length of about 280 m places a window of restriction for the ISDN rates and codes that concentrate their energy around about 360 kHz. Very high quality echo cancelers can remove this restriction. But it appears certain that for the rate contemplated for the (2B + D) rate at 160 kbit/s with 2B1Q or 4B3T code in the hybrid duplex mode, the transmission is not going to suffer by this restriction. Likewise, at the (2B + D) rate with 2B1Q code (actual transmission at 80 kbit/s) in the hybrid duplex mode of operation, the average bridge tap length of 280 m is unlikely to cause a major problem. However, only a detailed simulation study can predict the three eye openings in the 2B1Q code. From the survey data it is possible to determine the percentage of the loops with critical bridged taps at the proposed data rate. The code used for data transmission and mode of transmission (TCM or AEC) also play a significant role in the overall system architecture and its design.

5.4 THE MAJOR LIMITATIONS FOR LOOP DATA TRANSMISSION

The ideal subscriber loop for the transmission of ISDN data in a telephone network would be a uniform gauge wire between the individual subscribers and the central office. The ideal central office electrical environment would be completely free of all noise. The lines would have only information content. Each wire pair would be shielded or totally isolated from other wires, cables, power systems, lightning surges, and other electrical noise pollutants. The real networks are far removed from these ideal conditions, and we can classify these networks by the extent of imperfection.

There are essentially two characteristics that make the network physically imperfect: a) gauge changes, and b) bridged taps. Oxidation (short interruption) and water in the cables have been cited as some of the other impairments in the large loop plant of countries such as Japan, Canada, and the United States. No scientific or systematic approach has been attempted to deal with these problems except to avoid these loop sections for the ISDN services.

Open wires suffer from the climatic effects more severely. Water, frost, sleet, and ice interfere with the data-carrying capacity. Corrosion can cause a break in the continuity, especially where splices exist. Physical structures and towers can also alter the electrical characteristics. The proximity of power lines causes inductive coupling and large signals at power frequency (and its harmonics) may find their way in the data path. These impairments for the open wire influence the capability to handle higher speed data rates even though wire pairs are transposed to minimize the exposure to stray signal pickup. This pickup could be at power frequency or even at audio frequency in the form of crosstalk.

In the cable environments (in contrast to the open wire environments) copper or aluminum conductors are generally insulated with a plastic material or pulp. A bundle of the conductor pairs are twisted again to form the cable core. Different types of enclosing sheaths are also available. The extent of the twist is varied in different sections to give a randomized pick up from other wire pairs and also to cancel or decrease the crosstalk. The different wire sizes and the material changes also cause variations in the primary and secondary constants for transmission.

Further there are three electrical impairments in the telephone media: (1) crosstalk and (2) lightning surges, and (3) electro-magnetic pick-up. Pick-up arises from imbalance in the cable pair. Defective shielding of the cables and standard techniques to circumvent their effect exist in different countries. Residual effects of lightning surges can be grouped with central office impairment as impulsive noise.

The central offices are again far from the ideal. Impulse noise and surges corrupt the signal received from the subscriber. Further, some central offices tend to add on short bridged tap lengths on the wire pair as it is being scanned. The electrical characteristics of the cable do not materially change for voice transmission, but it has been shown that these change the data transmission characteristics enough to cause serious errors, especially in the Hybrid mode of transmission if adequate precaution is not taken.

The telephone loop plant environment has been used conventionally to carry voice frequency signals. The use of copper conductors to span the central office and the subscriber in order to establish speech signals was the original purpose of the loop plant. More recently, the capabilities of this existing network have been investigated to enhance the voice channels by carrier systems and to carry digital data. Subrates (such as 300 to 1200 and even the 2400 bits per second) well within the audio band can be handled with little or no conditioning of the telephone lines. However, it may still become necessary to check the transmission characteristics to meet the bit error requirements when the loop plant is being used for data services. The human ear, which is tolerant of the speech quality may still interpret the voice in spite of its quality, whereas the data transmission at even the low bit rates may fail the overall tolerance for the frequency of the occasional errors. In the rest of this section the limitations of the existing loop plant are discussed to carry the digital data at the maximum rates for ISDN requirements and yet satisfy the accuracy requirements of other conventional digital links.

Cable pairs in the loop plant have two important electrical constraints: the characteristic impedance and propagation constant. These two constants, called the secondary constants, can be derived from the four primary constants: the series resistance, the inductance, the shunt conductance, and the shunt capacitance per unit length. The two secondary constants are essential for computing the circuit performance of the cable pair in carrying the data for ISDN services. During the manufacture of the cables, strict control of the conductor diameter, the thickness, quality, and the consistency of the insulation are all essential to ascertain the uniformity of these four primary constants. A five percent manufacturing tolerance is still accepted as the standard range for the primary constants in the Western Electric facilities in the United States. The secondary constants thus have an implicit range of variation. A reasonably accurate characterization of the data transmission properties is now possible around the statistical averages of these four primary constants. However, the primary constants vary considerably with frequency and with temperature and the cable size. For example, a set of values for these four primary constants for the 26 American Wire Gauge (AWG) polyethylene insulated cable (PIC) at 70 kHz is 453.03 ohms per mile, 0.9546 millihenries per mile, 4.63 micromhos per mile, and 0.083 microfarads per mile. The same constant become 956.65 ohms per mile, 0.8381 millihenries per mile, 46.85 micromhos per mile, and 0.083 microfarads per mile at 1 mHz. Both sets of values are at 70° F. At 120° F and at 70 kHz, these constants become 502.51 ohms per mile, 0.9617 millihenries per mile, 4.643 micromhos per mile, and 0.083 microfarads per mile. Bare overhead open wire have different primary constants associated with them depending upon their size, pole spacing, etc. The ISDN loop plant may include a wide variety of transmission media. Special consideration is necessary to evaluate the type of media over which data can be transmitted at different rates.

5.4.1 Impairments Resulting from the Physical Design Rules

Most of the telephone loop plants have conventionally been designed using a maximum loop resistance design rule for providing access to the customer. This rule sets an outer limit for the dc loop resistance, thus permitting shorter loops to have finer conductor diameter.

In the United States, the value of the dc resistance has been set at 1300 ohms to facilitate the flow of supervisory signals, the ringing current, and the speech signals. This restriction permits loops as long as about 15000 ft of the 26 AWG (0.4039 mm) and 75000 ft of the 19 AWG (0.9119 mm) cable. In the United States and Canada four size (26, 24, 22, and 19 AWG) cables are used extensively. The two intermediate sizes (24 and 22 AWG) correspond to 0.5105 and 0.6426 mm diameter.

The European loop plant also has similar restrictions but the wire sizes can be finer especially in the United Kingdom, because of the shorter loop lengths. The metric size generally corresponds to 0.4, and 0.6 mm of copper conductor diameter in Italy and 0.32, 0.4, 0.5, 0.63 to 0.9 mm of both copper and aluminum in the United Kingdom. The Swedish loop plant has 0.4, 0.5, 0.6, and 0.7 mm wire diameters. The French, German, and Finnish (Norway) loop plants have generally used 0.4 and 0.6 mm cables, even though the German loop plant now appears to have 0.8 mm cables. The lengths of loops, based upon the resistance loop, varies in these countries depending upon the cable size and maximum resistance the network can tolerate.

The Japanese loop plant also uses 0.33, 0.4, 0.5, 0.65, and 0.9 mm cables with the intermediate sizes dominating the plant. The Australian loop plant has a selection of 0.32, 0.4, 0.51, 0.64, 0.9, and 1.2 mm diameter cables. However, the three most dominant sizes in the loop plant are the 0.4, 0.51, and 0.64 mm cables.

Further, the expansion of the loop plant has been rapid and few firm rules were used to make the loop plant uniform. Hence, it is very usual to see individual loops consisting of three or four individual sections of cable. The rapid growth of the telephone service, coupled with quick telephone service availability has prompted the telephone operating companies to construct loops consisting of wire pairs from different cables of the same size and also of different sizes. The feature in the loop plant is widely prevalent in most of the telephone plants established over the last few decades. For this reason, about 80 percent of the American loops, 55 percent of the Canadian loops, and 48 percent of the Japanese loop all have more than one section. The American and the Australian loops have the maximum average number of sections per loop. The newer telephone plants in other countries where the telephone plant is just being established, do not suffer by the gauge changes to the same extent.

The next dominant impairment is that of the bridged taps. The rapid expansion in the number of telephones has demanded more and more loops. One of the ways the operating companies have met this requirement is by tapping into an idle loop that was serving another customer in the vicinity of the new customer. The extension of the loop that served the older customer was left intact in the anticipation of serving him again if the loop returned to its older configuration. Further, while multiparty loops were being phased out, the connections to the secondary parties were left intact. Hence, in the history of the loop plant, events such as those mentioned, have led to a loop plant environment where loops have idle sections bridged onto the loop and left as taps.

It is not uncommon in the surveys of the older loop plants around the world to see that loop plants have four bridged taps in some of the loops. On a statistical basis, the results of such surveys indicate that about 87 percent of the American loops have more than one bridged tap and that every loop has an average of 1.64 bridged tap per loop. Eighty percent of the Canadian loops have at least one bridged tap. About 48 percent of the Japanese loops and 54 percent of the Australian loops have bridged taps. In Europe, about 20 percent of the Italian loop plants have bridged taps.

5.4.2 Impairments Resulting from Environmental Conditions

Oxidation, water in the cables, and intermittent and/or unidirectional contact also causes concern for the transmission of data in the loop plants of some national networks (United States, Canada, and Japan). Oxidation generally occurring at cable splices depends upon the material, exposure to the atmosphere, and the atmospheric conditions. In corrosive neighborhoods and in very old loop plants, this problem causes severe noise and interruption problems. However, the ISDN designers overlook this problem because the newer digital services are generally requested in newer urban and suburban regions where the loop plant is relatively new.

Water in the cables influences the capacitive component of the primary constants most profoundly. In the underground cables this problem can influence the bit error rate by closing the eye by an additional 5 to 7 percent. This effect differs from any other effects such as those resulting from temperature or bridged taps because the primary constants are significantly and unevenly altered. In the simulation studies for evaluating the eye opening with and without sections of loop being water logged, we have found the T1 carrier system at 1.5414 mHz still operated dependably in spite of the degradation. This is largely due to a slack in the design rules of this carrier system calling for repeaters every 6000 ft. Such an overdesign philosophy may not be feasible for the ISDN environment because the loops have to carry data bidirectionally between the central office and the subscriber at the higher speeds without any repeaters.

Short interruption resulting from oxidation and/or physical conditions make such loops unsuitable for ISDN applications. By and large when the operating companies become aware of the problem after a customer complaint, the cable or the wire pairs are avoided for data transmission.

5.4.3 Limitations Resulting from Electrical Interference

The major limitations of an electrical nature for the transmission of the ISDN data arises from the effects of crosstalk, impulse noise, electromagnetic pickup and lightning surges, and the effects of loading coils and echoes. The crosstalk and impulse noise problems in the loop plant are both acute and these have been researched adequately and discussed in this section and Section 5.4.4, especially for the ISDN applications.

Electromagnetic pickup from the proximity of power lines and substations can be a serious problem. Power line pickup can become a serious problem, especially if the subscriber telephone loop lines run parallel to the power lines. In the open wire environment, frequent transposition, which provides for equal exposure to the positive and negative polarity, tends to cancel out the induced signal. The problem of imbalance can still be severe. Crosstalk signals are usually cancelled out by the transposition. In the loop and local trunk and cables environment the twist in the cable curtails electromagnetic pickup and crosstalk.

Bare overhead wires are also susceptible to lightning surges. Direct hits are rare because such wires are not very high; reasonable protection from lightning may be expected from higher structures. However, in area with high atmospheric electrical activity, a ground wire is installed over the overhead wires, thus providing an umbrella for the open wires. Lightning arresters are sometimes essential, minimizing the damage and the length of the service interruption. Underground, unexposed, and shielded wire pairs rarely suffer from direct lightning hits, even though secondary pickup effects from surge voltages and currents may persist. Shielded cables may offer an effective guard against lightning and stray pickup signals from external sources.

Loading coils also curtail the digital capability of subscriber loops. The coils are a means of providing inductive loading of cable pairs. In the United States, these coils consist of two closely coupled windings on a high permeability magnetic material. The self and the mutual inductances are both harnessed in a series-aiding mode of connection, thus minimizing the size of weight of these coils. These coils have been traditionally used to reduce the attenuations and to make the impedance and delay more uniform throughout the voice frequency band. This effects are obtained at a considerable price of the same characteristics at frequencies beyond the voice spectral range of interest. For example, inductive loading of the 22 AWG cable forces the attenuation of the local cable pair at about 0.8 dB per mile throughout voice frequency band from 200 to 3200 Hz, whereas the nonloaded loop loss would vary from 0.8 dB to about 3.1 dB. However, above 3.5 kHz the loss of the loaded cable increases dramatically.

Loading coils are generally placed in long loops over 18,000 ft. (about 5.48 km). The standard distance of the first load coil is about 3000 ft. (0.914 km) from the central office and the spacing is about 6000 ft. (1.828 km) for the remaining coils. The presence of these coils is detrimental to ISDN applications because of the distortions they bring about in the transmission characteristics of the wire pairs at higher frequencies. Hence, loops considered for digital data transmission at rates near or over the voice frequency cutoff need special conditioning or need to be totally excluded. Removal of loading coils for ISDN applications is generally not economically feasible. Realistically for the ISDN applications, the problem is not severe because loops longer than 18,000 ft. tend to serve isolated customers. In addition, the concept of establishing the carrier serving area limits the length of loops to be within 12,000 ft. (3.657 km). When the need for ISDN services arises for a group of distant subscribers, then there is a good chance that these customers will be within a carrier serving area. For the subscribers near to the central office, nonloaded loops will be used and for the distant customers the digital information would be piped to a local digital center by a high-speed digital carrier system. It is envisioned that such digital centers will be distributed close to most customers and commercial centers such that the local transmission between the center and the customers can be accomplished by the standard central office and subscriber channel units being developed for the ISDN applications. This approach of the ISDN planning is consistent with the philosophy that the proposed data centers be strategically located in the United States.

Crosstalk Interference

Crosstalk generally refers to the phenomenon of the cross coupling between signals on one channel and another channel. The channels may contain transmission paths such as wire pairs or cable pairs. The coupling generally occurs through extraneous paths. This type of coupling causes an interference in the main signal because low-level induced signals from other channels couple into the main signal. When the coupling results from a series of uncorrelated random signals from a large number of other channels via low-level coupling coefficients, then the overall interference is like random noise. In rare instances, the interference can also become a very low-level discernible signal from one channel to another when the interference occurs via a fairly strong coupling coefficient. Crosstalk occurs in most communications channels because the main signal path cannot be totally isolated from all other paths, which also carry information or signals in their own paths.

Crosstalk studies have been conducted in great detail for different channel environments. The objective of these investigations is to curtail the interference from dominating the intelligibility of the conversations in voice frequency channels or from increasing the bit error rate in the data channels. However, in this chapter, we confine our attention to the particular way crosstalk effects the transmission of data for the ISDN services. In this context, the two important crosstalk interferences are: the near end crosstalk (NEXT) and the far end crosstalk (FEXT). At the present stage of ISDN-related crosstalk studies, the interference from other types of crosstalk, such as the interaction, direct and transverse crosstalks, do not cause serious limitations.

The telephone loop plant wire pairs dedicated to ISDN applications have been labeled Digital Subscriber Lines (DSL). This term (first introduced in Ref. 2) now refers to all digital lines in the telephone loop plant capable of carrying data in the range of 64 to 324 kb/s and used for that purpose. A preliminary discussion of crosstalk in the context of the DSL environments has been presented in Ref. 4. Here, we present the broad impact of crosstalk in three pertinent directions; (1) compatibility between the DSL system and other carrier systems as they coexist in the same cable and/or bundle within a cable, (2) the compatibility between the different DSL systems in the same cable and/or bundle within a cable, and (3) the effects of activity factor on crosstalk effects. The last consideration becomes important because ISDN services and the subsequent DSL systems have not become spread out enough to be of immediate concern, even though this could become an important issue as ISDN gains momentum.

A considerable amount of judgment becomes essential since different national environments have different admixture of carrier services in their loop plant. Further, different guidelines have been used during the evolution of the loop plant to its final form in different nations and different operating companies. Hence, we discuss these issues as they exist in the United States, since the loop plant seems to have experienced a rather explosive growth and the design rules have been less stringent to accommodate this growth. It is our impression that the conditions in most other environments are less severe and hostile to the systematic introduction of ISDN.

Crosstalk in the DSL systems has a three-fold influence: First, the DSL, being repeaterless for the entire span, has to transmit at the maximum level to ensure an adequate signal level at the receiver. Higher transmit signal levels of the DSL enhance the crosstalk interference into other systems especially at the receiver end of the other systems because their own signal level would have experienced a maximum attenuation. Thus, to ensure the accuracy of the received DSL signal and also minimize the crosstalk interference into other existing systems in the loop plant, we can assign a maximum distance limitation for the DSL. Once again this first consideration translates into different sets of design rules for different national and operating company environments.

Second, to be assured about the accuracy of the received DSL signal, the crosstalk power coupled into the DSL as a ratio of its own signal power has to be within statistically generated sets of ranges. These ranges are influenced by the bit error rate requirements one places on the DSL and upon the number and composition of the interfering systems. Hence, depending upon the loop environments, the percentage of loop population to be served by the DSL, and an evaluation of the existence of the disturbing systems, another set of design rules can be formulated. The set of rules from the second consideration compliments the rules from the first consideration. Impulse noise considerations also start to play a significant role in this aspect of the design rule. Their influence is discussed in Section 5.4.4.

Third, the self interference of one DSL crosstalking into another DSL also plays a significant role. Having to be accommodated in the same cable and/or same bundle, these systems can systematically destroy each other's signal. This problem assumes significant dimension because the telephone plant originally designed to carry voice band signals now carries signals in much higher bands of frequencies and at a higher amplitude. The coupling at these higher frequencies can become higher and thus become detrimental to the data communication at the ISDN rates. The cable characteristics and the cross coupling between the conductors plays a dominant role. The line code used influences the band where the signal energy is concentrated and the coupling varies with frequency. Hence, in the rest of the section one can see the need to specify the line code in the crosstalk calculations. The bipolar code is generally chosen because of its favored status in most telephone networks.

Compatibility Between Systems—NEXT

In this section we discuss the implications of introducing the high speed ISDN loop (DSL) in the American loop plant environment. Analog carrier systems are sometimes used; it becomes essential to determine whether the carrier systems can coexist with the DSL system in the same cable and bundle groups. This problem does not exist in other national or company networks where such carrier systems are not used.

Only near-end crosstalk interference (NEXT) needs to be considered when determining the compatibility of DSL's and other systems sharing the same facilities. (Certain pathological system configurations may lead to far-end crosstalk interference, however, these are ignored here.) The crosstalk noise power from one system to another can be written as

$$P_N = \int_{-\infty}^{\infty} S_T(f) \, L_{XT}(f) \, |H_R(f)|^2 \, df \tag{5.1}$$

where

$S_T(f)$ = power spectral density of the transmitted signal of the interfering system
$L_{XT}(f)$ = crosstalk loss of cable
$|H_R(f)|$ = gain versus frequency characteristic of the disturbed system.

Figures 5.11 and 5.12 show the power spectral densities of alternate bipolar and transmitter encoded duo-binary coding. These graphs are normalized to a fre-

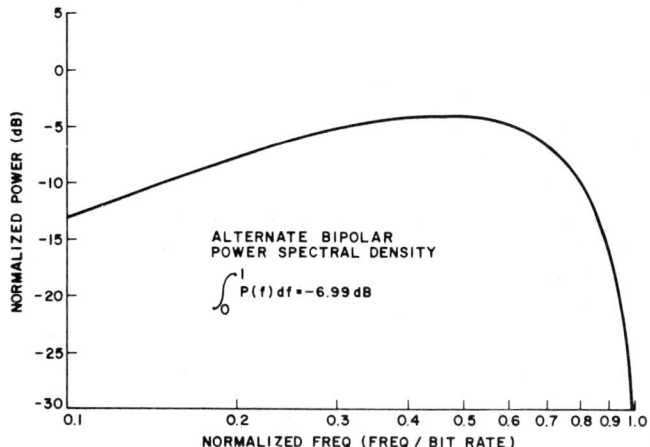

Reprinted with permission from: "A Tutorial on Two-Wire Digital Transmission in the Loop Plant" by S.V. Ahamed, P. Bohn, and N. Gottfried in *IEEE Transactions on Communications*. © 1981 IEEE.

Figure 5.11 Normalized power versus frequency for alternate bipolar signalling in the loop plant.

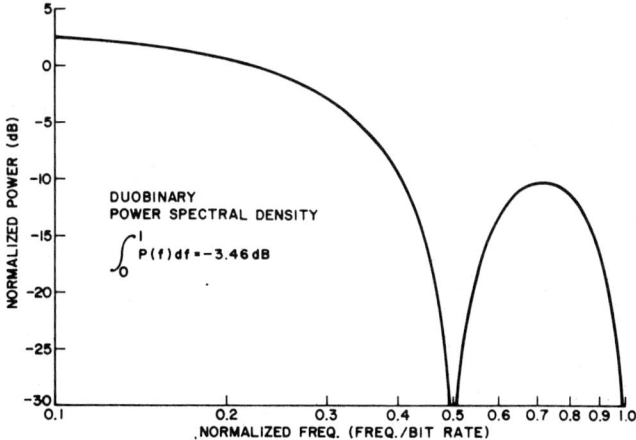

Reprinted with permission from: "A Tutorial on Two-Wire Digital Transmission in the Loop Plant" by S.V. Ahamed, P. Bohn, and N. Gottfried in *IEEE Transactions on Communications*. © 1981 IEEE.

Figure 5.12 Normalized power versus frequency for duobinary signalling in the loop plant.

quency of $1/F_B$ and normalized to a power level of A^2/F_B, where F_B is the bit rate on the transmission line and A is the peak amplitude of a rectangular pulse waveform. Three other encoding techniques have desirable properties and should be considered: delay modulation coding, transmitter encoded class 4 partial response coding, and nonreturn to zero (NRZ) coding, such as used to generate channel encoded duobinary of class-4 partial response coding. The power spectral densities shown in the figures are based on the assumption that the input data stream consists of a continuous random sequence of 1's and 0's. Since one cannot be guaranteed that the incoming data will always be random, this implies that some type of data scrambling is employed. Data scrambling satisfies two objectives. First, it generates a clock component in the power spectrum with high probability, and second, it eliminates discrete spectral components (arising from repetitive data sequences), which complicate the compatibility problems.

Figure 5.13 shows the pair-to-pair and 49 disturbers, 1 percent NEXT as a function of frequency which was used in this study (17, 18). The exact amplitude and shape of $H_R(f)$ depends on the particular service being analyzed.

The way in which (5.1) is used to evaluate compatibility between services is as follows. In the case of DSL interference into another service, $H_R(f)$ for the disturbed system at its maximum possible gain (or equivalently, its maximum system loss) must be determined, along with the maximum allowable noise power P_N that the disturbed system can tolerate. For analog systems, P_N will be dictated by idle channel noise requirements and for digital systems, by bit-error-rate requirements. Once the above quantities have been established, the amplitude of the DSL power spectral density may be calculated by use of Equation (5.1).

Fig. 5.14 represents an example of the results of the calculation described above. Shown is the maximum allowable transmit power of a DSL, which still allows compatibility with DDS (19) and the SLC-8™ system (20, 21) as an example of an analog carrier system. Only the calculations for coding techniques

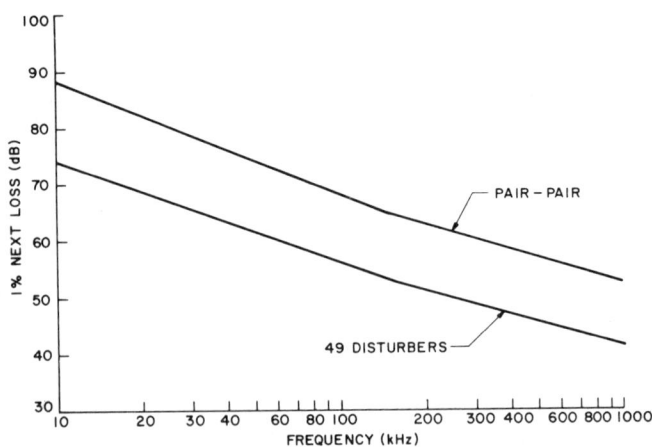

Reprinted with permission from: "A Tutorial on Two-Wire Digital Transmission in the Loop Plant" by S.V. Ahamed, P. Bohn, and N. Gottfried in *IEEE Transactions on Communications*. © 1981 IEEE.

Figure 5.13 1% NEXT loss versus frequency in the loop plant.

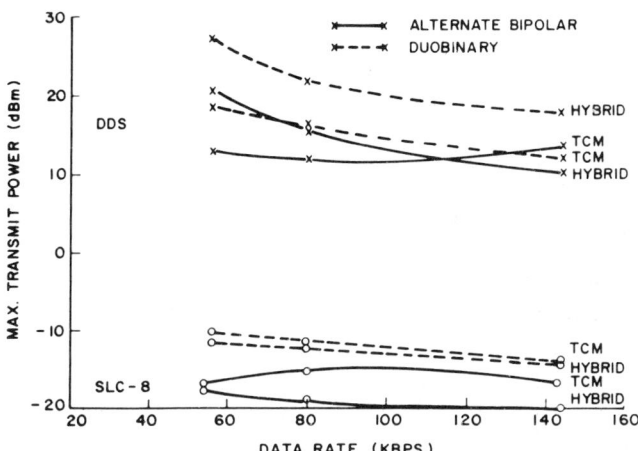

Figure 5.14 Maximum transmit power versus data rate in the loop plant.

Reprinted with permission from: "A Tutorial on Two-Wire Digital Transmission in the Loop Plant" by S.V. Ahamed, P. Bohn, and N. Gottfried in *IEEE Transactions on Communications*. © 1981 IEEE.

fall between the two shown in Figure 5.15. Two items of explanation are required of the figure. The first involves the labeling of the horizontal axis. The measure here is data rate, not line rate, which is why for each data rate there are two curves: one for the hybrid system in which the line rate equals the data rate; and one for the TCM system in which, in the present study, the line rate is 2.25 times the data rate. The second note of explanation involves the vertical axis, labeled maximum transmit power. This power is the maximum permissible average power of a random bit stream in a bandwidth equal to the line bit rate, that is,

$$P_{ave} = \frac{A^2}{R_L} \int_0^{FB} S_T(f)\,df \qquad (5.2)$$

where

$$R_L = 135\,\text{ohms}.$$

The square filtering function applied to (5.2) is somewhat arbitrary; however, the use of other more practical filtering functions leads to only minor variations of the results. In addition, in the case of TCM transmission, Fig. 5.14 applies only during the burst period; the long-term average power (that is, over several bursts) is lower by the amount 10 log C, where C is the time compression factor (in this case 2.25).

In the case of compatibility with analog carrier systems such as the SLC-8™ system, crosstalk statistics for 49 disturbers were used. This implies that in a

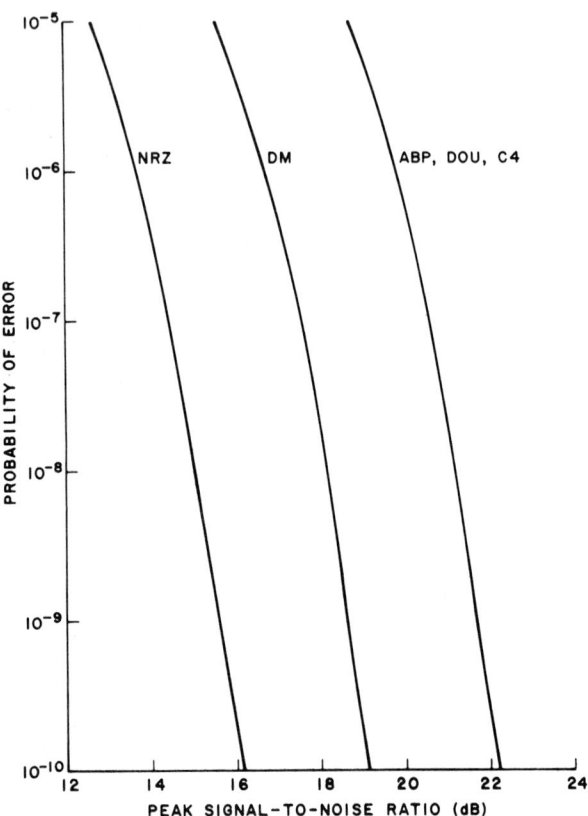

Reprinted with permission from: "A Tutorial on Two-Wire Digital Transmission in the Loop Plant" by S.V. Ahamed, P. Bohn, and N. Gottfried in *IEEE Transactions on Communications*. © 1981 IEEE.

Figure 5.15 Probability of error versus peak signal-to-noise ratio for different line codes.

particular cable group there is one SLC-8™ system and 49 disturbing DSL systems. In the DDS example, crosstalk statistics for 22 rather than 49 distributors were used for the burst mode frequencies. The reason for this is as follows. There are two modes in the DSLs in the same binder group that may be operated. In the first mode, the TCM transmitters bursts in a time-uncorrelated manner. This mode of operation results in NEXT interference from one DSL to another, since some systems will be in the receive mode while others are in the transmit mode. In this mode of transmission, at any instant of time, given N DSLs in a cable group, on the average only $N/2.25$ are transmitting. Then, rather than N crosstalks, there are, on the average, $N/2.25$ and the crosstalk statistics must be adjusted accordingly. In this case, going from 50 to 22 crosstalkers increases the 1 percent crosstalk loss by approximately 2 dB. If in this mode of operation, the self-interference is too large to tolerate, the system may be designed to operate in the alternative mode. In this mode, all DSLs in an office are synchronized to burst together. In this way, self-NEXT is eliminated. NEXT interference into other systems (DDS, T1, etc.) is little different from that caused by a continuously transmitting system. The slight advantage gained by bursting is obtained by the fact that during the burst the probability of causing an error in the interfered system is P_e and during the idle period it is zero. Thus, when considering bit-error rates, one may average the two intervals and, for a given BER, the burst mode may have a P_e larger by the compression factor C. However, for a C of 2.25, this allows one to decrease the required SNR by only about 0.2 dB for BER's on the order of 10^{-8}.

Crosstalk interference from other services into the DSLs is calculated in a similar manner as above. The interfering power spectral density is determined from a knowledge of the interfering systems. The receiver transfer function $H_R(f)$ is obtained from the DSL design. The maximum noise power is obtained by comparing the required signal to noise ratio for a given BER with the incoming signal level. The probability of error as a function of peak signal power-to-noise ratio for the various coding schemes is shown in Figure 5.15. Here, it is assumed that the interfering crosstalk noise has a Gaussian amplitude distribution. To determine the minimum allowable DSL receiver power, $H_R(f)$ is varied according to the receive power level until P_N has a value consistent with the SNR requirement.

Figure 5.16 shows example results of this calculation for the two cases of 49 SLC-8™ and of 49 T1 carrier systems interfering into one DSL. In both cases, a BER of 10^{-7} and 60 percent worst case eye opening margin have been assumed. In the SLC-8™ system case, only interference from the high group carriers was considered since these carriers are always colocated at the central office and are always on. In addition, only the receive levels for the TCM system are shown, since SLC-8™ interference into a hybrid system should not be a limiting factor.

As can be seen from Figures 5.14 and 5.16, crosstalk from analog carriers can severely limit the range of many DSL systems. This picture can be improved considerably by not allowing DSLs and analog carrier systems to operate in the same cable group. An additional margin of 10–15 dB can be allowed in both the maximum transmit power and minimum receive power by requiring cable binder unit separation.

Self-Induced Crosstalk Limitations

In this section, the limitations of the DSL system range that result from crosstalk from other DSL systems will be considered. Both far-end crosstalk (FEXT) and near-end crosstalk (NEXT) interference will be analyzed.

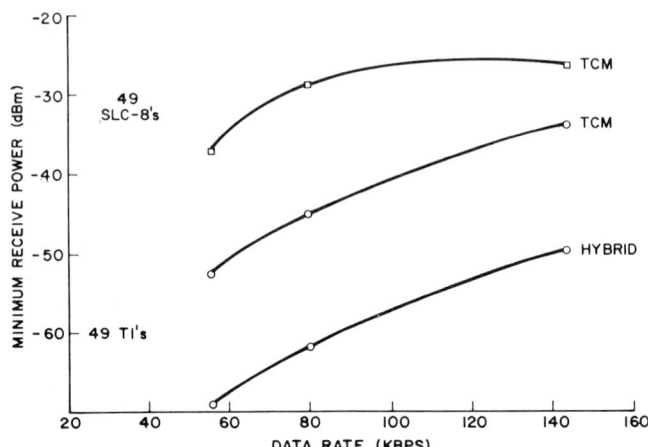

Reprinted with permission from: "A Tutorial on Two-Wire Digital Transmission in the Loop Plant" by S.V. Ahamed, P. Bohn, and N. Gottfried in *IEEE Transactions on Communications*. © 1981 IEEE.

Figure 5.16 Minimum received power versus data rate.

a) Far-End Crosstalk: The mode of FEXT interference to be analyzed here is that of nonequal level FEXT. It can be described with the aid of Figure 5.17 and 5.18 shows two DSLs sharing the same binder group in a cable. In the case of nonequal FEXT, two terminals in two different offices transmit at the same power: $P_{T1} = P_{T2} = P_T$. The receive power on system 1 is $P_T - \alpha L_C$, where α is the cable loss per unit length. The interfering power into system 1 at the receiver is given by

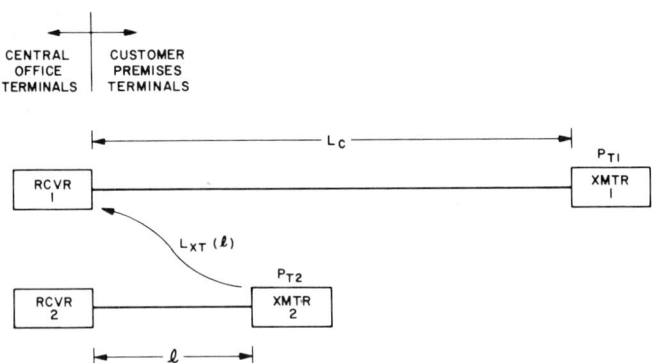

Reprinted with permission from: "A Tutorial on Two-Wire Digital Transmission in the Loop Plant" by S.V. Ahamed, P. Bohn, and N. Gottfried in *IEEE Transactions on Communications*. © 1981 IEEE.

Figure 5.17 Illustration of FEXT.

The Major Limitations for Loop Data Transmission

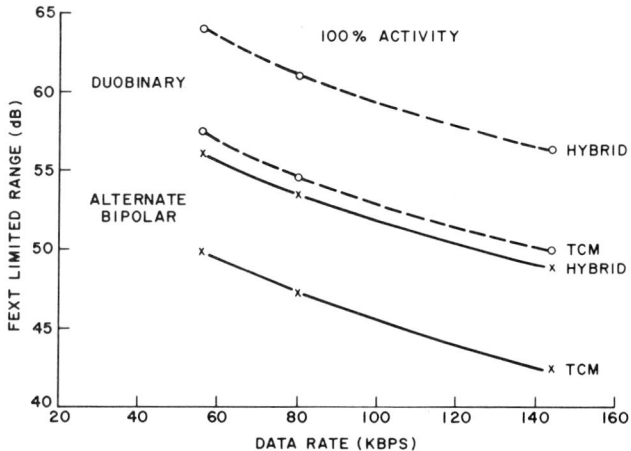

Reprinted with permission from: "A Tutorial on Two-Wire Digital Transmission in the Loop Plant" by S.V. Ahamed, P. Bohn, and N. Gottfried in *IEEE Transactions on Communications.* © 1981 IEEE.

Figure 5.18 FEXT-limited range versus data.

$$P_{XT} = P_T - L_{XT}(l) - \alpha l.$$

Now, the signal-to-noise ratio is

$$SNR(l) = (P_T - \alpha l_C) - P_{XT}$$
$$= L_{XT}(l) + \alpha l - \alpha L_C.$$

The general assumptions is that FEXT varies by 20 dB/decade in frequency and 10 dB/decade in length. Thus, one may write

$$L_{XT}(l,f) = K_0 - 20 \log f/f_0 - 10 \log l/l_0.$$

Now, one would like to determine the minimum *SNR* (*l*) and, therefore, substituting in the expression for $L_{XT}(l)$ and differentiating the result with respect to *l* gives

$$\frac{d}{dl} SNR(l) = \alpha - \frac{10}{l} \log e$$
$$= 0 \text{ when } l = \frac{10}{\alpha} \log e$$

and hence we obtain

$$SNR_{min} = -\alpha L_C + K_0' + 10 \log \alpha$$

where

$$K' = K_0 - 20 \log f/f_0 + 10 \log e - 10 \log(10 \log e)$$
$$= K_0 - 20 \log f/f_0 - 2.03 \text{ dB}.$$

By using the above analog with (5.1), one obtains the results shown in Figure 5.12. These results are based on $K_0 = 42.5$ dB at $f = 3.15$ MHz and $l_0 = 1$ kFT for the 1 percent pair-to-pair loss.

b) Near-End Crosstalk: Near-end crosstalk between DSLs is calculated exactly the same as in the case of NEXT between unlike systems. Example results are shown in Figure 5.19. These results assume that the TCM systems are operating in an unsynchronized manner. If all the TCM systems in a cable are synchronized and transmit together, then self-NEXT interference is eliminated. In addition, a 60 percent worse case eye opening is assumed. This is discussed further in Chapter 6, Section 6.3.2 of this volume.

Effect of Activity Factor on Crosstalk Loss

All of the results on crosstalk interference above assume 100 percent cable fill and 100 percent activity (all customers, or systems, operating continuously). Since the services offered by DSL systems can vary greatly, the proper activity factor can also vary greatly. The following generalization can be made regarding customer activity.

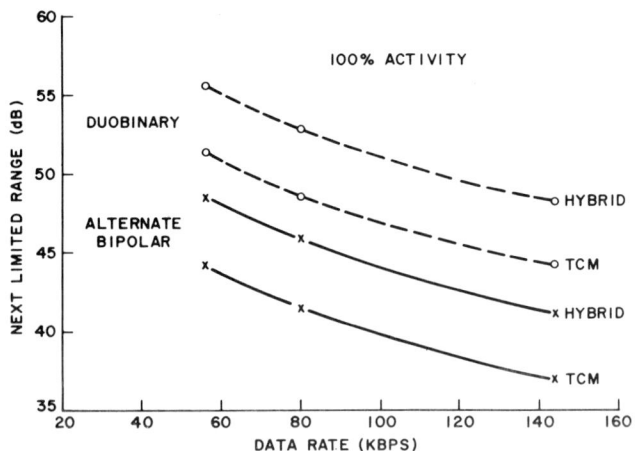

Reprinted with permission from: "A Tutorial on Two-Wire Digital Transmission in the Loop Plant" by S.V. Ahamed, P. Bohn, and N. Gottfried in *IEEE Transactions on Communications*. © 1981 IEEE.

Figure 5.19 NEXT-limited range versus data rate.

The crosstalk loss advantage gained by various activity factors is

Activity Factor	Loss Advantage
50 percent	1.5–2.0 dB
25 percent	4.0–4.5 dB
10 percent	7.0–7.5 dB.

Based upon the assumptions made with respect to the activity factor related to the DSL, the crosstalk limitations can be appropriately adjusted.

5.4.4 Impulse Noise

Impulse noise in central offices is of major concern in the transmission of ISDN data over the subscriber loops. The subscriber data is received at the central office after considerable attenuation and dispersion resulting from the loop transmission properties. The received signal is then processed and the information bits are extracted from the processed signal. The impulse noise in central offices cause an additional noisy signal to be superimposed atop the received signal, thus altering its spectral and amplitude characteristics. The processor (digital, analogue or even just a simple equalizer) can thus yield erroneous bits of information. The factors that dominate the extent of damage the impulse noise can cause can be summarized as: (1) relative strengths of the signal strength to the noise, (2) the spectral overlap in the received signal strength and the impulse noise events, and (3) the relative instant amplitude and the polarity of the noise event with respect to the timing and polarity of the information bit. For example, a positive impulse falling over a positive bit in a polar of a bipolar code will not lead to an error because the threshold detector will judge the combined signal to be a $+1$, whereas the other for the impulse or the information bit can lead to an error if the combined signal becomes less than the present threshold value.

The relative signal to noise ratio depends upon two factors: (1) the signal strength after attenuation and (2) the noise characteristics. The signal strength depends upon the line loss at the main spectral region where line code concentrates its energy. The noise characteristics depend upon the source of the impulse noise event and the nature of the central office. As discussed in Section 5.4.3, lightning surges, switching events, dialing pulses, etc. can cause significantly different noise properties in different central offices; the electronic switching systems being the quietest. However, the line attenuation depends upon the line length, the loop configuration, the loop make up, the line code, etc. Thus, it can be seen that the bit error caused by impulse noise depends upon a set of randomly distributed variables. Statistical calculations give an idea of the average number of bit errors in a prescribed duration or actual simulation studies indicate the extent of eye closure (see Section 6.3.2) that a particular event causes in a given loop if the event is present at a certain instant of time. From our simulation studies (2) at the (2B+D) rate with the bipolar code over all the loops at the 1973 Loop Survey of the predivestiture Bell System, a typical event recorded in an ESS central office, the following conclusions are recorded. Loops with about 30 dB loss at 72 kHz are practically unaffected by the event. Loops 30–40 dB of loss suffer to the extent that the s/n ratio at the received eye diagram falls from 7 dB to 3 dB. Loops over 40 dB of loss tend to suffer severely from central office impulse noise.

There are two ways of considering the impulse noise requirements of a DSL. First, given a BER requirement at a particular bit rate and a particular loop acceptance test, over how long a loop can one satisfactorily operate? Or, one may rephrase this: given the same BER requirement and specifying the loop length over which the DSL must operate, what must the loop acceptance test be? The first formulation implies that, although a large percentage of the loops are expected to pass the acceptance test, the system penetration is limited to those loops passing the test. The second formulation suggests that marketing requirements indicate that this system must have a particular range and the question to be answered is: what percentage of the loops have sufficiently low impulse noise to satisfy the range requirements? It is this last formulation that would be desirable; however, there is insufficient information on impulse noise in the loop plant to realistically solve the problem. Therefore, the solution to the first formulation has been undertaken here.

The loop acceptance test used as a basis here is the same test used for a 56 kbit/s DDS; that is, using a Western Electric 6F Impulse Noise Measure Test Set with a 497F filter, there can be no more than seven counts at a threshold level of 44 dBrn in a 15 min interval. In addition, a BER requirement of 10^{-7} will continue to be assumed here. The following assumptions will be made.

a) The long-term number of noise peaks that exceed a given threshold (in dBm) will increase by a factor of 10 for each 10 dB decrease in the given threshold (19).

b) The noise count varies as the square of the receive filter (voltage) bandwidth (19).

c) For a burst DSL system with a time compression factor C, only N/C impulses affect the system, where N is the number of impulses counted by the noise measurement set in a specified interval of time.

Assumptions a) and b) may be combined to obtain

$$N(BW_2, TH_2) = N(BW_1, TH_1) \times \left(\frac{BW_2}{BW_1}\right)^2 \times 10^{(TH_1 - TH_2)/10} \qquad (5.3)$$

BW represents the bandwidth and TH represents the threshold level.

Also it has been shown (19) that (for an alternate bipolar line code) given a noise impulse exceeding the threshold, the probability that it will cause an error is 3/16. This and (5.3) may be combined with the fact that the bandwidth of the 497F is 25 kHz to translate the DDS acceptance test to the minimum DSL receive level consistent with the BER requirement. The results of such calculations are shown in Figure 5.20.

5.5 EVOLVING TRENDS IN ISDN

The major forces presently shaping the ISDN architecture, the protocols, the configuration, and the semiconductor chips will exert a major influence on the ISDN features of next decade. However, a few trends have emerged and are gaining steady momentum.

First, the general consensus in the United States favor the (2B + D) rate (i.e., two times the data rate of 64 kbit/s and a low rate data channel at 16 kbit/s together with an overhead of 16 kbit/s yields the information rate of 160 kbit/s).

Second, the time compression multiplexing mode that has been adopted for CSDC Customer Switched Digital Capability introduced by the predivestiture Bell System and some of the operating companies is being questioned seriously. This technique has definite advantages associated with it. However, the adaptive echo canceler (AEC) techniques are gaining steady prominence as more and more functional semiconductor chips tailored for this device are becoming avail-

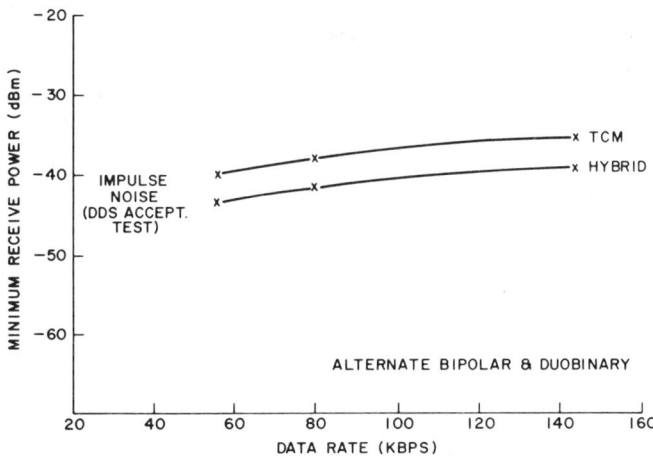

Reprinted with permission from: "A Tutorial on Two-Wire Digital Transmission in the Loop Plant" by S.V. Ahamed, P. Bohn, and N. Gottfried in *IEEE Transactions on Communications.* © 1981 IEEE.

Figure 5.20 Minimum receive power versus data rate.

able. Guarded optimism is well in order for a gradual introduction of the AEC mode of ISDN data transmission. In the past, other major national networks such as those in Canada, Japan, Italy, and France to some extent had experimented with the TCM techniques. After the predivestiture Bell System adapted the TCM mode, many other countries followed suit and it still appears that for the CSDC rate of 56 kbit/s, the choice was sound and readily implemented.

Third, the choice of line codes is again under serious scrutiny. The bipolar code which had been favored by T1 carrier systems, and adopted for the CSDC applications seems to be losing its popularity. The recent investigations are directed towards the 2B1Q, and its variation, the partial response codes, and other more bandwidth efficient codes. The answers have not been definite but the new semiconductor chips capable of cheaply implementing the algorithms necessary for the more band efficient codes are surfacing. The preliminary crosstalk sensitivity figures indicate about 6 dB preference for these codes. The impulse noise studies are not complete but the intuitive preference for the newer codes arises because of the narrower bandwidth. The central office noise characteristics also become important in the code selection.

Fourth, the conventional equalizer designed in (2), seemingly adequate for the CSDC applications in the TCM mode, operating at 64 kbit/s (56 kbit/s customer information rate and an overhead of 8 kbit/s) needs a detailed study to justify whether such a device will be adequate for the proposed ISDN rates and with the echo canceler type of technology. Decision feed back equalizers, using tapped delay lines and FIR filters, provide the delicate balance between the equalizer functions and the echo canceler functions. Only a more comprehensive simulation study provides definite answers to the combined effect of the equalizer and the echo canceler in view of the loop characterization of the particular national and/or local network.

Fifth, the impedance matching networks have been simple fixed resistors. Our preliminary studies indicate that changing the terminating impedance from a resistor of 120 ohms to about 127 ohms can reduce the average reflections of the loop population by about 5 dB at the central office side at 80 kHz in the American loop environment. Similar results are not evident at the subscriber side because of a greater dispersion of the image impedance (because of localized bridged taps). The average reflection at the subscriber can be about 6 dB higher than a similar reflection at the central office side.

Finally, the digital signal processing of the received wave shapes to extract the digital and timing information at the subscriber is also being considered for microprocessor-based channel units. It is not evident if the custom designed chips will not suffice. However, with easier IC fabrication techniques and the rapid turn around of the chips, the ISDN designers will have ample crucial decisions to make in the next few years. Suboptimal system architecture and component decisions at this stage can have severe ramifications in the future.

The recent and more active participation of CCITT and T1D1 committees in unifying and standardization of the ISDN protocol, interface and networking is benefiting the steady transition towards the ISDN and the vendors of the ISDN components. Some national networks have found it economically favorable to ignore the recommendation of these committees in view of the peculiarities and the networks. The implications of these choices have to be appreciated and comprehended by the network administrators.

These trends have also opened new areas of investigations for the ISDN system designers. The answers are highly variable from network to network, and from urban to suburban loops. It is to be noted that these early findings are based upon the literature currently available and biased by our studies of the American loop plant. Similar studies and findings would facilitate the optimum choice of ISDN elements and components for other networks.

5.6 REFERENCES

1. Bell Telephone Laboratories, "Transmission Systems for Communications," 5th edition, 1982.
2. S. V. Ahamed, "Simulation and Design Studies of the Digital Subscriber Lines," *Bell System Technical Journal*, Vol. 61, 1982, pp. 1003–1077.
3. F. T. Andrews, R. J. Haas, "The Role of Digital Switching in the Evolution of the Subscriber Loop," *Proceedings of the Sixth International Symposium on Subscriber Loops and Services*, 1984, pp. 29–33.
4. S. V. Ahamed, P. P. Bohn, N. L. Gottfried, "A Tutorial on Two-Wire Digital Transmission in the Loop Plant," *IEEE Trans. on Comm.*, Vol. COM. 29, 1981, pp. 1554–1564.
5. B. S. Bosik and S. V. Kartalopoulos, "Time Compression Multiplexing System for a Circuit Switch Digital Capability," *IEEE Trans. on Comm.*, Vol. COM. 30, 1982, pp. 2046–2052.
6. D. H. Morgan and A. J. Schepis, "The Fiber SLC® Carrier System—An Innovative Application of the Lightwave Technology for the Loop Plant," *Proceedings of the Sixth International Symposium on Subscriber Loops and Services*, 1984, pp. 35–39.
7. P. P. Bohn, C. A. Brackett, M. J. Buckler, T. N. Rao, "The Fiber SLC Carrier System," *Bell Laboratories Technical Journal*, 1984, Vol. 63, pp. 2389–2416.
8. S. A. McRoy, J. H. Miller, J. B. Truesdale, and R. W. Van Slooten, "Integrating with the 5ESS Switching System," *Bell Laboratories Technical Journal*, 1984, Vol. 63, pp. 2417–2437.
9. I. Dorros, Keynote Address. "The ISDN—A Challenge and Opportunity for the 80's," *IEEE International Conference on Communications*, Denver, Colorado, Vol. 1 of 4, pp. 17.0.1–17.0.5.
10. R. E. Buss and G. R. Leopold, "Data-Voice Multiplexer Technologies for Packet Switched Networks," *Telephony*, June 1986, pp. 56.
11. The 1986 National Communications Forum, Seminar NET-23 "ISDN—The Reality I," and Seminar NET-24 "ISDN—The Reality II," October 1, 1986, Chicago, IL.

12. J. A. A. Le Guillou, F. Marcel, and A. J. Schwartz, "PRANA at the Age of Four: Multiservice Loops Reach Out," *IEEE Trans. on Comm.*, Vol. 30, No. 9, 1982, pp. 2185–2210.
13. M. Dupire, C. Ab der Halden, and J. Brooke, "L'Installion D'Abonne du Reseau Multiservice de Biarritz," *ISSLS 84, The Sixth International Symposium on Subscriber Loops and Services*, Nice, France, October 1–5, 1984.
14. L. M. Manhire, "Physical and Transmission Characteristics of Customer Loop Plant," *Bell System Technical Journal*, Vol. 57, 1978.
15. S. V. Ahamed, ibid., 1982, part II.
16. S. V. Ahamed and R. R. P. Singh, "Physical and Transmission Characteristics of Subscriber Loops for ISDN Services," *1986 International Conference on Communications*, Toronto, June 21–24, 1986.
17. T. F. McIntosh, "Experimental Verification of Theoretical Low Frequency Crosstalk Models," in *Proc. 27th Int. Wire. Cable Symp.*, Nov. 14–16, 1978.
18. S. H. Lin, "Statistical Behavior of Multipair Crosstalk," *Bell Syst. Tech. J.*, Vol. 59, pp. 955–974, July–Aug. 1980.
19. E. C. Bender, J. G. Kneuer, and W. J. Lawless, "Digital Data System: Local Distribution System," *Bell Syst. Tech. J.*, Vol. 54, pp. 919–942, May–June 1975.
20. D. H. Morgan, "Expected Crosstalk Performance of Analog Multichannel Subscriber Carrier Systems," *IEEE Trans. Commun.*, Vol. COM-23, pp. 240–245, Feb. 1975.
21. R. K. Even, R. A. McDonald, and H. Seidel, "Digital Transmission Capability of the Loop Plant," in *Proc. Int. Commun. Conf.*, 1979.

Chapter 6

DIGITAL SUBSCRIBER SYSTEMS

6.1 RECENT GROWTH OF DIGITAL SUBSCRIBER SYSTEMS

The digital (r)evolution has had an impact in the loop plant environment. Even though the digital carrier systems (1) have been used over the last two decades and are well trenched in the loop plant, digital subscriber systems are paving the path towards making enhanced customer services a reality. In this chapter, the digital subscriber systems that are essential to implement the integrated services digital network are discussed.

6.1.1 ISDN and Digital Subscriber Systems

The implementation of ISDN calls for high-speed data links between the central office or a digital distribution center and subscribers. The data rates of these links were introduced in Section 5.1.3. The modes for transmission were introduced in Section 5.1.4. The entire system, which handles the high-speed data over these links and offers digital connectivity between the central office or the digital distribution center and the subscriber, is called the Digital Subscriber System (DSS). It consists of the incoming data encoder, the input filter (if any), the scrambler (if any), the transmitter, the subscriber loop, the receiver hybrid (if any), the line terminating network, the line equalizer, the echo canceller (if any), the timing recovery circuit (at the subscriber end), the threshold detection circuits, the equalizer and echo canceller adaptive circuits, the descrambling circuit, the error detection, testing and data realignment circuits, and decoding circuits.

Each one of these components accomplishes a specific function, even though the more recent customized IC chips and components may tackle more than one function at a time. From a functional standpoint, the incoming data encoder receives the data to be transmitted and generates the encoded version of the same. For example, for the bipolar (AMI) code, the encoder provides the alternate mark inversion necessary, and for the 4B3T code, the encoder provides the specific 3-bit binary sequence for every four bits received, by looking up the conversion table. The data scrambler scrambles the incoming sequence of data and provides a new sequence of the pulse sequence to be transmitted on the line. The need for scrambling generally arises in order to assist the timing recovery circuits at the subscriber end by ascertaining a minimum density of ones in the received data sequence, and to provide minimal privacy in the data communication. The wave shape encoder (if any) provides a particular wave shape for every pulse to be transmitted. The input filter provides the output wave shaping before the signal is put upon the subscriber line. The hybrid is generally used with the adaptive echo canceller system and the hybrid provides a directional isolation so that the transmitted data finds its access to the digital subscriber line, and the received data finds its access to the receiver without being severely contaminated by the effects of the transmitted data. Total isolation being impossible in the practical environment, the hybrid provides the first stage of isolation. The echo cancellers in the AEC environment provide a second degree of isolation and clean up the received signal for the binary decoding by a threshold detector. The line equalizer is generally necessary to reconstruct the received pulses. An ideal equalizer perfectly compensates for the line loss over the entire band of frequencies which have significant spectral power distribution of the specific code and/or wave shape. The threshold detector differentiates between the various levels of the pulses.

For the three-level ternary (such as the bipolar) code the three levels to be detected are $-1, 0$, and $+1$ and a two-level threshold detector will suffice. For a multilevel code, additional thresholds are necessary. The timing recovery circuits at the subscriber terminal recover the clock at which the data was transmitted from the central office. This recovered clock is used to keep the subscriber clock in synchronism with the central office clock and to determine the instant at which the threshold detector should scan the pulse height to detect the data. The timing information is also important in supplying the shift register clock for tapped delay lines (if any) in echo cancellers and/or decision feed back equalizers. The descrambler recovers the original data from the scrambled sequence of received bits. The error correction circuit checks for transmission errors and corrects the errors if an error correcting code was used to encode the transmitted data. Finally, the data realignment circuits becomes necessary if there is misalignment of the recovered data. The protocol and overhead bits contain adequate information to check the data alignment. The data alignment circuit is activated only if a misalignment is detected.

The details of the overall design of the entire DSS need considerable optimization and a study of the interaction between the various components. In this section, we present only the concepts associated in the preliminary design of the these digital subscriber systems, which are essential in the implementation of the ISDN.

6.1.2 Developments in DSS Design

Digital Subscriber Systems (DSS) incorporate the Digital Subscriber Line (DSL) and the terminating channel units. The digital subscriber line provides the digital path between the central office and the subscriber, and the terminating channel units provide the termination at the central office and at the subscriber. Both the system and component aspects are necessary to facilitate the design of these systems. As indicated in Chapter 5, the optimal design depends upon a large number of external parameters such as the loop environment, the line rate, the line code, etc. However, the design data, as it has been published for the American loop environment, is available and some of the basic findings are condensed in this chapter.

In perspective, we present the DSS study in the predivestiture Bell System. A systematic study of the DSS was undertaken in AT&T Bell Laboratories during the late 1970s and early 1980s. The United States, Japan, and the major European countries had been contemplating the feasibility of ISDN. Many basic questions regarding component design and its realizability remained to be investigated. In the early phase of the study, devices tended to be inadequately defined, and their functions were vague. To ascertain the device feasibility in context of the functions, a variety of filter circuits, equalizers, balancing networks, echo cancellers, timing recovery circuits, scramblers, encoders, and decoders have been isolated and investigated. Limited experimental component design study to provide an intuitive basis for the combined effect of all the components on the system performance has proven to be an impossible task. This became exaggerated by the fact that the components themselves had to be designed and optimized in context of the ISDN requirements. Nevertheless, an enormous amount of experimental work was also undertaken.

In most experimental investigations, laboratory studies can become expensive, time consuming, and often yield only a subset of the answers sought. Further device optimization, which needs a fairly exhaustive range of variation on a number design parameters based on experimental studies, can become uneconomical, if not impossible. Hence, a dedicated simulation system to initiate the extended simulation studies and computer aided design of the DSS can be a viable choice, and the limited experimental facility can serve to confirm the prediction of the simulation studies.

The loop survey data and the cable characterization are held as the invariable external parameters. The effects of every other design, environmental, the device parameter is treated as a variable during the study. Typical of such design parameters have been the modes of transmission (Hybrid or the AEC and TCM), the design of bandpass filters (if any), the equalizers, the echo cancellers, timing recovery circuits, etc. The results of such design studies of DSS and a condensed description of the simulation facility are presented in this chapter.

6.2 DATA TRANSMISSION SYSTEMS AND THEIR COMPONENTS

The five systems used for ISDN data transmission are: (a) the Time Compression Multiplexing (TCM) system; (b) the Adaptive Echo Canceller (AEC) or the Hybrid System with adaptive echo cancellation; (c) the Frequency Division Multiplexing (FDM) system; (d) the four-wire system; and (e) the hybrid impedance balance system. The principles of each were discussed Section 5.4.1. For reasons discussed, the two competing, 2-wire systems that have received considerable attention are the TCM and AEC systems. These two systems and their components are discussed next.

6.2.1 The TCM System and Its Components

In the TCM system, the intervals for transmission and reception are isolated in time. The separation of the time intervals, during which the compressed data is burst onto the loop and during which the compressed data is received from the loop, simplifies the task of maintaining little or no interference of the received data from the transmitted data. The data at the channel units (transmitter and receiver) is received uniformly from the data source and the line. However, the data at the transmitter has to be received in a buffer or a series of small buffers (2, 3) from the source and transmitted at a much higher rate over the line. Conversely, the data from the line is received into a buffer or a series of small buffers at a much higher rate from the line and uniformly emptied to the receiver. Hence, changing the data rate becomes an integral part of the TCM systems. A series of geometrically decreasing arrays of shift registers is preferable to on long shift register because the first bit of arriving data does not sit and wait for the last bit to arrive, but instead gets onto the line and to the other end. This minimizes the delay in transmission. For the digital voice, this feature is desirable in minimizing the delay of any residual echo. For data applications, the overall transmission time is reduced. This concept has been implemented in (4).

The enhancement of the rate will have to be in excess of twice the data rate. The reason for the rate enhancement is based upon three additional timing considerations and discussed next as two independent reasons. The first reason based upon the first consideration, can be viewed as being independent of the second reason which is based upon the second and the third considerations.

The first timing consideration is based upon the fact that the transmitter receives its own echos for a finite duration, say t_e, (See section 6.3.2) after the transmission is complete. Thus, the transmitter cannot be forced into the receive mode as soon as the transmit cycle is complete. The transmitted signal echos received are too strong in relation to the received signal and these echos can become 60 dB stronger than the received signal thus drowning the information completely (5).

The second consideration is based upon the duration, say t_t, for the pulse to travel down the subscriber line and the third consideration is based upon the settling duration, say t_s, for the tail of the last received pulse at the receiver. These durations are discussed further.

Both the two durations t_t and t_s, are necessary at the central office and at the subscriber as they receive the data. The channel unit at each end of the loop serve as transmitter and as receiver of the burst data and the switching of the functions have to take place at an appropriate instant of time. The TCM principle is depicted in Figure 6.1 and the functioning of the TCM system is quite simple because the change of the channel unit functioning can simply be accomplished by a transmit/receive (T/R) switch shown in Figure 6.2. The loop is forced into an idle period while both the transmitter and the receiver are quiet for a duration known as the guard interval.

Consider the cyclic functions for the channel units in the TCM system at the central office side. The transmitter bursts the data into the channel for t_b seconds. The last bit of data, after it has been bursted, takes a finite while, known as the transit time (t_t seconds), to travel to the subscriber. The receiver at the subscriber is active for this finite while after the transmitter at the central office has been quiet to permit the last bit transmitted to be received. The short duration known as transient or settling time (t_s seconds) will also become necessary between the instants the receiver at the subscriber is turned off and the instant at which the transmitter is turned on in the same channel unit to permit any circuit and system transient in the local channel unit to subside. The transient time is generally very small and sometimes ignored in view of other considerations to follow. Thus, the minimum duration between the instants transmission ends at the central office

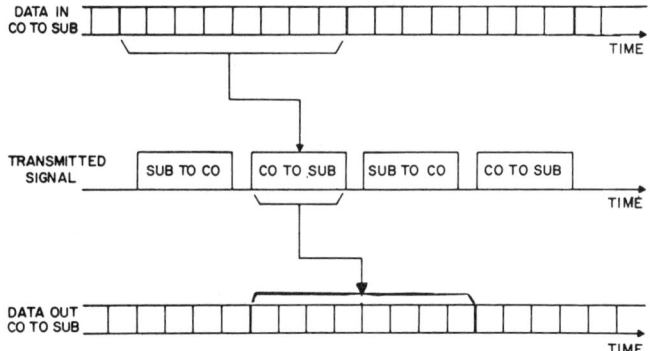

Reprinted with permission from: "A Tutorial on Two-Wire Digital Transmission in the Loop Plant" by S.V. Ahamed, P. Bohn, and N. Gottfried in *IEEE Transactions on Communications*. © 1981 IEEE.

Figure 6.1 Principle of digital transmission under the Time Compression Multiplexing (TCM) mode.

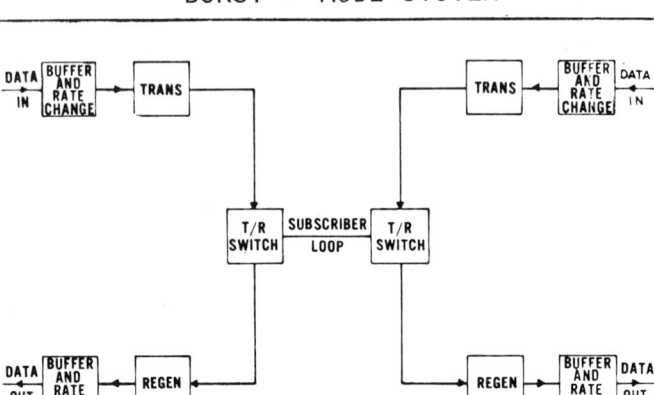

Reprinted with permission from: "A Tutorial on Two-Wire Digital Transmission in the Loop Plant" by S.V. Ahamed, P. Bohn, and N. Gottfried in *IEEE Transactions on Communications*. © 1981 IEEE.

Figure 6.2 Components of the TCM system with transmit/receive (T/R) switches.

and the transmission begins at the subscriber is $(t_t + t_s)$ seconds. It is the travel time plus the transient time. This interval is the overhead required to burst one block of data. The events at the subscriber are essentially the same as those at the central office but the transmit and receive times are interleaved to facilitate the data transmission. The overhead remains the same.

The minimum duration before the channel unit can be forced from the transmit mode to the receive mode is called the guard interval. The requirement on this guard interval due to the first timing consideration (i.e., for the transmission line echoes to subside) play a more decisive role. From the investigations (Section 6.3.2.F) an interval of as many as 12 to 14 pulse periods may be necessary to ensure that the relative echo amplitude is 60 dB below the received signal for a line rate of 144 kbit/s in the American loop environment. This duration exceeds the transit and transient time and thus dominates the second and third timing consideration. In this discussion, we have assumed no additional delay by the rate changing or buffering circuits, and it is possible to achieve this by using circuits presented in (2). If these circuits are not used and conventional shift-in shift-out registers are used, the overhead may increase and the arrangement of the buffering circuits becomes important (5).

The loop configuration, the nature of discontinuities, the bridged tap configuration of the worst loop, and the data rate influence the guard interval due to the echos at the transmitter. The duration for the data bits to travel down the

Data Transmission Systems and Their Components

longest line, the transient time, and the data rates all influence the guard interval of the second type. In the American loop plant, the worst echos subside to within 50 dB of the received signal at 144 kbit/s data rate in 10 pulse periods and to within 60 dB in 12 to 14 pulse periods.

By the same token, the maximum duration to travel down a 18,000 ft. (5.48 km) of 26 AWG cable pair is about 33 microsec or about 4.7 pulse periods. If an additional transient period of two pulse periods is granted, the overhead is about 6.7 pulse periods, thus making the minimum guard interval about seven pulse periods. Hence, the guard interval from the first set of requirement (echos to subside) overrides and about 12 pulse periods is the minimal requirement for the worst loop conditions. In reality, for the American TCM, Circuit Switched Digital Capability system (4), the guard interval is chosen as 16 pulse periods, which is in excess of either of the two requirements. The burst rate is about 2.25 times the information rate with a 200 bit block transmitted at 144 kbit/s over the subscriber telephone lines. The burst repeat interval is 3 ms with 432 pulse periods at 144 kbit/s comprising two 200 pulse periods for the two bursts, and a 16 pulse period of the guard interval in either direction of transmission.

6.2.2 The Adaptive Echo Canceller (AEC) System and Its Components

A hybrid (Figure 6.3) in the AEC system initially maintains some directionality for the transmitted signal to find its way into a subscriber loop and for the received signal to find its way into the receiver. If the hybrid has ideal directional properties, then all of the transmitted signal will go out onto the subscriber loop

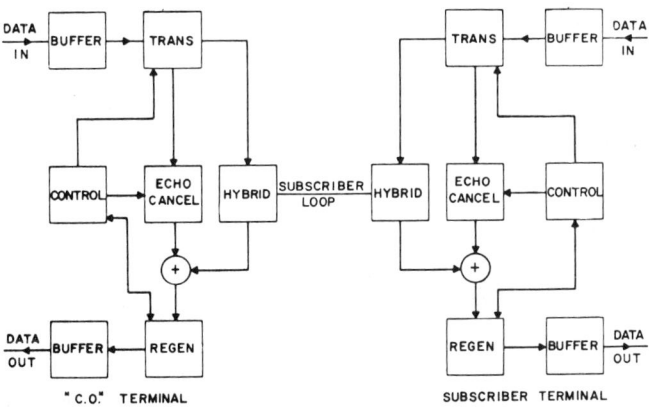

Reprinted with permission from: "A Tutorial on Two-Wire Digital Transmission in the Loop Plant" by S.V. Ahamed, P. Bohn, and N. Gottfried in *IEEE Transactions on Communications*. © 1981 IEEE.

Figure 6.3 Components of the hybrid system with echo cancellation of the AEC system.

and all the received signal will go into the receiver. However, these hybrid devices being far from ideal, cannot prevent the stronger transmitted signal (which can be 40 dB stronger) from contaminating the weaker received signal. There is a good electrical reason for this. The hybrid devices have a network called the balancing network whose characteristics are to match those of the subscriber line very closely. If the network had its impedance at all the harmonic frequencies of the received pulses exactly the same as that of the line, then the reflection from the hybrid would be zero. The extent of the mismatch in the impedance gets transformed into the extent of the reflected signal from the hybrid.

The AEC system isolates the transmitted and received signal by maintaining a delicately balanced echo cancellation circuit which computes the magnitude and phase of the reflected signal from the sequence of pulses that have been already transmitted. This signal is systematically subtracted out of the composite signal received, thus yielding only the received signal had there been no reflections. It is here that the need for an accurately functioning echo cancellation circuit arises. This circuit, capable of generating and cancelling the contaminating echo, is called the echo canceller and is depicted in Figure 6.3.

The requirements on the type of circuitry can be defined as follows. If the received signal suffers an attenuation of 40 dB at center band of a bipolar code, and the echo in the worst case in two times as loud as the transmitted signal (or 6 dB louder), then a 46 dB echo canceller will make the magnitude of the resulting echo signal as loud as the received signal. If the noise of the echo is to be 15 dB lower than the received signal, then a 61 dB echo canceller is necessary. In this simplified discussion, it is assumed that the line equalizer treats the signal and the echo noise after cancellation in the same way. If the equalizer has variable spectral amplifications and the spectral contents of the signal and noise are different, then the overall signal-to-noise ratio at the detection circuitry is going to vary after equalization or the regeneration of the pulses.

The next consideration is that the echo cancellation circuitry has to perform a ceaseless function as long as transmission is maintained. Further, from an earlier training sequence, at the slant-up time, a general purpose canceller circuit has to be adapted to be functional for a particular loop. This calls for a learning process (see the Control box in figure 6.3) in which the prior sequence of transmitted pulses gives rise to a echo canceller signal over a finite duration of time during which the reflections are expected. The learning process is called the adaptation and the most frequent way of obtaining the canceller output is by a tapped delay line. The tap weights are learned earlier and continuously updated as long as the transmission is in effect. If there is no correlation between the bits transmitted, the adaptation algorithm functions in order to minimize the overall difference between the actual echo and the echo canceller output. Many techniques for implementing the echo canceller algorithms are available. The learning period for the echo canceller circuit is called the adaptation time and one of the processes for reaching the final tap weights is called the convergence process. Since the echo canceller output cannot exactly equal the echo at all times, the tap weights have a tendency to oscillate about their final settings. The extent of this oscillation is generally very small and at the peak excursion of the oscillations, the net echo canceller output is still sufficiently close to actual echo so that the signal-to-noise ratio at the detection circuit is still within the prescribed limits.

The parameters indicating the performance of an echo canceller depend upon the algorithm used in computing the magnitude and the phase of the echo that the canceller is attempting to cancel. However, at the outset, a few parameters become important. First, consider the bit rate. Generally, the echo cancellers function with a tapped delay line, multipliers, and adders. As the bit rate starts to increase, the time available to perform arithmetic functions becomes less. At a certain rate, one can visualize that the integrated circuit multipliers used in the canceller chip may not function dependably. This is especially the case if a high degree of accuracy is needed. Thus, the number of the bits necessary to preserve the accuracy of the canceller signal becomes the second parameter to consider in context with the dynamic range of operation expected from the conceller. The third parameter of interest is the immunity that the device offers to noise in the received signal. In Section 5.4.3 the extent and nature of electrical noise to the DSL is discussed. It now becomes imperative that the echo canceller be immune to the noise and its variations, yet be effective against genuine echos over its entire dynamic range. Another significant parameter is the start up time for the canceller. Generally, a training sequence is necessary to cold start the AEC systems; this duration can be quite long. Some of the older algorithms can require up to 10 seconds for the adaptive mechanisms to converge. Yet the more recent simulation studies have required only two to three seconds.

Many newer configurations of the echo canceller systems include two levels of cancellation. A coarse fixed parameter canceller based upon the statistical patterns of echo can be used to reduce the echo level by an order of magnitude and then an highly sensitive adaptive circuit to decontaminate the received signal further. Wave form coding is also used in the transmitter in the more recent systems. For such systems, additional information about the details of the derived wave form is essential and an input to the adaptive echo canceller becomes necessary. One such termination at the subscriber is shown in Figure 6.4. This circuit also incorporates the equalizer in two stages. The first stage of equalization is based upon the analogue automatic gain control (AGC) equalizer tracking the height of the reconstructed pulses. The second stage is based upon an adaptive decision feed back equalizer (DFE) for the compensation of reflections from bridged taps. Between these arrangements of the echo canceller and the equalizer, the AEC systems can be designed to function satisfactorily over a dominant population of the nonloaded loops. The entire system has to be designed as single

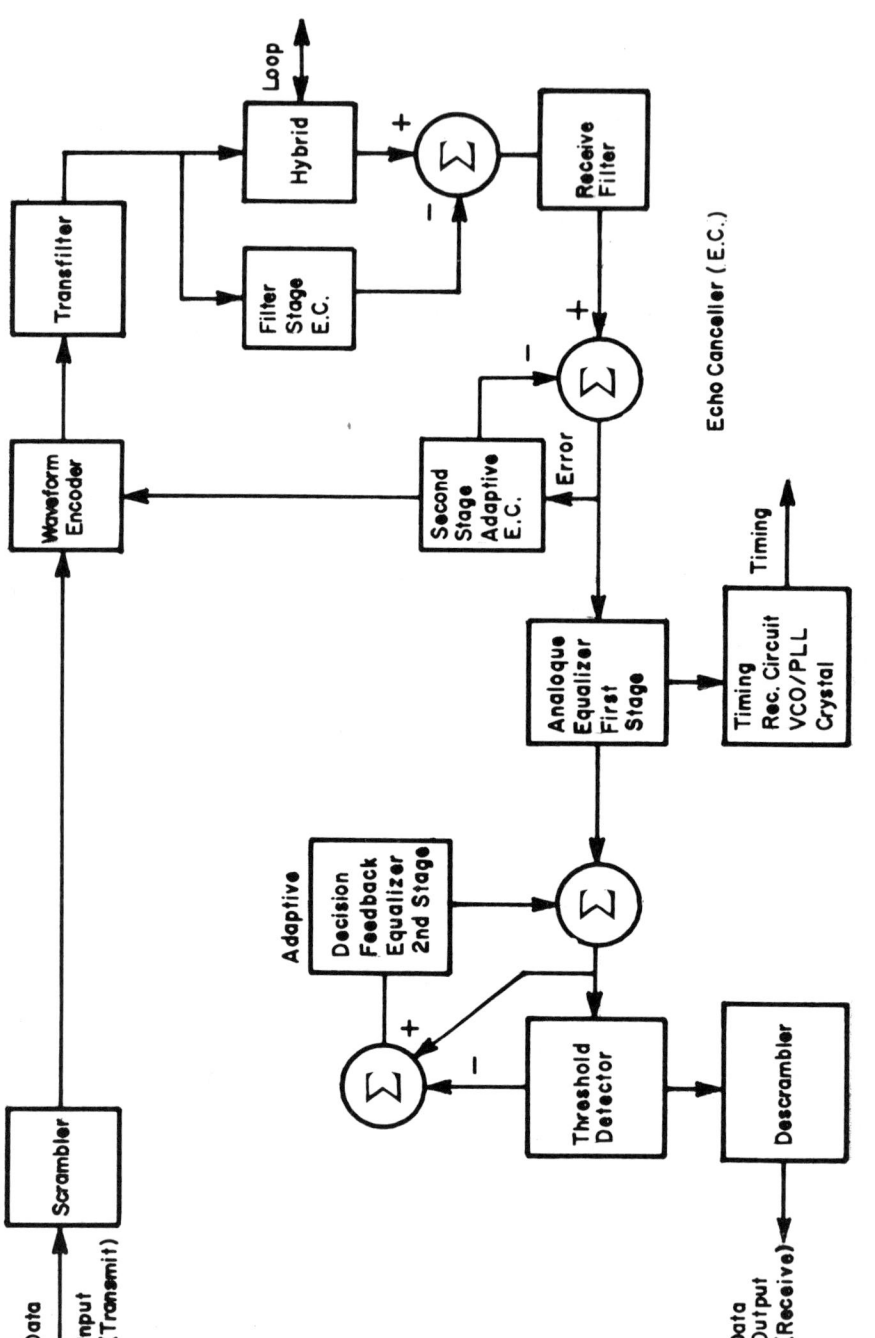

Figure 6.4 A typical subscriber unit configuration of AEC with two stages for echo canceller and equalizer.

entity in the loop environment because of the complex interrelation between the components. For example, the length of echo canceller tapped delay line can be controlled by the transmit and received filters characteristics, and the convergence (or for that matter, the divergence) time of the canceller depends upon the tap setting algorithm and the central office impulse characteristics, etc.

6.3 DESIGN AND IMPLEMENTATION OF DIGITAL SUBSCRIBER SYSTEMS

There are two basic decisions in the implementation of the DSS at any given line rate. First, the mode of transmission (Section 6.2), and second, the choice of line code. These two critical choices in the design of the DSS form the two focal points. Around these points, the design of components and other peripheral systems can be based. Almost all of the algorithms for implementation, and designs of the devices are critically influenced. The technological progress determines the realizability and cost of the component devices and the economics of these new services will be the eventual justification for ISDN.

The development of a computational facility is also another critical step in the design of DSS. Simulation plays a dominant role. The number and the range of the secondary design parameters is too large for intuition or experience to cast a valuable insight. Computer-aided design techniques do prevail and intelligent choices can be made by man-machine interaction. The discrete steps in one such design procedure used for the CSDC (Customer Switched Digital Capability) at 64 kilo bits per second are discussed here. Such procedures are equally applicable for the (2B+D), ISDN system, and component design.

The development of design rules is also critical in the implementation of the DSS. At this stage, after the choice of the ISDN data rate, the mode of data transmission, the system and component design, it becomes essential to study the loop conditions detrimental to the eventual deployment of the newly designed DSS. Such detrimental loop conditions can be studied by the simulation facility and the design rules are developed to systematically exclude the ISDN services to be offered on such loops. An example of the detrimental loop make up occurs when a critical length coarse gauge bridged tap exists on a loop. Open-ended bridged taps whose length causes a strong reflection about one pulse period after the main pulse has reached the receiver, causes a maximum intersymbol interference if no special precautions are taken. (See footnote in Section 5.3.2.2). Such an interference causes a strong possibility of error in the recovery of information. The study of such conditions or combinations of such conditions and the extent of the severity of possible errors is presented in Section 6.3.3.

6.3.1 The Choice of Line Codes

The coding of binary information to a predetermined pulse pattern for the ease and accuracy of digital transmission and the fabrication of the devices has been investigated thoroughly. Results from such studies facilitate the selection of a particular code for ISDN application. For the transmission of digital data in the T1 carrier systems and the Circuit Switched Digital Capability (a forerunner to the introduction of the ISDN services in the United States) the bipolar code has been implemented.

Code selection is based on several important considerations. The most important of these are as follows: the NEXT (Near End Crosstalk) advantage, the impulse noise immunity, the required receiver bandwidth, the ease of clock recovery, the presence of dc component, the error propagation, and finally, the ease of implementation. Three distinct codes will be used to exemplify these considerations. First, bipolar code (BP) has been chosen as a reference because of its simplicity and robustness. Its primary advantages are the lack of dc component, spectral peak at half the baud rate, the ease of implementation, and the lack of error propagation. The relative disadvantages of not offering clear channel clock recovery can be offset by scrambling the data stream. Simulation studies and experimental results have confirmed the system performance for this type of coding.

Second, partial response coding (e.g., duobinary and modified duobinary) also results in a three-level eye diagram. These codes can offer better crosstalk and impulse noise immunity by confining the transmitted energy to a lower frequency band (0-1/2 bit rate), thereby allowing a higher transmitter power because of the reduced crosstalk effects at lower frequency. These codes, however, are more demanding than the bipolar code in their equalizer/regenerator requirements. Duobinary requires passing the dc component of the signal (or dc restoration otherwise). Modified duobinary requires very precise timing recovery because of the horizontal eye closure it produces.

Third, nonreturn to zero codes have also received interest for digital transmission. As in the case of the duobinary code, eye diagrams without dc restoration necessitate precise low frequency equalization for satisfactory performance. The advantages offered in crosstalk and impulse noise immunity again must be weighed against the additional design complexity necessary to achieve satisfactory performance. Application limitations are presented in Table 6.1

More recently, the interest in the 4B3T code for the ISDN applications has been growing. There is a consideration reduction in the bandwidth requirement for this code, thus making it attractive from the crosstalk considerations. It is estimated that this code, in conjunction with a perfect equalizer and echo canceller, will offer a 6 dB NEXT advantage over the bipolar code. The simulations to evaluate the eye degradation of this codes and the robustness against bridged taps has to determined by detailed simulation studies. We have not studied this code in enough detail to comment on the ease of clock recovery or the extent of the clock jitter.

Variations of the 4B3T codes (such as DI43 and MM43) are also under review for ISDN applications. The more effective utilization of the bandwidth is the

Table 6.1
Choice of line codes and influence on the applications limitations

APPLICATION LIMITATIONS		
COMPATIBILITY CONSIDERATIONS ⇨	TRANSMISSION LIMITATIONS ⇨	APPLICATION LIMITATIONS
SDL XTLK INTO OTHER SERVICES	MAX XMT PWR.	DISTANCE LIMITS
XTLK FROM OTHER SERVICES IMPULSE NOISE	MIN RCV PWR	% LOOPS SERVED
SELF INTERFERENCE NEXT, FEXT	SYSTEM RANGE	

main attraction and a detailed simulation of the system for eye openings (Section 6.3.2) in the real subscriber environment is necessary before these newer codes can be adopted for the DSS. The final choice of the 2B1Q code by the Bell Operating Companies in the United States indicates the need for continued simulation studies in the particular loop environment of the individual operating companies.

6.3.2 The Role of Simulation

Simulation results and their experimental verification become essential in guaranteeing the dependability of the selected system. Exhaustive experimental studies and verification can be virtually ruled out as a possible approach because of the wide disparity of the loop plant environment and the possible variation of the engineering design. In the early phases of the study, neither the structure of the system nor the functional boundaries of the components are firm. It is here that the innate flexibility in the software becomes essential, if the simulation system is to become valuable as a design tool.

In the ISDN system and component design, extensive computer simulation can be a valuable in establishing the system performance. Data bases, such as the 1973 and the 1983 Loop Surveys (6, 7) provide the range of possible loop configurations to carry the high-speed data. Most of the important modes of transmission, such as the TCM, hybrid, four wire, and adaptive balance systems, may be initially explored to choose the most feasible and economically viable system in context with the choice and the state of IC technology. Generally, it is also desirable to model the system components individually to study the effects of varying their designs and optimizing them. This feature becomes important because of the variations in the IC technology (CMOS, MOS, hybrid, etc.) chosen to implement these devices for the system.

Simulation systems can be hosted on any computer. The ease of building such a system, the availability of the hardware, and its data base management features (in view of extremely large amounts of numerical processing of data and complex algebraic functions necessary to generate the results) become crucial in the choice of the hardware and the software. In Figure 6.5, we depict the block diagram of one software organization for the simulation of the DSS. More recently, powerful microcomputers are viable contenders to the minicomputers and main frames conventionally used in such large scale simulations. Mass storage and data base management techniques also become essential. Extended graphics plays the critical role of extracting and presenting pertinent and crucial data in its perspective. In these large system design and optimization procedures, the designer-computer interface can be streamlined by special purpose software dedicated to the study of the ISDN systems.

Typical initial data bases housed in the simulation facility would be the loop survey data, cable characteristics, temperature data of different geographical networks, urban suburban loop statistics, etc. Primary derived data bases would be physical and spectral domain characterizations, Fourier excitation functions at different bit rates, line codes, impulse noise characterization of different central offices, crosstalk data for the different cables and bundles, etc. Secondary databases would be the time domain characterizations, eye diagram data, scatter plot data, worst loop tabulations under different conditions, etc. It can be seen that the memory and mass storage requirements can become overbearing if no systematic data management techniques are used. The system software designer and user insight become critical factors at this stage. The following steps have proven effective in the design of the digital subscriber systems components.

(1) *Frequency Domain Simulations*—Spectral domain analysis is carried by the *ABCD* matrix representation of each section in the loop and each of its bridged taps. These matrices are systematically condensed into a single *ABCD* matrix for the entire loop. From this composite *ABCD* representation, the transfer functions are extracted and combined with the transfer functions of the other system components such as filters, equalizers, transformers, etc. System transfer functions thus generated contain both the attenuation and phase delay information. Hence, at any particular frequency, the system performance can be entirely specified by the transfer function.

(2) *Time Domain Simulations*—For time domain simulations, we have computationally forced the system into a limit cycle condition (6). This condition exists when a periodic repetitive excitation function is imposed upon the system and the boundary conditions at the beginning and the end of the period containing the repetitive excitation are identical. In relation to the pulse duration for data transmission, the repetitive period is made arbitrarily long such that a pseudorandom sequence of data can be embedded within the period.

The simulation capability permits any number of predefined pulse periods

Design and Implementation of Digital Subscriber Systems

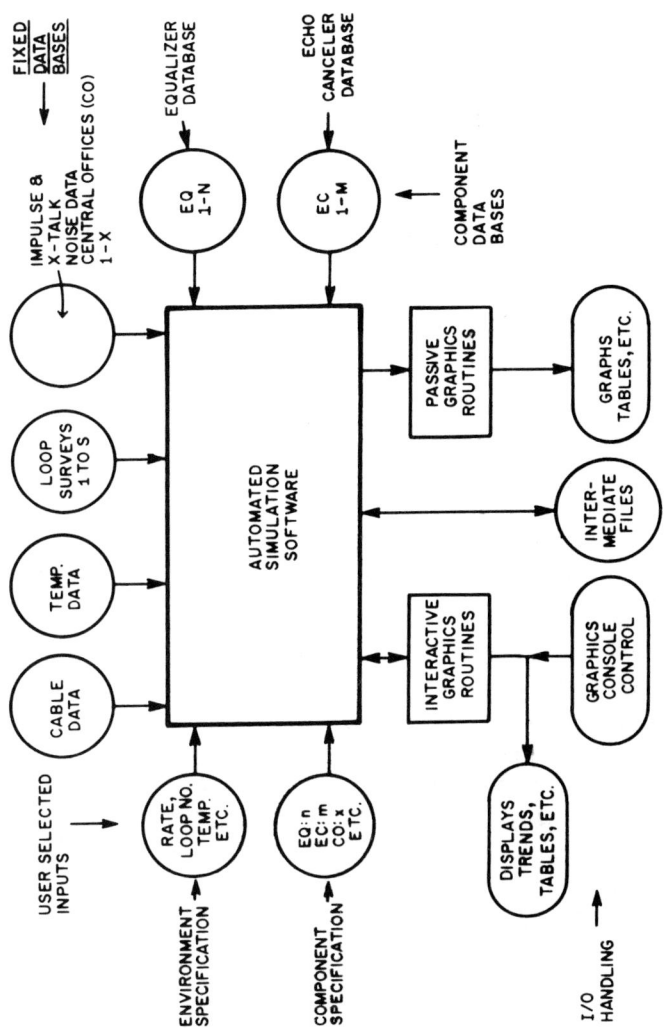

Figure 6.5 Block diagram of the simulation software.

to be contained within the repetitive period, thus enabling the long sequences of data (in its entirely) to be resolved into Fourier components. Next, the transfer function of the entire system (including filters, equalizers, echo cancellers, terminating impedances) after determining the composite ABCD parameters of the entire loop (including bridged taps, loading coils, gauge discontinuities, etc.) is computed in the frequency domain. Conversion back to the time domain is accomplished by performing the inverse Fourier transformation.

(3) *Total Simulation Capability: Use and Results*—The simulation facility organized around the databases offers significant potential to study the effects of changing the design of the system or the design of components and studying the incremental effects of changing parameters to optimize their values. In the design of a system and in the design of components (filters, equalizers, terminating impedances) both the spectral and the time domain simulation facilities are extensively used. In studying the changes of pulse widths (10–100 percent duty cycle) and in the effects of different codes (NRZ, WAL-2, AMI, multilevel, etc.), the time domain simulation facility is extensively used.

The simulation capability also encompasses an elaborate graphic software system. Results obtained from the facility are numerous and diversified. Hence, a systematic method of illustrating and cataloging the results becomes necessary. Individually tailored software to generate specific displays is also coded, altered, and updated to suit design requirements. In the following, some of the key results are summarized.

   a) *Typical Spectral Domain Loop Survey Results*—In this section the image impedances of the loops in certain segments of the 1973 Loop Survey are presented. For data transmission, the initial study has been confined to 831 of the 1098 loops listed in the survey. This segment of the loop population has no loading coils and loop length is under 18,000 ft. Two sets of results are presented. In Figure 6.6(a) and (b) the image impedance of the loop from the central office and from the subscriber are plotted at 32 kHz (e.g., for the data rate of 64 kbits with BP coding). The bridged taps are stripped and the resistive component is plotted on the x axis, with the imaginary component on the y axis. Each point on the plots corresponds to one of the 831 loops constituting the truncated 1973 Loop Survey. If the composition of all the loops was a single uniform gauge, all the points on this Figure 6.6(a) and (b) would overlap at one single point. The extent by which the scatter plot deviates from a single point indicates the range over which a cable impedance matching circuit would have to operate. Further, in the hybrid mode, the extent of reflections that occur for any particular loop would be indicative of how far the point

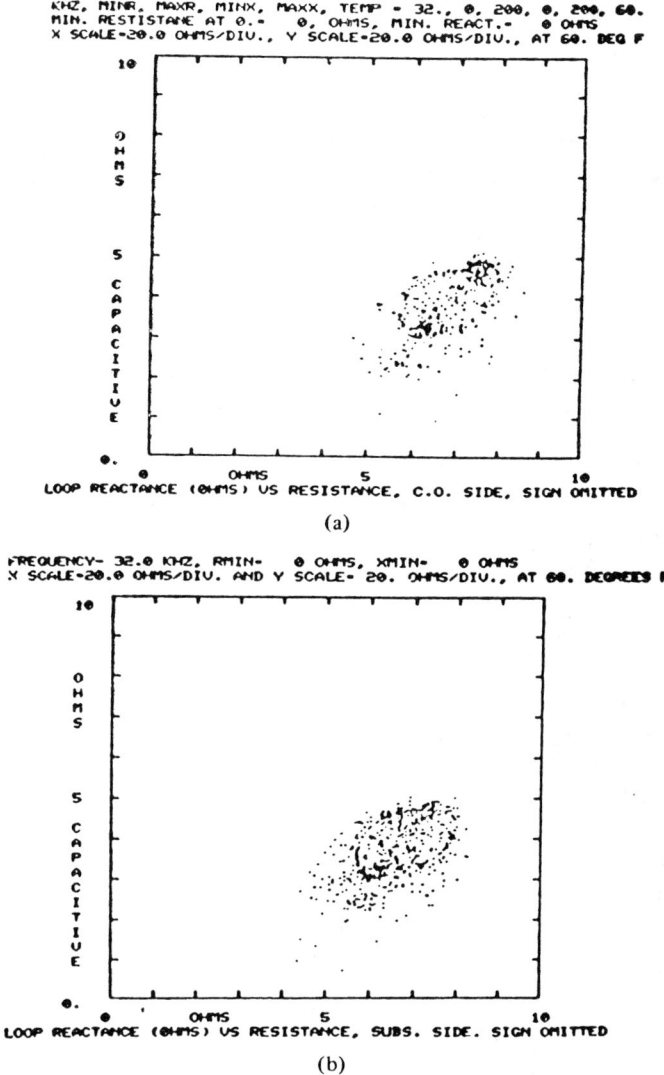

Reprinted with permission from: "A Tutorial on Two-Wire Digital Transmission in the Loop Plant" by S.V. Ahamed, P. Bohn, and N. Gottfried in *IEEE Transactions on Communications*. © 1981 IEEE.

Figure 6.6 (a) Scatter plot relating loop resistance and reactance at 32 kHz for bipolar code data transmission at a line rate of 64 kbit/s (central office side, all the bridged taps are removed) at 60°F. (b) Scatter plot relating loop resistance and reactance at 32 kHz (subscriber side, all the bridged taps are removed) at 60°F.

for that loop lies from the actual terminating impedance existing in the system. In the TCM mode, the severity and duration of reflection at the end of burst of data would also be indicated by the actual location of the point for the loop from the terminating impedance. In the AEC mode, the echo canceller quality necessary for the suppression of reflection can be deduced. In the TMC mode, the extent of guard space necessary between the bursts can be deduced.

Similar figures for the truncated loop population when the bridged taps are *not* removed are shown in Figure 6.7(a) and (b). The extent of the change in image impedance because of bridged taps becomes obvious by comparing Figures 6.6 and 6.7.

b) *Typical Time Domain Results*—Time domain results are obtained as pulse patterns or eye diagrams. Eye diagrams are the superposition of individual received pulses in the same time slot. The received digital sequence is invariably compared with the transmitted sequence for accuracy of transmission. For a typical loop, the received pulse pattern is shown in Figure 6.8(a). The loop in this case consists of four sections of cable: 10,669 (3.25 km) ft., #26 AWG; 2418 ft. (.737 km), #22 AWG; and 4875 ft. (1.486 km), #24 AWG with a 1300 ft. (.396 km), #26 AWG bridged tap located at the junction of the #22 and #24 gauges. The eye diagram is shown in Figure 6.8(b). For these simulations, the TCM mode of transmission is assumed at 144 kbit/s corresponding to any effective bidirectional rate of 64 kbit/s. Eye diagrams contain significant information determining the quality of transmission over a single loop. However, in representing the performance of a system over all of the loops in the population, data must be extracted from the eye diagrams to provide a quantitative evaluation. Hence, we have extracted seven statistics for each eye for each loop in the population and the quality of eyes generated by a proposed system can be condensed in terms of these seven statistics for all of the loops.

(4) *Eye Statistics and Scatter Plots*—Figure 6.9 shows a typical eye diagram of the seven eye statistics (the average positive eye height, the top eye thickness, the top eye opening, the central eye thickness, the bottom eye opening, the bottom eye thickness, and finally the average bottom eye height). When all of these parameters at normalized to average eye height of unity, then the eye statistics can be cross-compared for all the loops. Generally, four scatter plots are necessary to convey the seven eye statistics for the eyes in the two directions of the data transmission. Typical scatter plots depicting these statistics are shown in Figure 6.10(a)–(d) for 64 kbit/s (burst mode system operating at 144/kbits) with bridged taps intact. In Figure 6.10(a) and (b), the central office to subscriber transmission eyes are depicted. The

Design and Implementation of Digital Subscriber Systems

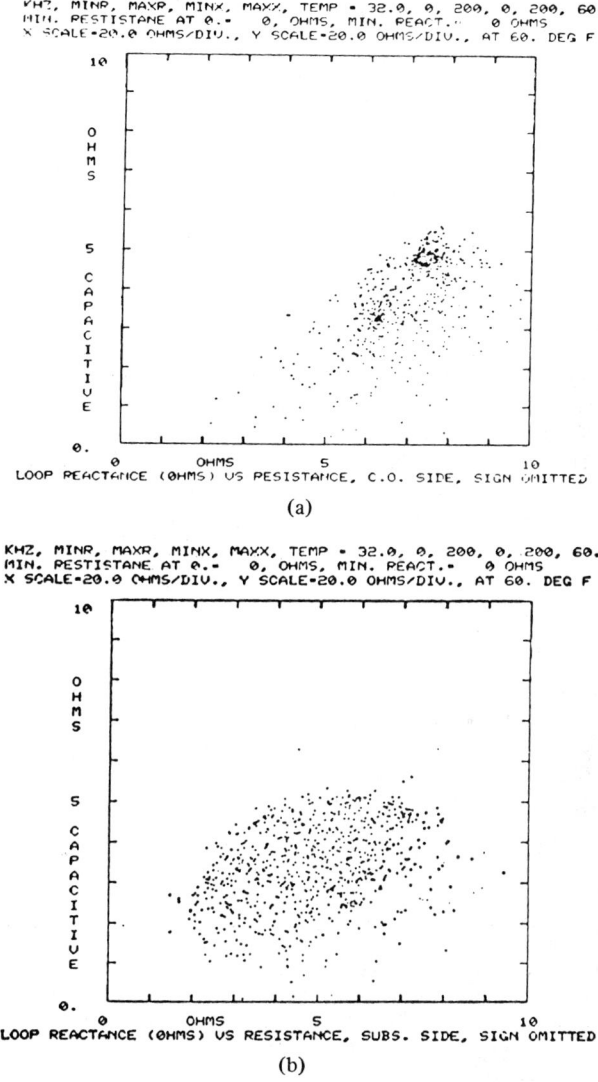

Reprinted with permission from: "A Tutorial on Two-Wire Digital Transmission in the Loop Plant" by S.V. Ahamed, P. Bohn, and N. Gottfried in *IEEE Transactions on Communications*. © 1981 IEEE.

Figure 6.7 (a) Scatter plot relating loop resistance and reactance at 32 kHz (central office side, bridged taps are left intact) at 60°F. (b) Scatter plot relating loop reactance to loop resistance at 32 kHz (subscriber side, bridged taps are left intact) at 60°F.

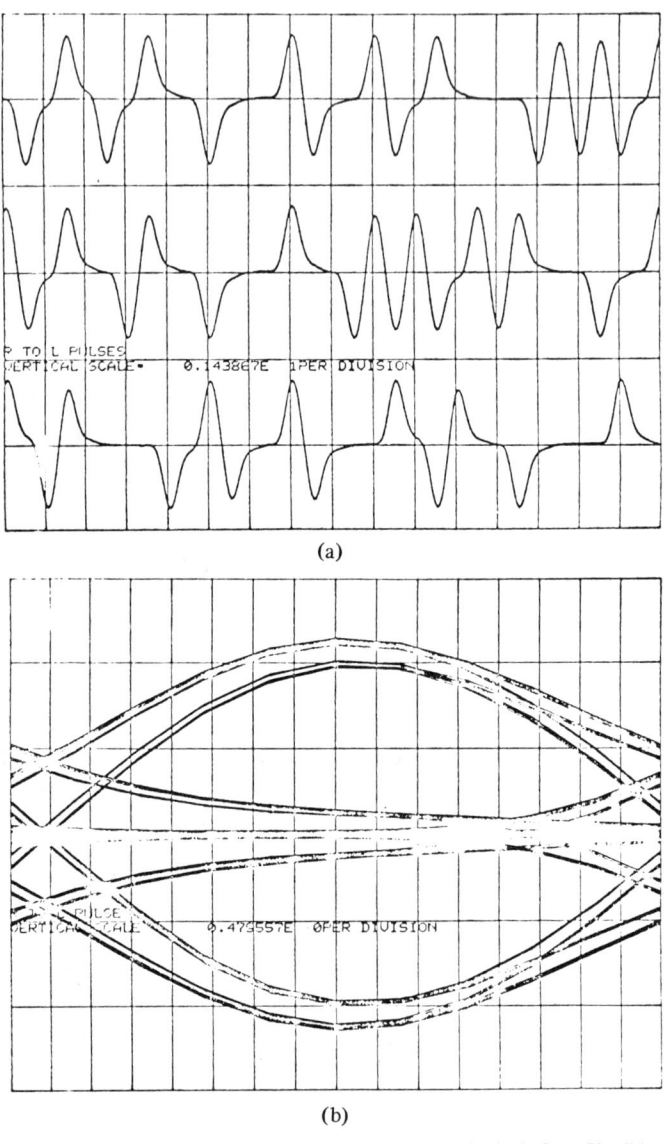

Reprinted with permission from: "A Tutorial on Two-Wire Digital Transmission in the Loop Plant" by S.V. Ahamed, P. Bohn, and N. Gottfried in *IEEE Transactions on Communications*. © 1981 IEEE.

Figure 6.8 (a) Received data stream at the subscriber end with AMI coding as data is transmitted at a line rate of 144 kbits. The loop configuration is a typical loop chosen from the 1973 Loop Survey. (b) Eye diagram generated at the subscriber end from the pulse pattern of (a).

proportions of eye opening to the average eye height are depicted in Figure 6.10(a). For a perfectly open eye, the points would be at the ±0.5 level. The extent to which these points deviate from these indicates the imperfection of the system. In Figure 6.10(b) the proportions of eye thicknesses (top, central, and bottom) in relation to average eye height are plotted for the same 64 kbit/s system. Figures 6.10(c) and (d) convey similar statistics in the opposite direction of data transmission.

(5) *Effects of Impulse Noise Events*—The simulation software is segmented adequately to permit flexibility to move freely from spectral to time and time to spectral domains. This capability permits the study of central office impulse noise events on the reception of data and on the eye closure.

The limit cycle approach of system simulation is appropriate under these conditions. The spectral components of the received signal are obtained and transformed to the time domain. The impulse noise event is added to the signal in the time domain and the spectral components are again resolved for the composite signal. Next, the effects of equalizer, filter, echo canceller, etc., are included. The entire pulse sequence and the event are then transformed into the time domain.

The results of one such simulation are shown as an eye diagram on a typical 12,865 ft. (3.92 km) loop. Figure 6.11(a) depicts the eye diagram at the

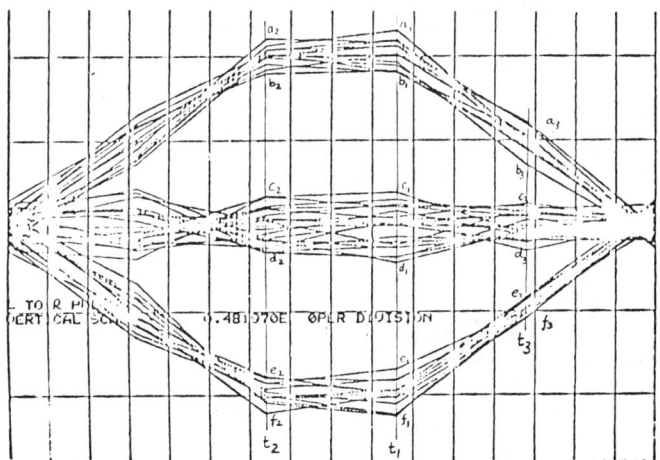

Reprinted with permission from: "A Tutorial on Two-Wire Digital Transmission in the Loop Plant" by S.V. Ahamed, P. Bohn, and N. Gottfried in *IEEE Transactions on Communications*. © 1981 IEEE.

Figure 6.9 Extraction of eye statistics from eye diagrams at t_1 t_2 and t_3 indicate the scanning instants selected and a_ib_i, b_ic_i, c_id_i, d_ie_i, e_if_i, $(a_i + b_i)/2$, $(e_i + f_i)/2$ indicate the data clusters.

central office and effects of the event are indicated. The event in the time domain is shown in Figure 6.11(b). The signal strength of the event at the central office is about 10 dB below the transmitted signal. The location of the event in the limit cycle also influences the error in detection; and to study the maximum number of errors the event can cause, it becomes necessary to let the time frame of the event slide in the limit cycle duration at the central office incoming lines. This is accomplished by varying the location of the event in Figure 6.11(b) relative to the start of the limit cycle.

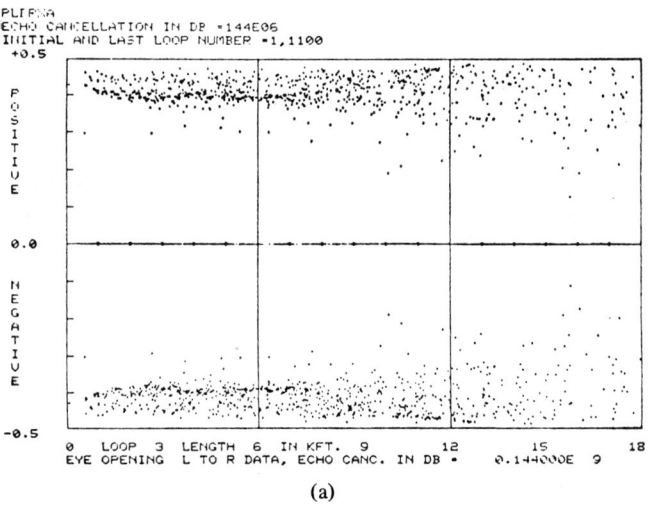

(a)

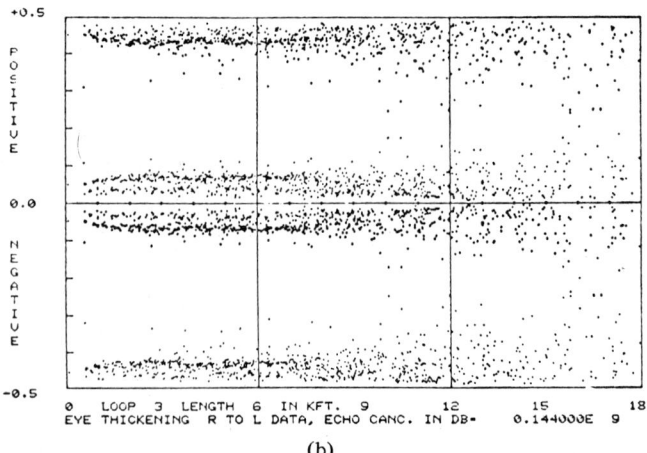

(b)

Design and Implementation of Digital Subscriber Systems 297

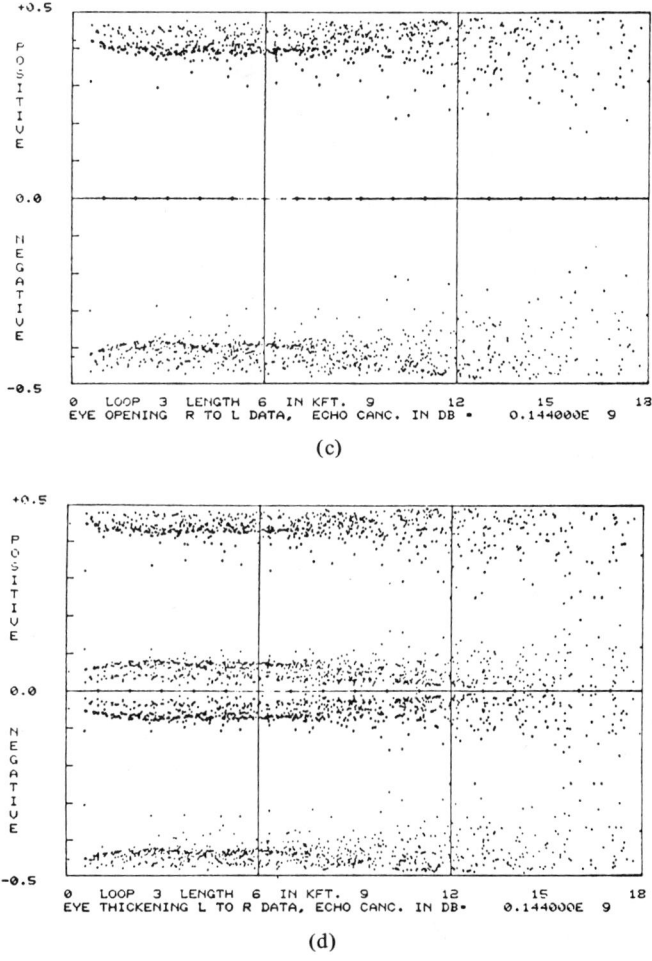

Reprinted with permission from: "A Tutorial on Two-Wire Digital Transmission in the Loop Plant" by S.V. Ahamed, P. Bohn, and N. Gottfried in *IEEE Transactions on Communications*. © 1981 IEEE.

Figure 6.10 (a) Typical eye opening scatter plot (at the subscriber end) as data at a line rate of 144 kbit/s is transmitted through all the loops in the 1973 Loop Survey with the bridged taps left intact. (b) Eye thickening scatter plot (at the subscriber end) as data at a line rate of 144 kbit/s is transmitted through all the loops in the 1973 Loop Survey with the bridged taps left intact. (c) Eye opening scatter plot at the central office end. (d) Eye thickening scatter plot at the central office end.

298   **Digital Subscriber Systems**

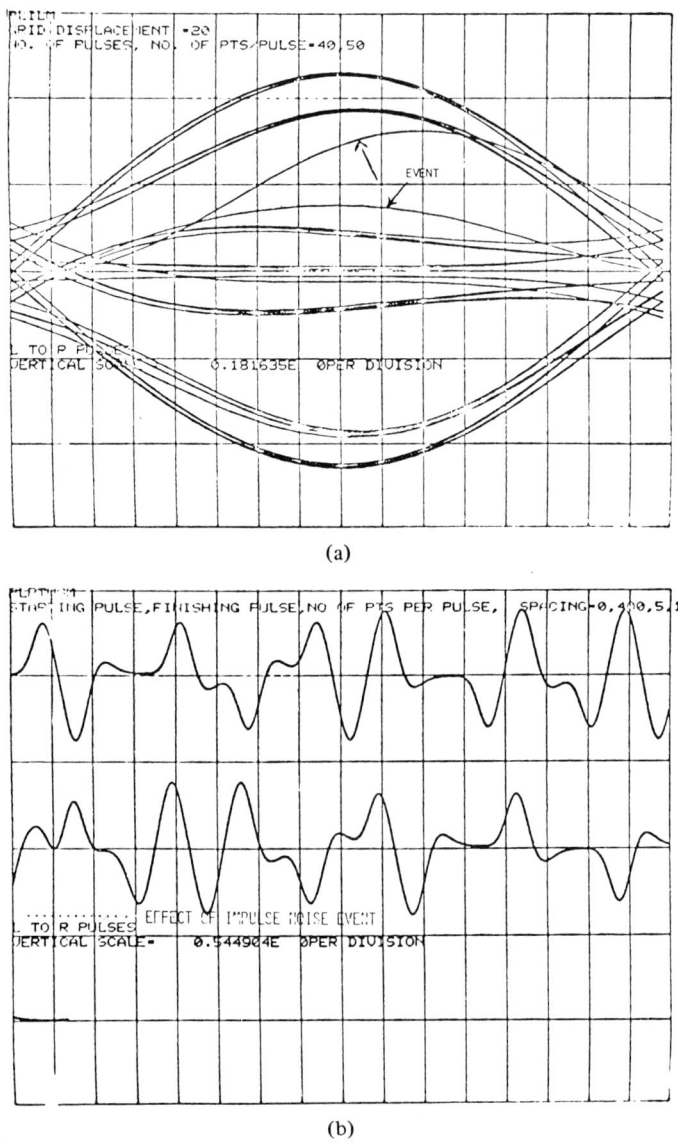

Reprinted with permission from: "A Tutorial on Two-Wire Digital Transmission in the Loop Plant" by S.V. Ahamed, P. Bohn, and N. Gottfried in *IEEE Transactions on Communications*. © 1981 IEEE.

Figure 6.11   (a) Eye diagram with an impulse noise event at the central office.
                        (b) Data stream and the impulse noise event in the time domain.

(6) *Guard Space Investigations for the TCM Mode*—In the time compression multiplexing mode, the cable loss is higher than that in the hybrid mode because of the increase rate of transmission during the burst. The reflections during the burst do not play a significant role since the receiver is inactive. However, as the receiver is activated, it becomes necessary to ascertain that the residual reflections from the burst will not be detrimental to the reception of far-end data. These reflections die down rapidly after the end of the burst but lingering effects prevail because of mismatch and bridged tap effects.

In this section, we use the simulation facility to relate the magnitude of the reflected signal to the received signal as the time between transmission and reception is prolonged to accommodate the necessary guard space. The relatively easy access to the data base is again utilized in this simulation. Typical results are plotted for the loops in the truncated data base in Figure 6.12(a) and (b) for the two directions of transmission at a burst rate of 144 kbit/s. In this figure, time (in pulse durations) is plotted after the burst. The relative magnitude of the reflected pulse to the received pulse (in dB) is plotted along the y axis. Each point in each of the 18 time slots pertains to the statistic of a loop for that time slot. There are 831 points in each time slot corresponding to the number of the loops in the truncated 1973 Loop Survey data base.

6.3.3 Design Rules For Loop Selection

Design rules help the operating companies and field engineers choose appropriate loops for ISDN digital transmission. These design rules are derived from a series of simulation studies and from systematically isolating the conditions which close the received eye diagrams. Some of the loop configurations are detrimental because of the presence of critical length bridged taps. For example, one bridged tap of coarse gauge, whose length is quarter wave length at peak energy concentration of the chosen code, causes the reflected wave to appear at the next scanning instant, thus creating a maximal intersymbol interference. If an optimized or a decision feedback equalizer is not used, the design rule calls for the elimination of such loops as possible contenders for data transmission.

The rules become complicated and cumbersome to evolve when the various gauge discontinuities, sizes, and numbers of bridge taps are involved. Hence, the development of design rules is perhaps the last step in the DSS implementation. From our simulation studies for the CSDC applications and the evolution of the design rules, we infer that high loss loops need a considerably more elaborate set of design rules, and that loops with even one bridged tap near the critical length can cause a complete eye closure as the length approaches to within 300 ft. of the critical length with #19 AWG bridged tap. The #19 AWG wire is coarse with a wire diameter of 0.0359 inches or 0.9119 mm and has an

300 Digital Subscriber Systems

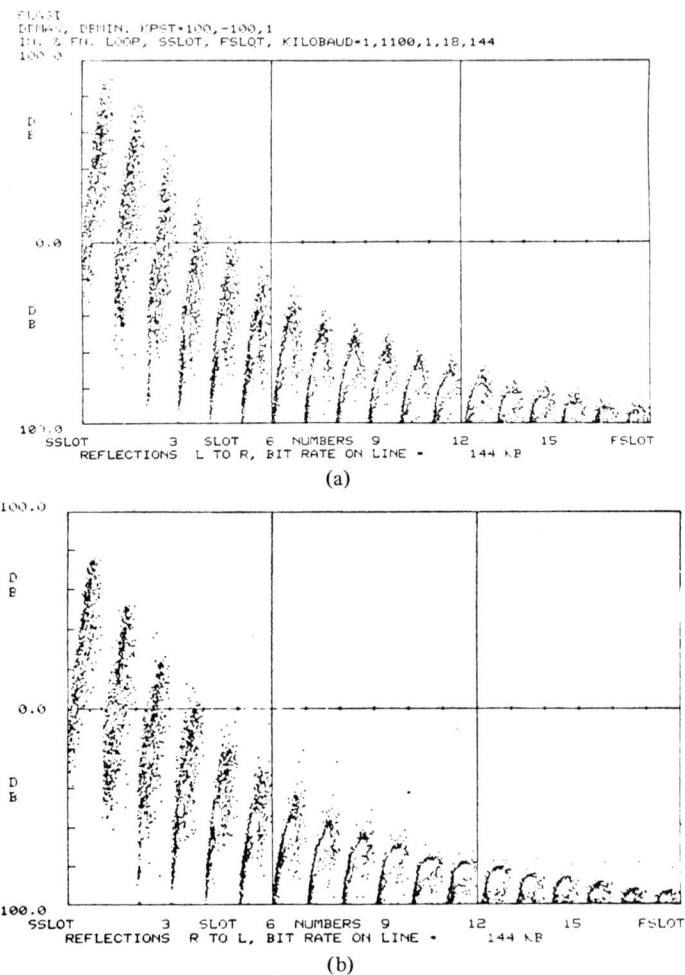

Reprinted with permission from: "A Tutorial on Two-Wire Digital Transmission in the Loop Plant" by S.V. Ahamed, P. Bohn, and N. Gottfried in *IEEE Transactions on Communications*. © 1981 IEEE.

Figure 6.12 (a) Reflections from the loops at the end of a burst (in TCM mode of transmission) of a data stream at 144 kbit/s plotted in dB against time in pulse intervals (central office end). (b) Reflections from the loops at the end of a burst (in the TCM mode of transmission) of a data stream at 144 kbit/s plotted in dB against time in pulse intervals (subscriber end).

attenuation of only 5.4 dB per mile (about 4.4 dB per km) at 80 kHz. By the same set of simulations, we have determined that the eye closure can go from 10 percent to 50 percent as the length of the #26 AWG (fine wire with a diameter of 0.0159 inches or 0.4039 mm and thus a higher attenuation of about 8 dB per mile or about 5 dB per km) bridged tap passes through its critical length.

6.4 THE COMPONENT OPTIMIZATION PROCEDURES

The DDS components depend upon the mode of transmission. Subscriber loops terminate in channel units and the channel unit accomplishes all of the functions necessary for the overall use of that subscriber line. A typical set of the functions may be supervisory and control functions, data transmission functions, fault localization and diagnostic functions. However, in connection with the transmission aspects of the DDS, line coding, signal isolation of the received data (i.e., TCM rate changing, and data buffering; or hybrid echo cancellation, network impedance matching) equalization, filtering, timing recovery, decoding of received data, fault detection, error recovery dominate the design considerations. In this section, we discuss the design optimization of equalizers and some aspects of filters and timing recovery circuits. Individually optimized components may or may not prove to be ideal for the overall DDS system. The component optimization has to attempted on the global system. The techniques we propose here provide only the initial designs rather than the final version of the components.

6.4.1 Analysis and Design Optimization of Equalizers

A digital system equalizer attempts to undo the effects of cable attenuation and restore the pulse shape at the receive. Pulse shape, however, results from a composite of the amplitudes of many frequency components, their phase relationships, and their harmonic relationships. Hence, a perfect pulse restoration at all bit rates entails perfect responses at all frequencies. In essence, an inverse loop circuit constitutes a perfect equalizer circuit. However, loops bit rates and coding algorithms are each numerous and uncorrelated, thus causing additional elements of complexity in the design of equalizers. Even though the results presented here are for the bipolar code, it should be realized that for other codes, the region of spectral emphasis for matching cable loss and equalizer gain will be dictated by the bandwidth requirement of the particular code.

Four equalizer designs are analyzed and evaluated for their application for ISDN digital transmission in the range of 56–144 kbit/s. The first design is for the line rate of 144 kbit/s transmission. The second design is specifically for a line rate of 324 ($= 2.25 \times 144$) kbit/s in the TCM mode. The remaining two are intended for a range of TCM applications in the region of 144 (2.25×64) kbit/s to 324 (2.25×144) kbit/s. Realizations of the designs are discussed and their actual performance are computed by a series of simulation programs discussed in Section 6.3.2. To prevent cable mismatch effects (which are indeed present in a hybrid system) from overshadowing the influence of equalization, we present simulation results utilizing continuous unidirectional transmission, thus focusing on the performance of the equalizer. Once optimized, the equalizers can be utilized in either a hybrid or a TCM mode system. It is the information rate which differs. The eye opening scatter plots obtained by analyzing the system performance on actual loops taken from the 1973 Loop Survey are used as the ultimate criteria to judge equalizer quality.

6.4.1.1 Three Section, 18000 ft. (5.48 km) Equalizer

Circuit Configuration and Representation

Each of the three sections has a five pole, five zero circuit. The location of these poles and zeros are obtained by the solution of five simultaneous equations, which are formulated to yield exact equalization of six thousand feet of #22 AWG cable at five discrete frequencies (1, 2, 10, 50, 100 kHz). Further, the transfer function of this RC circuit is controlled by the gain of an operational amplifier and this gain is adjusted by yield equalization of loops or loop sections under 6000 ft. (1.82 km) cable. Thus, loops of any length under 18,000 ft. (5.48 km) can be equalized by cascading one, two, or three sections of this equalizer and appropriately adjusting the gains on the three sections of the equalizer. Variation in temperature is converted to a variation in effective loop length and the eye height is maintained by the gains of the three sections of the equalizer.

The performance evaluation of this equalizer with composite cables is limited to a study of eye openings as the composition of the cables is changed gradually for an overall loop length of 18,000 ft. or 5.48 km. Two major simulations are performed. In the first simulation, a loop compromises #22 and #24 AWG cables with the maximum overall length. At one extreme the loop is totally composed of #22 AWG cables; conversely, at the other extreme the loop consists of #24 AWG cable. These two cases are presented at the left and right sides of the x-axis in Figure 6.13. As the distance along the x-axis is increased, the #22 AWG cable is shortened and the length of #24 AWG is cable is increased. The eye openings are plotted on the x-axis. The two eyes generated for alternate bipolar coding yield two data points for each loop composition.

Similar results for #19 and #26 cable compositions are shown in Figure 6.14. The simulations are carried out at 144 kbit/s. The performance of the equalizer and the quality of the filter are both reflected by the eye opening. The first filter at the transmitting end has the 3 dB point located at half the bit rate. The second identical filter is located at the receiver end. This pair of filters essentially yields a pulse shape with the amplitude rising from zero steadily reaching a peak and diminishing to zero pulse period after is peak.

In Figure 6.13 and 6.14 a pair of points along each vertical line represents an independent simulation for the appropriate composition of cables. Eye diagrams are internally generated and the seven eye statistics of the eye (the average positive height, the top eye thickness, the top eye opening, and the central negative height) are extracted. For each loop, two dots representing the positive

The Component Optimization Procedures 303

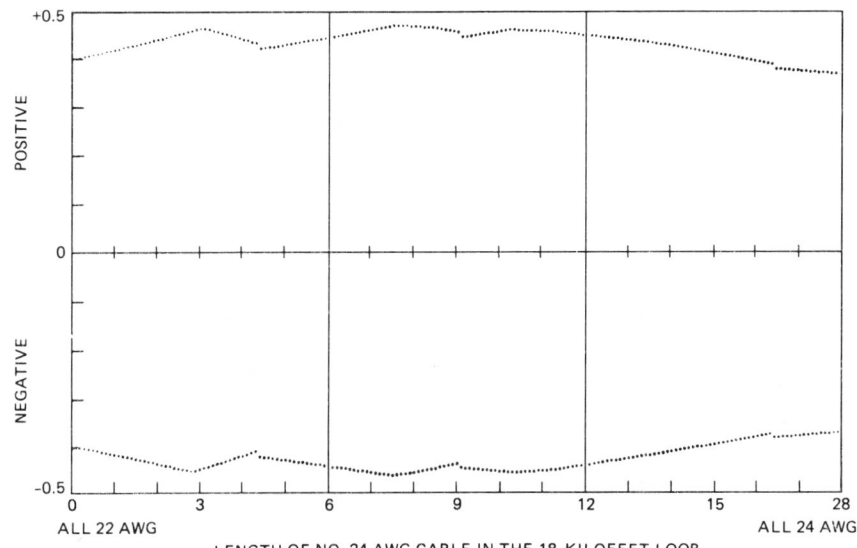

Reprinted with permission from the *AT&T Technical Journal.* Copyright 1982 AT&T.

Figure 6.13 Eye openings with Nos. 22 and 24 AWG composite cable loops having a total length of 18 kft (144 kb/s).

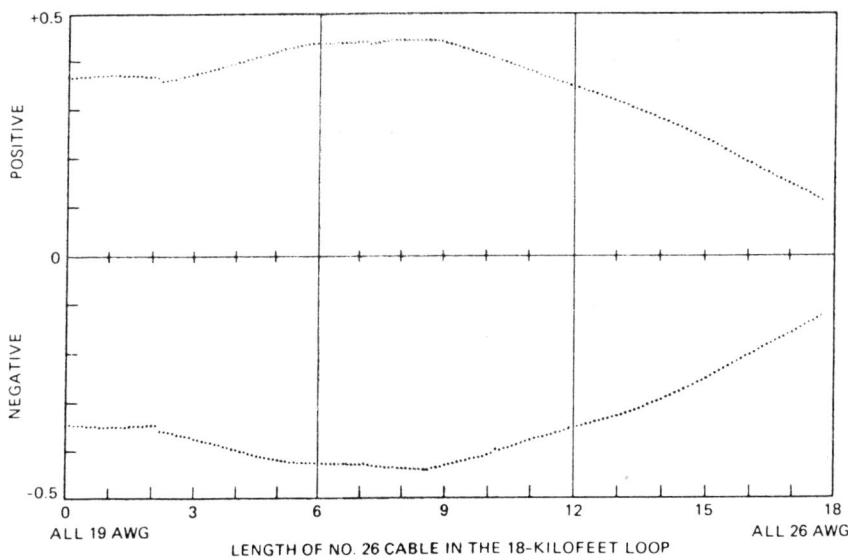

Reprinted with permission from the *AT&T Technical Journal.* Copyright 1982 AT&T.

Figure 6.14 Eye openings with Nos. 19 and 26 AWG composite cable loops having a total length of 18 kft (144 kb/s).

and negative eye opening are plotted. From Figure 6.13 and 6.14, it can be expected that the equalizer compensates 18,000 ft. (5.48 km) of #19, #22, #24, and #26 AWG cables with 70, 80, 70 and zero percent eye openings, respectively. With only #26 AWG loops, independent simulations indicate that this equalizer provides an eye opening of about 40 percent with 12,000 ft. (3.75 km) loops. It is also interesting to note that the eye openings in either direction of transmission are roughly the same because the effects of reflections (as they occur in a hybrid system) are eliminated, thus emphasizing only equalizer and filter performance. Proper filter performance is assured by 100 percent eye opening at zero cable length.

Equalizer Performance with the 1973 Survey Loops

When the equalizer is used at 144 kbit/s the performance of the equalizer may be examined by studying the scatter plots of Figure 6.15a and b. In Figure 6.15a, each loop is represented by two points along each vertical line. The eye opening of the top eye (generated by the transmission of alternate bipolar code) as a fraction of the overall average eye height is represented by a point in the top half of Figure 6.15a. Similarly, the eye opening of the negative eye is represented in the lower half of Figure 6.15a. All of the 831 nonloaded loops 5.2 of the loop data base have been simulated and the eye openings plotted against the length of the loops.

In Figure 6.15b each loop is represented by four points. The top eye thickness (one point), the central eye thickness (two points about the x-axis), and the bottom eye thickness (one point) are each represented as a fraction of the overall average eye height. All of these points are located along a vertical line at the length of the loop.

6.4.1.2 Four Section, 20,000 Ft. (6.1 km) Equalizer

The range and spectrum of equalization are both enhanced for this design. As opposed to the former design, a "staged turn-on" algorithm has been used for more effective equalization. In this mode, the four sections are turned on one by one as the range of each section is exhausted by the loop attenuation. Further, the equalization is also forced out to a frequency of 250 kHz, thus increasing the total maximum gain from 60 dB for the former design to about 72 dB. Each section of this six pole, six zero equalizer has exact equalization at 1, 2, 10, 50, 100, and 250 kHz for a 5000' #22 AWG loop.

Equalization Data for Loop Attenuation

The performance of this equalizer has been studied at three frequencies: 56, 144, and 324 kHz. The attenuation of the 831, 1973 Survey Loops, in the truncated data base are computed at these three frequencies. The results are accumulated

The Component Optimization Procedures

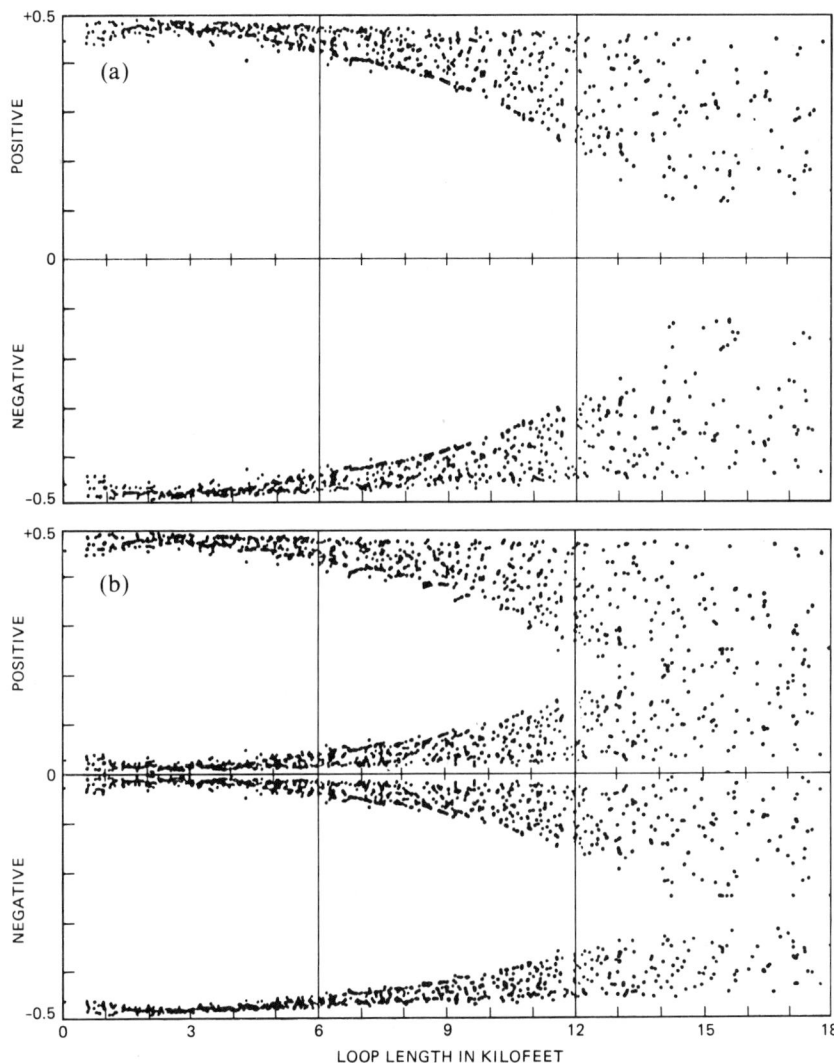

Reprinted with permission from the *AT&T Technical Journal*. Copyright 1982 AT&T.

Figure 6.15 Eye data scatter plot obtained by the simulation of 1973 Survey Loops with the three-section, 18,000-ft equalizer at 144 kb/s. (a) Eye opening. (b) Eye thickness.

in an interim database. Next, the loop attenuations are plotted against the equivalent length of each of the composite loops in terms of #22 AWG cable in Figures 6.16a, 6.17a, and 6.18a for the three frequencies at 56, 144, and 324 kHz, respectively. The equivalent length is calculated as the length of the #22 AWG cable, which has the same loss as the composite loop at about 200 kHz.

The gauge-length conversion numbers (Table 6.2) are used for determining the total equivalent loop length. The lengths of the different cables in this table are computed to have the same loss as 1,000 feet (0.3048 km) of #22 AWG cable. These scatter plots represent approximate straight lines, since the loss is directly proportional to the length.

Next, the equalization gain of this particular equalizer for each of the loops at the three corresponding frequencies is computed and accumulated in another database. The equalizer gains for all the loops are also plotted against the equivalent length of the #22 AWG cable at these frequencies in Figures 6.16b, 6.17b, and 6.18b.

If the equalizer is functioning satisfactorily, the points in Figures 6.16, 6.17, and 6.18 should also be colinear. The nonlinearity of the lines in Figures 6.17b, and 6.18b indicates the inability of the equalizer circuitry to track the line loss in the 50–80 dB range. The mismatch in the equalizer performance is indicated by the lack of a perfect overlap of the lines in Figures 6.16a and b, in Figures 6.17a and b, and in Figures 6.18a and b. A perfect equalizer should match the line loss perfectly at all of the harmonic frequencies and for all loop lengths. The extent of its imperfection is depicted by the imperfection of these straight lines.

The 1973 Loop Survey Eye Diagram Performance

Simulations at 324 kbit/s (the TCM rate corresponding to 144 kbit/s bidirectional hybrid rate) and at 126 kbit/s (the TCM burst rate corresponding to 56 kbit/s bidirectional hybrid rate) are presented to evaluate the performance of this equalizer. In Figure 6.19, the eye openings are depicted as a scatter plot for the 1973 Survey Loops when transmitting data at 324 kbit/s in the TCM mode. The scatter plot corresponds to CO (central office) to subscriber data transmission. A similar scatter plot is depicted in Figure 6.20 when the system is excited with 126 kbit/s in the TCM mode.

Table 6.2
Gauge Conversion Numbers at 108 kHz

Gauge # 1 kft	Equivalent kft of 22 AWG
19	0.658
22	1.00
24	1.409
26	1.923

NOTE: These gauge conversion numbers are held constant for all frequencies. Slight improvement in accuracy of results occurs if these numbers are optimized for each frequency independently.

The Component Optimization Procedures

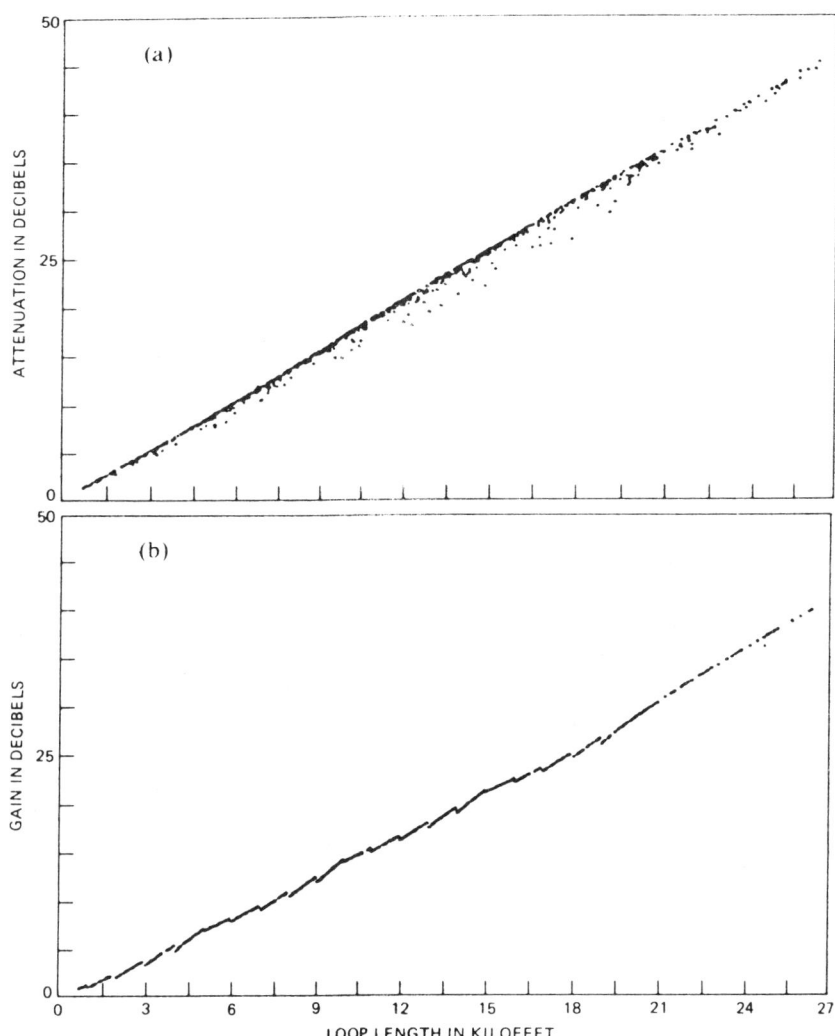

Reprinted with permission from the *AT&T Technical Journal*. Copyright 1982 AT&T.

Figure 6.16 Comparison of loop attenuations and equalizer gains for 1973 loop survey at 56 kHz plotted against the loop length expressed in equivalent kft of No. 22 AWG cable. Scale on the y-axis is 5 dB/division, and on the x-axis it is 1.5 kft/division. (a) Loop attenuation. (b) Equalizer gain.

6.4.1.3 Limited Range, Single Pole Equalizer

In the previous two designs, the equalizer gain exactly compensates the cable loss at specific frequencies (five for the first, six for the second design) and at specific cable lengths (6,000 feet of #22 AWG and 5,000 feet of #22 AWG cables). Variations as a result of length are accomplished by changing the gain

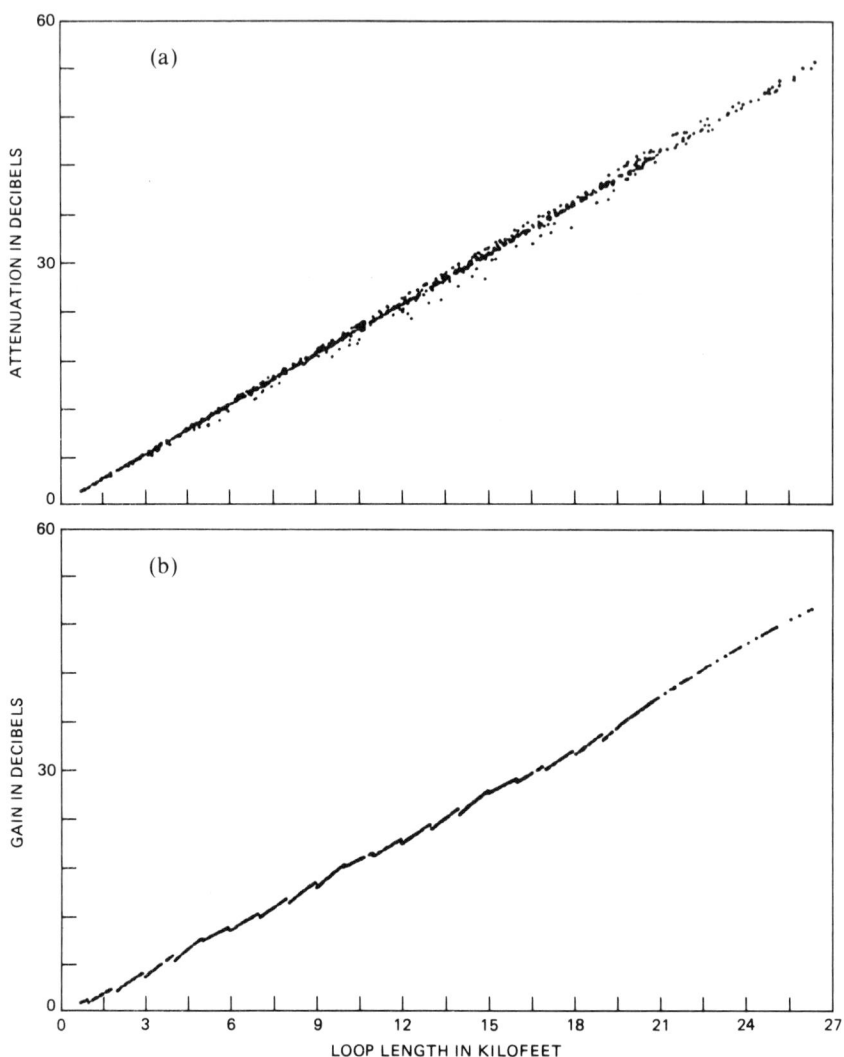

Reprinted with permission from the *AT&T Technical Journal*. Copyright 1982 AT&T.

Figure 6.17 Comparison of loop attenuations and equalizer gains for various loops at 144 kHz. The y-axis scale is 6 dB/division. (a) Loop attenuation. (b) Equalizer gain.

control on the operational amplifiers. Equalizers reported in Sections 6.4.1.1 and 6.4.1.2 are designed by broadly matching the shape of the cable loss frequency domain. Overall gain changes to accommodate different cable length are adjusted by changing diode resistances, which in turn adjust the gain of operational amplifier without drastically changing the frequency characteristics. A

The Component Optimization Procedures

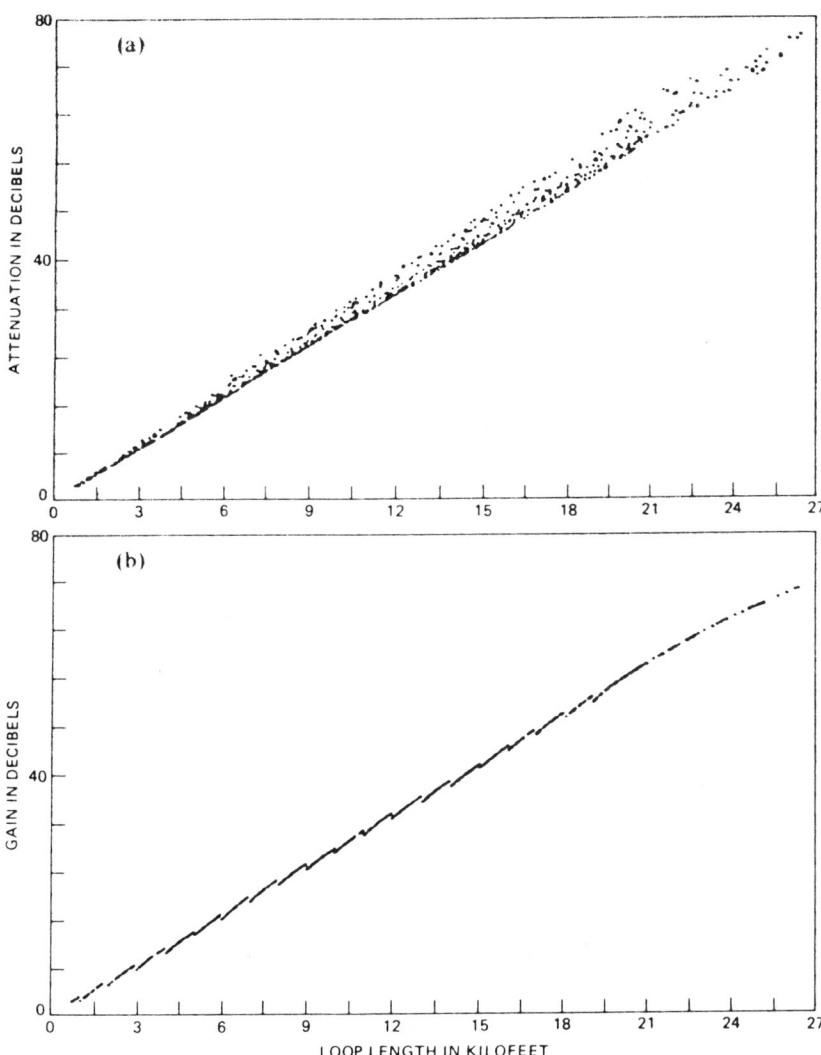

Reprinted with permission from the *AT&T Technical Journal*. Copyright 1982 AT&T.

Figure 6.18 Comparison of loop attenuation and equalizer gains for various loops at 324 kHz. The y-axis scale is 8 dB/division. (a) Loop attenuation. (b) Equalizer gain.

single pole and adjustable zero type of transfer function is used to generate a suitable equalization range over 6 to 36 dB of cable loss. When these poles and zeroes are redistributed, a corresponding range of equalization may be obtained for these lower frequency applications. The application considered here is for a 324 (= 2.25 × 144) kbit/s TCM mode data transmission. Cable attenuation

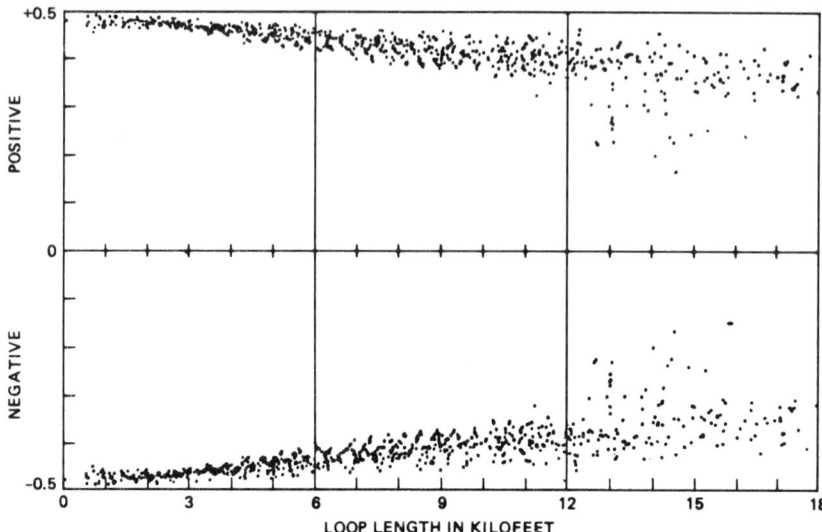

Reprinted with permission from the *AT&T Technical Journal*. Copyright 1982 AT&T.

Figure 6.19 Eye opening scatter plot at 126 kb/s with 20,000-ft, four-section equalizer in the TCM mode of transmission. Subscriber side data are given.

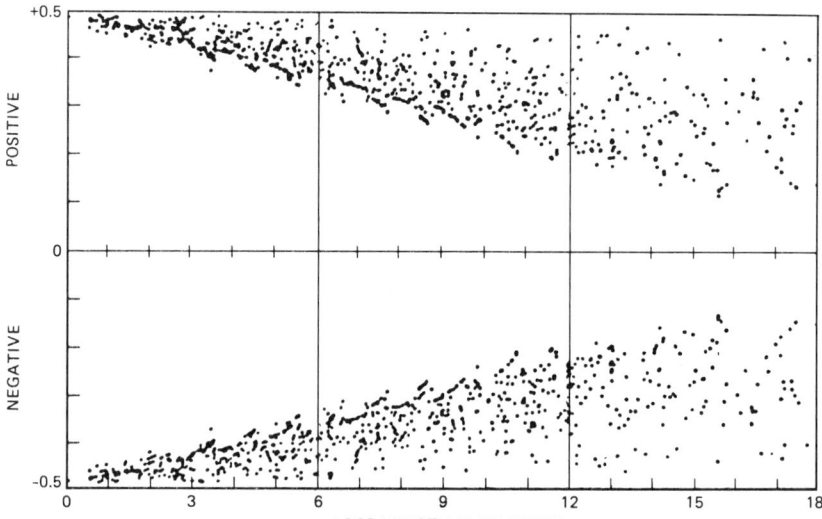

Reprinted with permission from the *AT&T Technical Journal*. Copyright 1982 AT&T.

Figure 6.20 Eye opening scatter plot at 126 kb/s with 20,000 ft, four section equalizer in the TCM mode of transmission. Subscriber side data are given.

and equalizer gains at the half-bit rate (162 kHz) are both plotted in Figure 6.21. The dotted lines indicate the cable attenuations of the #19, #22, #24, #26 AWG cables. The full lines indicate the corresponding equalizer gain. The chief limiting factor is the range of equalization. The equalizer saturating at about 36 dB cannot compensate for cable losses beyond this limit. Hence, the design of this particular type of equalizer has been abandoned for the 324 kbit/s TCM mode of bidirectional data transmission. A limited study of eye opening data has indicated that 30–40 percent of the loops would carry bidirectional data with a 60 percent eye opening at 324 kbit/s TCM mode of data transmission.

6.4.1.4 Enhanced Range, Wider Frequency Optimized Equalizer

The design philosophy for this equalizer is the same as that for the former equalizer, however, two cascaded gain amplifiers are used in this case. The first has a low frequency zero and two higher frequency poles while the second has a low frequency zero, another right half plane zero and a pair of complex poles. However, optimization by the incremental displacement of poles and zero is essential to follow the loop attenuation through the spectral band of interest.

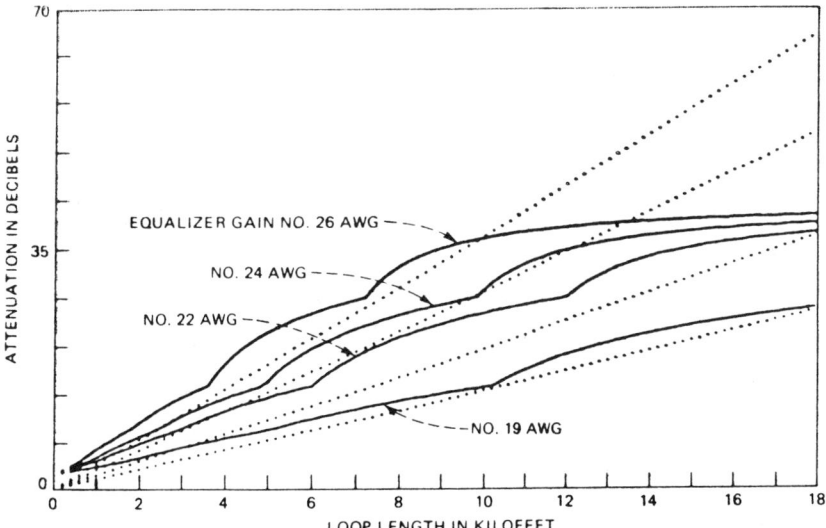

Reprinted with permission from the *AT&T Technical Journal*. Copyright 1982 AT&T.

Figure 6.21 Cable losses (dotted lines) and equalizer gains of a limited range, single-pole equalizer plotted for a 324-kbit/s TCM design. Loop length (1 kft/division) is plotted on the x-axis and loop loss (7 dB/division) is plotted on the y-axis.

Optimization Strategy

An optimization approach in both the frequency and distance domains is simultaneously used to attempt to match the equalizer gain against the cable loss. First, half-bit rate frequency cable attenuation lines are generated for #19, #22, #24 and #26 AWG cables. Then the gauge conversion factors (See Table 6.1) at this frequency are used as a basis of all the gauge conversions. The maximum equivalent loop length (as extracted from the Loop Survey) is then determined to correspond to about 16,700 feet of #26 AWG cable. Next, #26 AWG cable losses at 5, 10, 50, 100, 150, 200, 250 kHz are generated by the primary characteristics of the #26 AWG cable. The poles and zeros of the equalizer are given incremental changes such that the equalizer gain and the cable losses are almost colinear at about half-bit rate frequency over the span of one-third to two-thirds of maximum cable length (see Figures 6.22a and b). The first step of optimization is more or less easily achieved; however, in the second step of shifting the R and C's of the equalizer circuit such that noncollinearity of the equalizer gain and the cable loss at the low frequency end of the spectrum has to be weighed against the noncollinearity of the two at the higher end of the spectrum.

The optimization becomes complicated by the fact that the low frequency equalizer gain displays an intersection with the cable loss going from a lower gain to a higher gain as the length of the cable is increased. On the other hand, the intersection at the higher frequency follows an opposite type of characteristic changing from higher gain to lower gain as the length increases. Thus, a large number of trials is necessary to achieve a satisfactory compromise.

The effect of this optimization implies one distinct effect on the single pulse response. At lower cable lengths there is a distinct overshoot because of the higher frequency over equalization and the lower frequency under equalization. Conversely, at longer cable lengths there is a tail left behind in the single pulse response because of lower frequency over equalization and higher frequency under equalization. But there is a certain range of lengths at which the single pulse response approaches perfection. This perfection is about two-thirds the maximum cable length. Thus, the scatter plot generated by this type of equalizer shows near 100 percent open eyes (to be discussed next) in the central one-third length of the loops. Further, the performance is consistently good over a range of 126 to 324 kbit/s.

Selected Loop Performance:

Case 1. Very Short Loop Response

At 55 ft. (17 m) of #24 AWG cable the eye diagram is imperfect. The cable distortion component can be ruled out this minimal length. However, in our

The Component Optimization Procedures

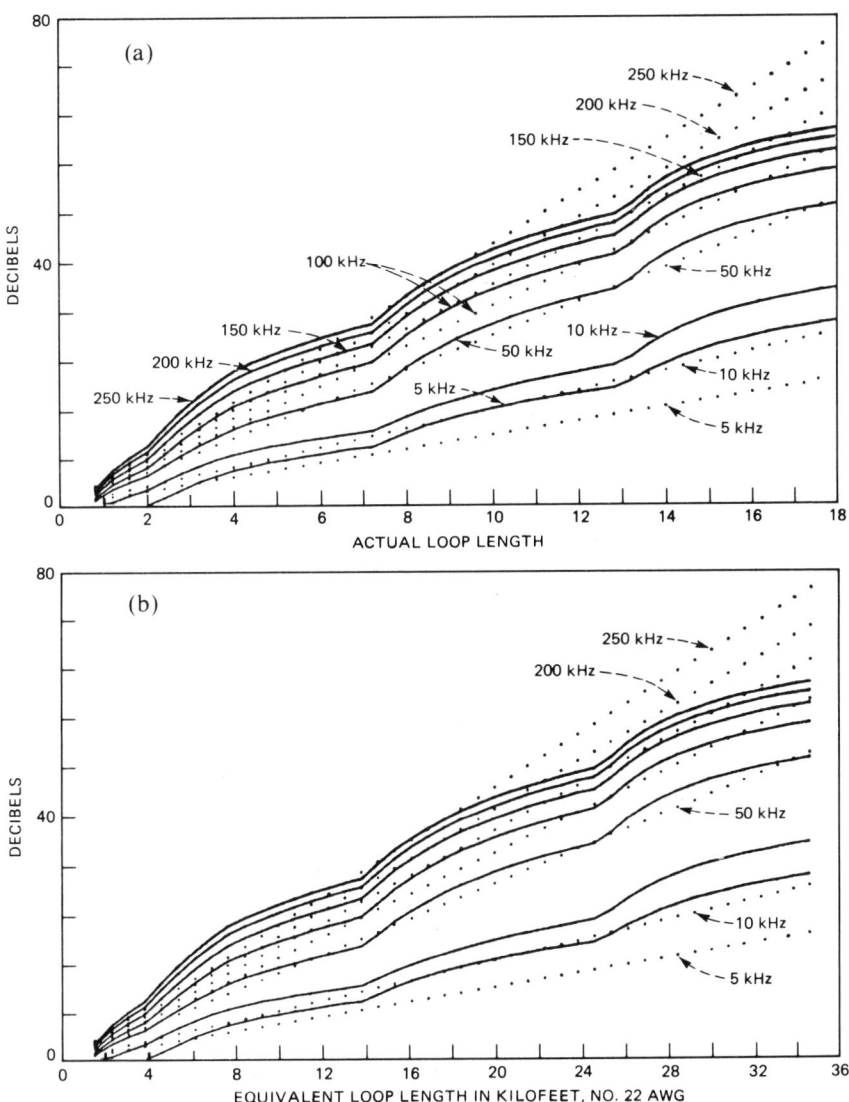

Reprinted with permission from the *AT&T Technical Journal*. Copyright 1982 AT&T.

Figure 6.22 Comparison of cable loss (dashed lines) and equalizer gains of the enhanced range, wider frequency, optimized equalizer at different frequencies for the No. 26 AWG cable. The y-axis scale is 8 dB/division. (a) Plot against the actual loop length (the x-axis scale is 1kft/division). (b) Plot against the equivalent loop length in terms of No. 22 AWG cable (the x-axis scale is 2 kft/division).

effort to reduce the circuit complexity, we have eliminated the transmit filter and optimized the pulse width at about 40 percent of the pulse period. The harmonics of this pulse shape are affected by the receiver filter and by the very slight effect of the equalizer. This has been understood to be the reason for the slight imperfection of the eye diagram shown in Figure 6.23 at 324 kbit/s operation.

Case 2.   Loops Around 6,000 Ft. (1.86 km)

Two loop responses around this length are presented in Figure 6.24 at 324 kbit/s. A seven-section loop with an overall length of 6,093 ft. (1.86 km) is chosen at 60 F. The eye diagram is relatively similar from both directions of transmission.

Case 3.   Loops Around 12,000 Ft. (3.66 km)

Two loops around 12,000 ft. (3.66 km) depict the response of the equalizer. In Figure 6.25a the eye closure is about 18.7 percent at 324 kbit/s. This loop is also composed of seven sections as detailed in the figure caption and is the worst loop in that region of overall loop lengths. The best cable performance also takes

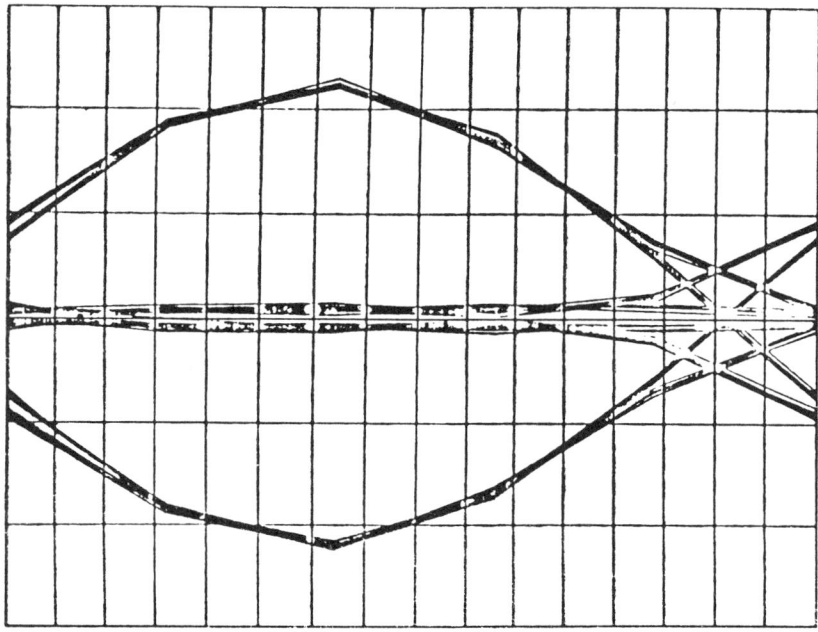

Reprinted with permission from the *AT&T Technical Journal*. Copyright 1982 AT&T.

Figure 6.23   Eye diagram showing the effect of finite-pulse width with 55 ft. of No. 24 AWG cable at 324 kbit/s.

The Component Optimization Procedures

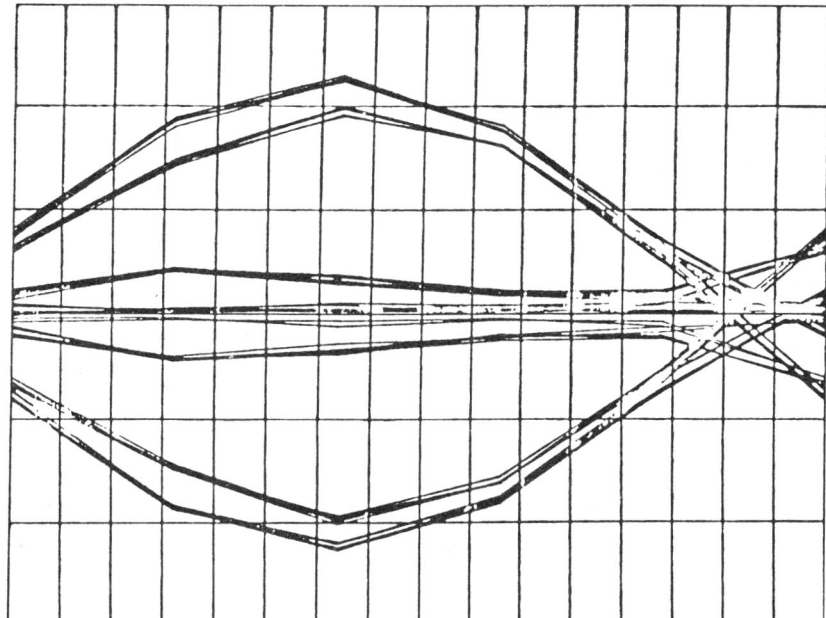

Reprinted with permission from the *AT&T Technical Journal*. Copyright 1982 AT&T.

Figure 6.24 Seven-section loop (88 ft., No. 19 AWG; 3,571 ft., No. 22 AWG; 17 ft., No. 26 AWG; 310 ft., No. 22 AWG; 19 ft., No. 24 AWG; 1,299 ft., No. 26 AWG; and 789 ft., No. 24 AWG) response at 324-kbit/s TCM excitation.

place around this length with a single section loop with a 11,998 ft. (3.66 km) of #26 AWG cable as shown in Figure 6.25b. The eye closure is only 3 percent and the response is consistent with the optimization strategy of the equalizer as discussed in this section.

Case 4. Loops Around 18,000 Ft. (5.48 km)

The longest loop eye diagram with the three section loop of length 17,962 ft. (about 5.48 km) is shown in Figure 6.26. Here the eye closures become 36.7 percent (Central Office to subscriber side) and 36.6 percent (subscriber to Central Office side) for 324 kbit/s data transmission. In Figure 6.27 the worst loop response is depicted. The eye closures are about 43 and 42 percent for the two sides of transmission and one can visualize the presence of long tails left behind single pulses in these two eye diagrams.

The 1973 Loop Survey Eye Diagram Performance

The loops in the truncated loop survey data base have been simulated at four different TCM frequencies (126, 162, 216, and 324 kbit/s to correspond to full

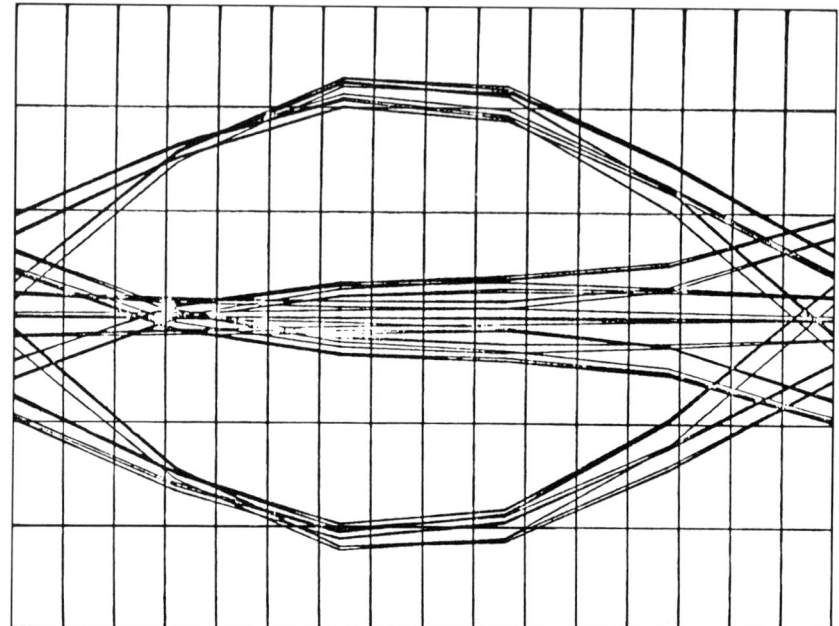

Reprinted with permission from the *AT&T Technical Journal*. Copyright 1982 AT&T.

Figure 6.25a Seven section (3450 ft., No. 26 AWG; 261 ft., No. 24 AWG; 50 ft., No. 26 AWG; 313 ft., No. 22 AWG; 2477 ft., No. 24 AWG; 5329 ft., No. 22 AWG; and 200 ft., No. 24 AWG) loop response at 324 kb/s TCM excitation.

duplex bidirectional rates of 56, 72, 96 and 144 kbit/s). The scatter plots of eye openings are plotted in Figures 6.28, 6.29, 6.30, and 6.31. All of these scatter plots exhibit a pattern yielding the best eye openings in the area of the 8,000 to 13,000 feet rather than at about zero length. This particular performance is due to the equalizer optimization strategy.

In conclusion, the design of the equalizer has a profound effect on the capability of the loops to transmit the digital signal. Here we have compared the various design strategies and studied their effects. An equalizer design using the presently available configurations was optimized for a center frequency of 81 kHz to correspond to 162 kbit/s TCM mode. However, a reoptimization of this design at 126 and 324 kbit/s was deemed unnecessary because of its satisfactory performance. The optimization procedure described here should prove valuable for the design and implementation of any equalizer configuration, provided its maximum gain is adequate for the attenuation expected from the longest of the loops. Generally, bridged taps limit the maximum range of equalization, and the optimization strategy present in this paper may be employed for more severe loop conditions.

The Component Optimization Procedures

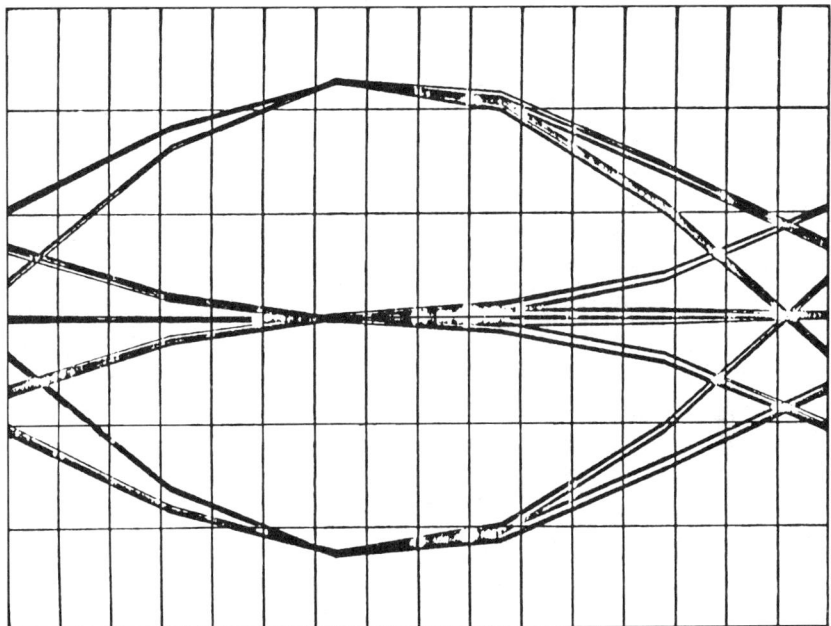

Reprinted with permission from the *AT&T Technical Journal*. Copyright 1982 AT&T.

Figure 6.25b Eye diagram for a single-section loop consisting of 11,998 ft. of No. 26 AWG cable at 324 kbit/s.

All of the simulation results presented here are for the bipolar (AMI) code. However, it can be visualized that the design philosophy and the optimization techniques are universal to the extent that the region of spectral emphasis (Figures 6.21 and 6.22) will be dictated by the bandwidth requirements of the particular code and the bit rate.

6.4.2 Filters For Digital Subscriber Systems

Filters in DSS for the proposed digital network play numerous roles. First, the transmit filter at the transmitter ascertains that energy of the digital information entering into the subscriber network is within the necessary spectral band for the particular code in use. Limiting this bandwidth helps the minimize crosstalk into other systems. One of the systems most susceptible is the analogue carrier system. The Bell Operating Companies in the United States use the analogue Subscriber Loop Carrier (especially the SLC®-1 and SLC®-8 systems). The effects of sharing the same cable and/or bundle with the ISDN digital lines causes serious crosstalk problems. For reasons of this type it becomes necessary to curtail the energy transmitted in the bandwidth outside the region of interest and the ISDN filters play an important role here in limiting the bandwidth at the transmitter. A certain amount of control can be effected by changing the wave

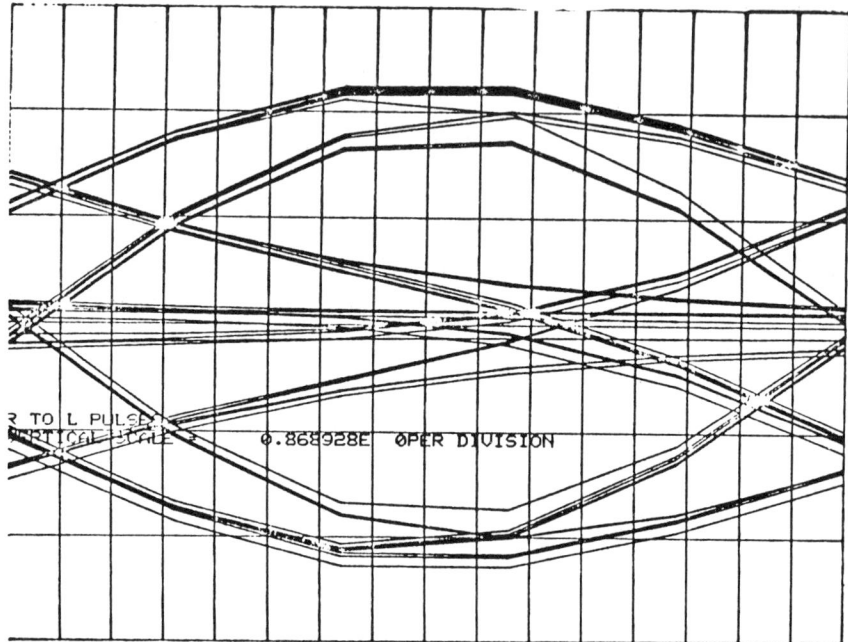

Reprinted with permission from the *AT&T Technical Journal*. Copyright 1982 AT&T.

Figure 6.26 Eye diagram for longest loop (10,669 ft., #26 AWG; 2,418 ft., #22 AWG; and 4,875 ft., #24 AWG) response. The total length is 17,962 ft. and the excitation is 324 kbit/s. TCM mode.

shape at the transmitter and by eliminating the need for the transmit filter. However, a considerable amount of pulse shape optimization, in conjunction with the loop characterization, is necessary to select the transmitted pulse shape.

Second, the filter at the receiver performs one critical function. It limits the energy of interest, over the bandwidth, that is necessary for the equalizer to reconstruct the pulses. The noise in the loop plant, (especially the crosstalk and impulse noise) can be spread beyond what is necessary for the recovery of information. The receive filter thus prevents this noise energy from causing errors in the receiver. If appropriate band limiting by the receive filter is done in conjunction with the equalizer characteristics, the accurate recovery of the received information can be enhanced.

Additional filtering to recover the clock at the subscriber end is also essential. The clock recovery circuit (see section 6.4.3) generally has a high Q (ratio of reactance to the resistance) phase locked loop. The resonant frequency of this phase locked loop circuit is at the bit rate or a subharmonic of the bit rate, thus a fairly stable sine wave is generated at the output and the clock is thus recovered. The type of filters used depends upon the circuit implementation. Digital filters

The Component Optimization Procedures

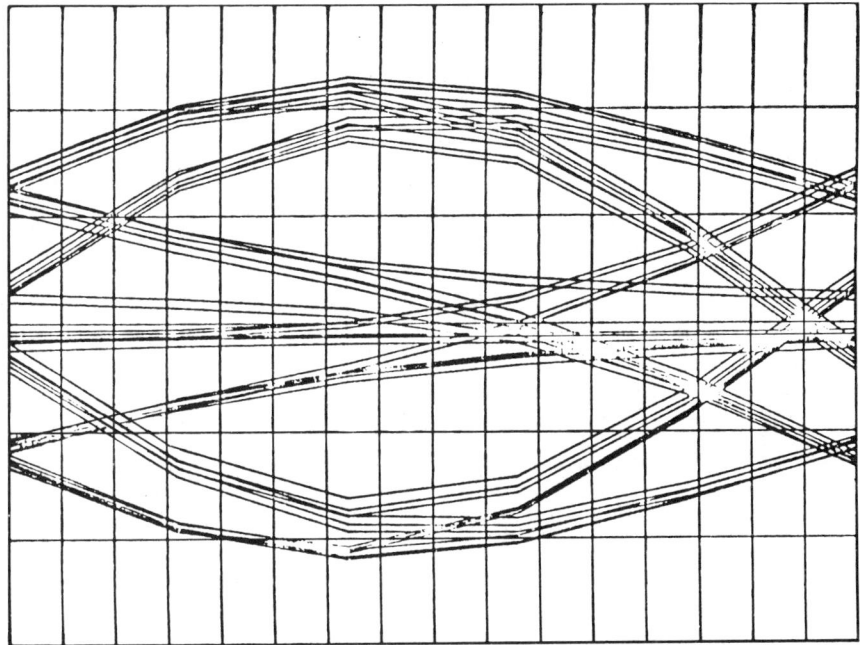

Reprinted with permission from the *AT&T Technical Journal*. Copyright 1982 AT&T.

Figure 6.27 Worst eye opening for an eight-section (7,789 ft., No. 26 AWG;; 2,998 ft., No. 24 AWG; 2,564 ft., No. 26 AWG; 25 ft., No. 24 AWG; 957 ft., No. 26 AWG; 2,120 ft., No. 24 AWG; 141 ft., No. 26 AWG; and 720 ft., No. 24 AWG), 17,314 ft., loop at 324 kbit/s.

are also considered and offer the usual advantage of being integrated with the rest of the digital circuits.

In the CSDC (Circuit Switched Digital Capability) introduced by the predivestiture Bell System, an additional high-pass filtering was introduced to limit the bandwidth to be above the voice frequency band (300 to 3200 Hz). This final transmit filter was included to prevent possible excessive crosstalk into the plain old telephone service. The guidelines of the network spectrum management policy influence the need for these additional filters. The national network guidelines dominate the decisions regarding the spectral limits over which energy can be transmitted in the digital subscriber systems.

6.4.3 Timing Recovery Circuits

The recovery of symbol timing at the subscriber is essential to keep the local transmit clock in synchronism with the central office clock. The timing information is recovered from the received data after equalization. One of the standard techniques for the recovery of the clock is to have a high Q (the ratio of the

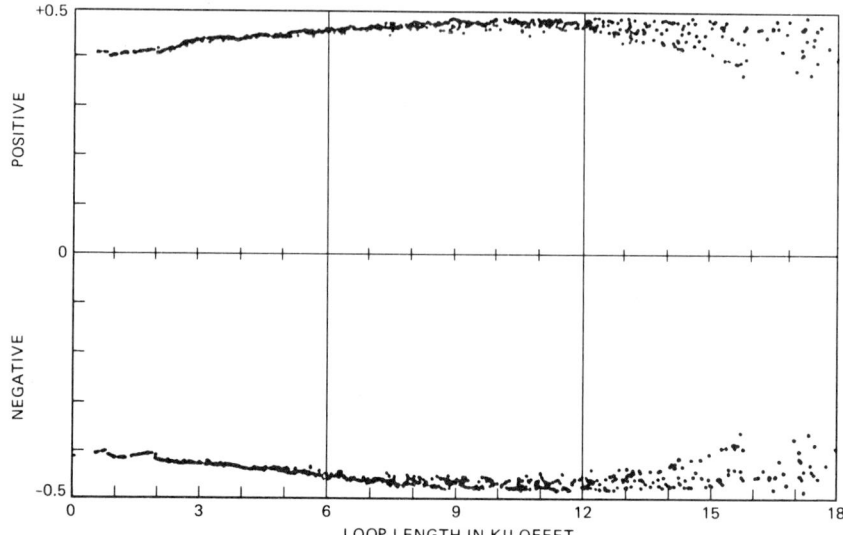

Reprinted with permission from the *AT&T Technical Journal*. Copyright 1982 AT&T.

Figure 6.28 Eye opening scatter plot for the 1973 survey loops at 126 kbit/s mode. (Effective rate is 56 kbit/s.)

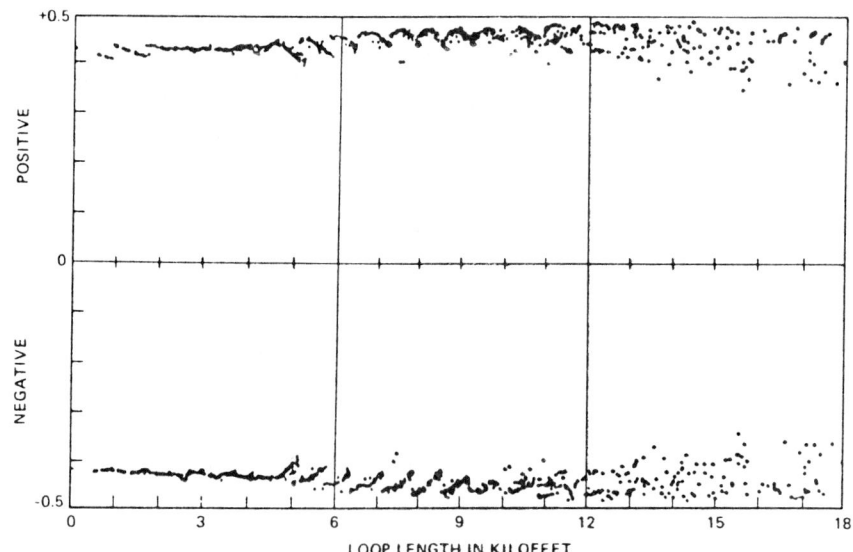

Reprinted with permission from the *AT&T Technical Journal*. Copyright 1982 AT&T.

Figure 6.29 Eye opening scatter plot for the 1973 survey loops at 162 kbit/s mode. (Effective rate is 72 kbit/s.)

The Component Optimization Procedures 321

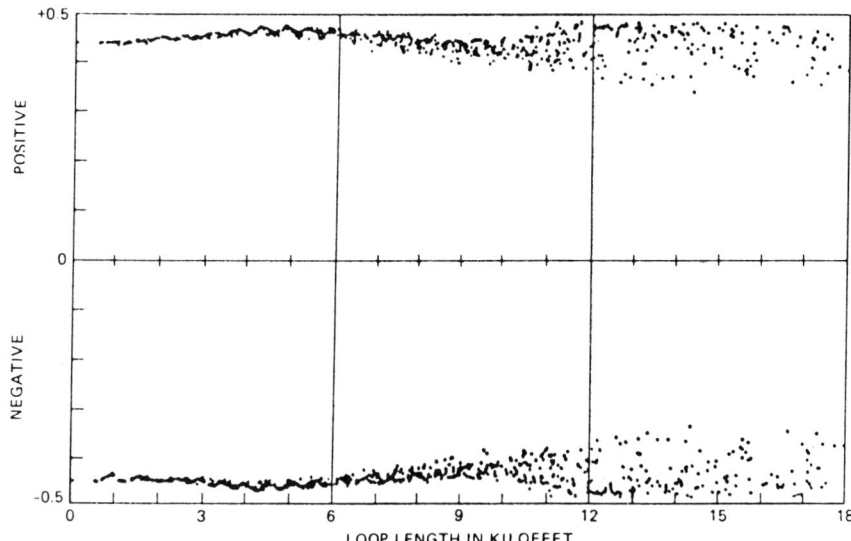

Reprinted with permission from the *AT&T Technical Journal*. Copyright 1982 AT&T.

Figure 6.30 Eye opening scatter plot for the 1973 survey loops at 216 kbit/s TCM mode. (Effective rate is 96 kbit/s.)

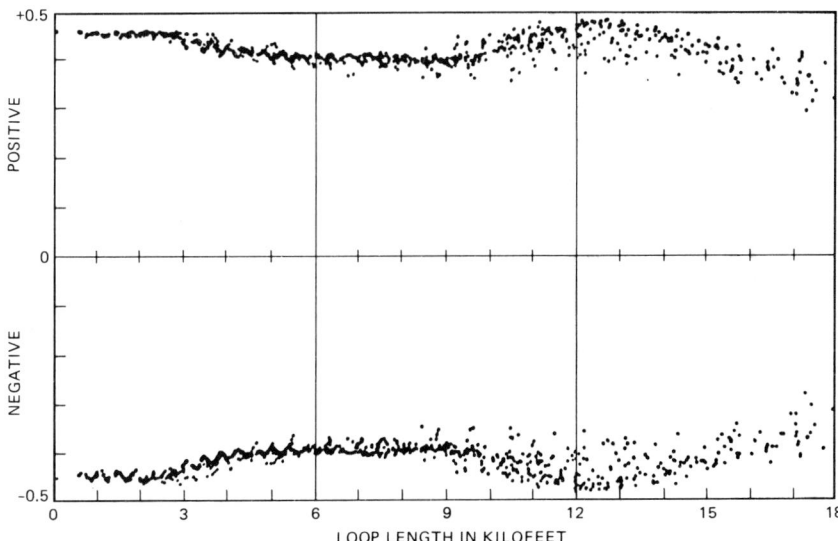

Reprinted with permission from the *AT&T Technical Journal*. Copyright 1982 AT&T.

Figure 6.31 Eye opening scatter plot for the 1973 survey loops at 324 kbit/s. (Effective rate is 144 kbit/s.)

reactance to the resistance) tank circuit resonant with the central office clock and to use it as a filter and thus recover the original timing at which the central office data was transmitted. Phase locked loops with high Q circuits are used for this purpose. Generally, the clock recovery circuit generally has a detector and/or clipper to exaggerate the harmonics before they are injected into the phase locked loops. The output of the loop is now at the clock frequency and the zero crossing of the output wave give the receiver clock.

In the design of components, it is essential to know that the receiver clock is going to maintain in synchronism under the worst of the conditions. One such condition exists when the loops become very long and the received signal is highly attenuated and distorted. The second condition occurs when the concentration of binary data diminishes and there are no zero crossing to excite the phase locked loop. This problem is handled by scrambling the transmitted data and descrambling it at the receiver. Thus, the presence of binary data is ascertained. The chance of not receiving any data at the receiver occurs only if the data stream to be sent bears a definite logical relation with the scrambling word. This is seen as an extremely remote possibility. The third condition occurs if the Q of the tank circuit is too low. Optimal design of the receiver calls for the right choice of Q.

The extent to which the zero crossing of the output of the phase locked loop wanders from a fixed instant of time is called the phase jitter. It is desirable to maintain the jitter as low as possible. The reason for holding the jitter to a minimum is to hold the scanning instant (at which the threshold detector operates) as close to the peak of the received signal as possible. The scanning instant is thus derived from the zero crossings of the output of the phase locked loop. The threshold for detection is derived as half the average height of the received data stream.

The simulation software (section 6.3.2) can be programmed to follow the functioning of the timing recovery circuit accurately. The results are shown in Figures 6.32 to 6.36. In Figure 6.32, the eye diagram after equalization is depicted. For the accurate recovery of the digital information, the scanning instant has to fall in the open region of the eye diagram. The extent of the variation of the scanning instant cannot be so large that the threshold detector (which senses the received data as $+1, 0, -1$) is likely to scan during the closed duration in the diagram depicted in the first one third of the pulse period. The pulse period in Figure 6.32 is the width of the figure.

The simulation for the functioning of the timing recovery circuit is done for the TCM mode of data transmission at 144 ($=2.25 \times 64$) kbit/s to yield an information rate of 64 kbit/s. The purpose of this particular study is to identify that (a) the subscriber clock will not fall out of synchronism with the central office clock during the transmission of the burst from the subscriber, and (b) the phase jitter of the recovered clock is not so excessive during the reception of data at the subscriber that an error in the data recovery is possible.

The Component Optimization Procedures

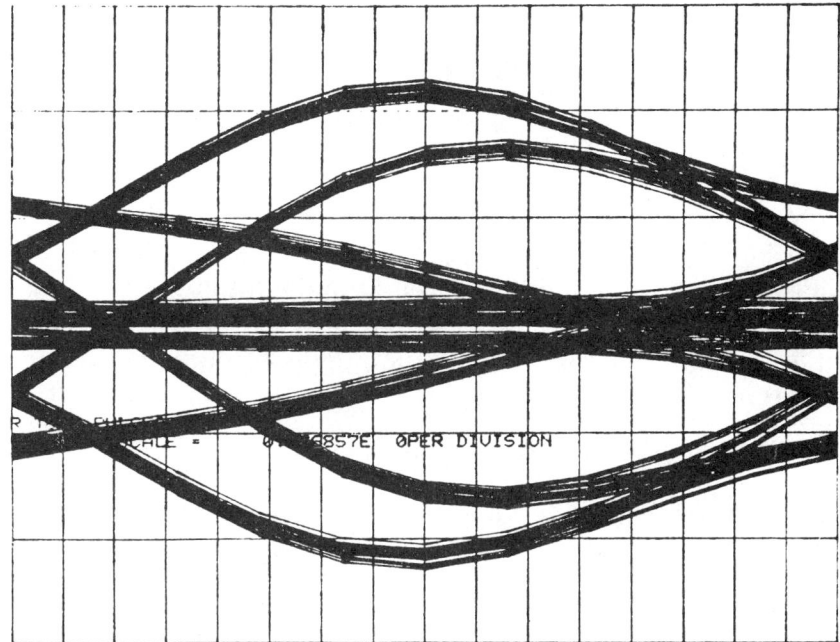

Figure 6.32 Eye diagram generated at the output of the equalizer for a typical loop.

The function of the timing recovery circuit during the transmit duration at the subscriber end is depicted in Figure 6.33. The end of the received burst period is depicted at the start of the first line (AB) of the sine wave at the top of this figure. The end of the burst occurs during the middle of the third line (EF). The gradual decay of the amplitude of the output of the phase locked loop is evident. However, stability of the recovered clock depends upon the constancy of the zero crossings of the sine wave. To depict this constancy a super position of the output from the phase locked loop is shown in Figure 6.34 and in Figure 6.35. It can be seen that the interval over which the sine wave crosses the zero voltage level is very narrow (about one two hundredth of the pulse period) and thus the jitter of the recovered clock is about 0.9 (180.0/200.0) degrees during the transmit time from the subscriber. Hence, the requirement (a) above is satisfied by the timing recovery circuit.

The next requirement calls for a stable clock during the reception of data. The jitter during this phase of operation is depicted by the width of the vertical line near the center of the eye in Figure 6.36. The width of this line depicts the extent of the variation in the zero crossing of the sine wave during the data reception phase. The scanning instant is generally chosen to be coincident with the peak of the eye height. Figure 6.36 depicts the location of the scanning instant and its jitter, thus ascertaining that the requirement (b) is satisfied.

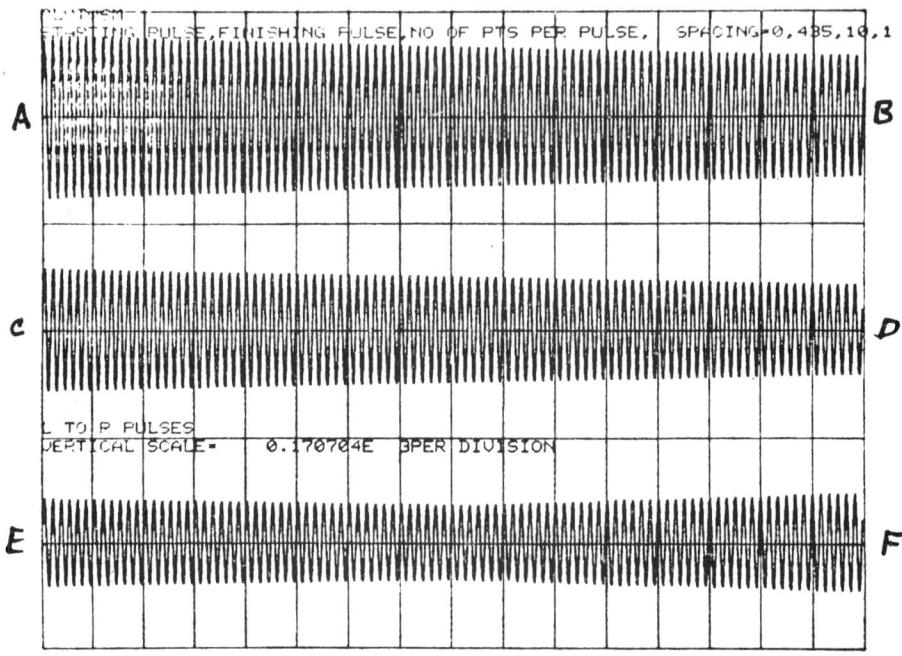

Figure 6.33 Voltage output from the Phase Locked Loop (PLL) during the bursting period of the TCM transmission.

Even though the threshold detector will perform satisfactorily under these conditions, one additional result becomes evident. The maximum eye opening may not be coincident with the instant at which the maximum eye height is detected. This is especially the case for long loops with bridged taps where the eye diagram is likely to be most distorted. From our early study a phase lag of about 10–15 degrees significantly enhanced the effective eye opening at the scanning instant for the worst loops, but did not significantly degrade the eye opening for the average and good loops. In the circuit switched digital capability offered by certain operating companies in the United States (see Section 5.2.1) environment, the eye openings were sufficient enough not to warrant the need for this 10–15 degree phase lag of the scanning instant, but this feature may be exploited to reduce the possible data transmission errors.

6.5 GLOBAL OVERVIEW OF THE STATUS OF THE DSS

The entire DSS consists of many individual components with predesigned characteristics. Most systems to be used in a particular subscriber loop environment

Global Overview of the Status of the DDS

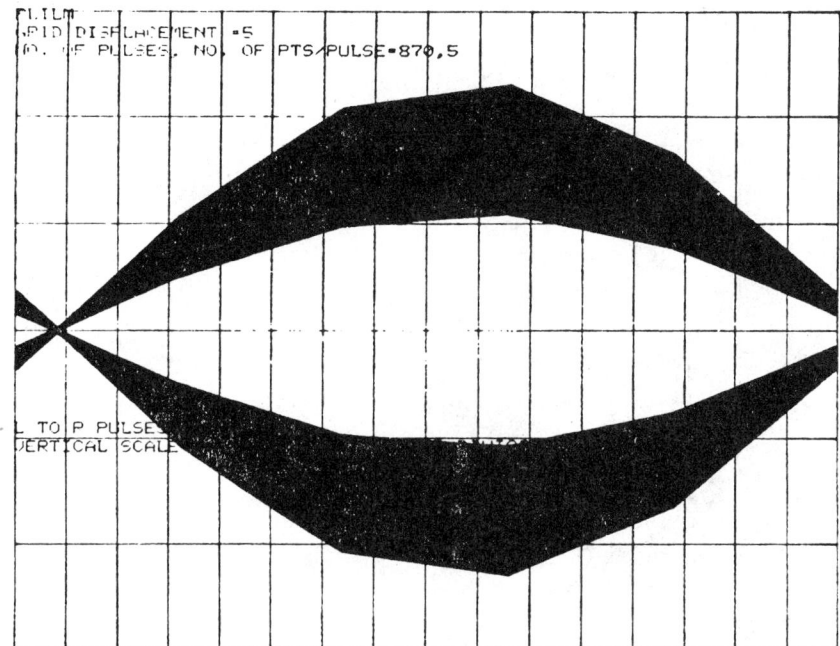

Figure 6.34 Output from the PLL during the received period of the TCM transmission. This output is in the form of an eye diagram to indicate the phase jitter at the zero crossings.

need some optimization. It is not evident at this stage whether one set of components will satisfy all loop environments. Hence, different national and network designers are experimenting with the different rates, modes, codes, component designs, and alternatives in context of their own environments in order to establish the suitability of the design parameters for their local conditions. In this section, we summarize the findings in as we find it in the published data. It is the object that these findings in different networks may facilitate a unified approach to the evolving Global ISDN of the future.

6.5.1 TCM Systems

In the United States, the CSDC (the Circuit Switched Digital Capability, a forerunner to the total ISDN capability) focuses on delivering 56 kbit/s and 1.33 kbit/s to the customer. Here most of the technical solutions have been achieved for the rather severe subscriber loop conditions (average of 2.3 gauge discontinuities/loop; 1.64 bridged tap/loop; 87 percent loops bridged taped; 23.2 dB loss at 72 kHz, 29.7 dB at 162 kHz; $(29 - j\,0.9)$ to $(188 - j\,112)$ ohms at 72

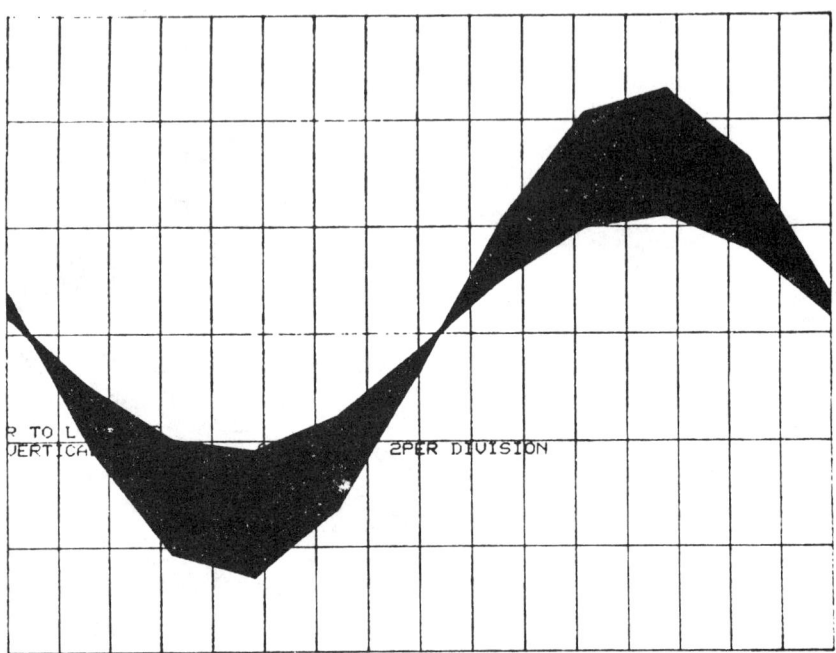

Figure 6.35 The zero crossing width from the negative to positive transition of the PLL output indicates the phase jitter.

kHz; (13.3 + j 88) to (242 − j 108) ohms at 162 kHz subscriber side image impedance variations). These solutions, even though satisfactory for the CSDC objectives, may pose severe questions for the eventual ISDN capability at 144 kbit/s. An additional and rather exhaustive study for the higher bit rate may well be warranted. The ESS, stored program offices are less prone to severe impulse noise events as they exist in the crossbar (XBAR) offices. Hence, the ISDN range has two limitations one for XBAR at a loss of 40 dB where the frequency and intensity of the impulse noise events can cause the bit error rate (BER) to exceed the permitted value. The second (higher) limitation for the quieter, ESS offices is at 45 dB range at 72 kHz because of the crosstalk to and from other systems including identical systems. The penetration is predicted at 88 percent in XBAR and 93 percent in the ESS central office environments. The burst rate is at 144 kbit/s sent a burst frequency of 333.3 Hz. Burst synchronization eliminates the self NEXT range limitation. Total integration of the system as one HIC (Hybrid Integrated Circuit) package is achieved (4).

In the two Japanese systems (8, 9); one system at 88 kbit/s (64 information + 16 information + 8 signalling) and the other system at 68 kbit/s (64 information + 4 signalling); the loop conditions are less severe. All the design

Global Overview of the Status of the DDS 327

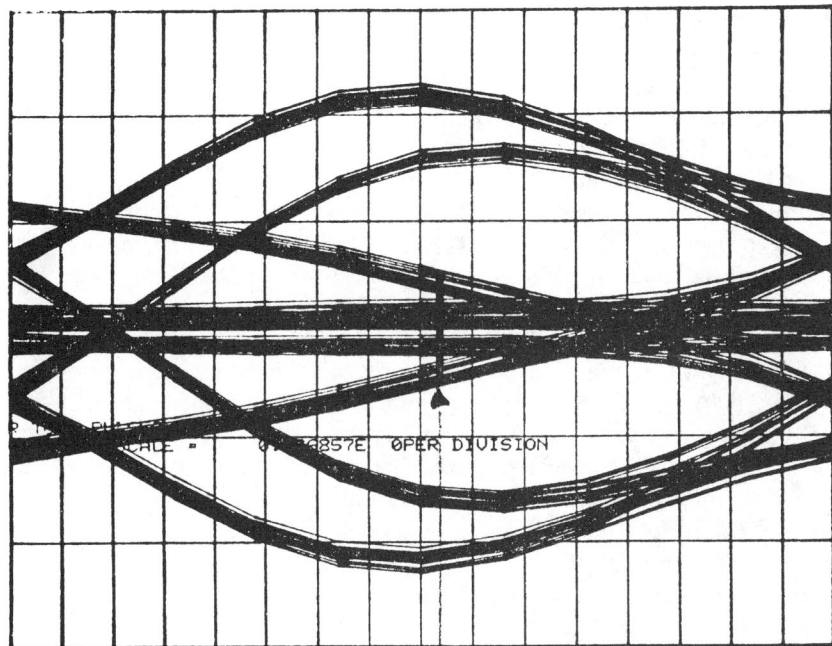

Figure 6.36 The extent of the phase jitter at the scanning instant when the eye height reaches the maximum value. The vertical line width denotes the uncertainty due to the phase jitter. It can be seen that the Timing Recovery Circuit function at the subscriber does not cause errors for this particular loop.

requirements are adequately handled by one CMOS, IC and one supporting TTL, MSI circuits. The 7 km range objective is seen to be satisfied by achieving a maximum equalizer gaining about 42 dB at 100 kHz. The penetration is 99 percent and is consistent with FEXT and impulse noise (Yokohamo Central Office) limitations. The first TCM systems uses a 233 bit burst at 200 kbit/s with a burst frequency of 400 Hz. The second system uses a 145 bit burst at 160 kbit/s with a burst frequency of 500 Hz. By and large, the TCM systems have a firm foothold in the Japanese telephone network. The subscriber loop environment in this network is conducive to the effective use of the inherent features of the TCM technology. The devices and the interface boards are becoming more directed towards the TCM systems. Having command of the high quality semiconductor technology, the Japanese vendors of the ISDN integrated circuit chips and interface boards can sway the favor toward these TCM systems by creating a desirable economic climate for the eventual deployment of the TCM systems in a considerable fraction of the ISDN markets.

The Canadian study (10) encompasses three possible rates at 72, 144, and 288 kbit/s information rate, even though the primary emphasis at this stage appears to be at 72 and 144 kbit/s. The subscriber loop conditions are akin to the predivestiture Bell System loop environment. The 45 dB limit at 80 kHz corresponds to 4.2 km of 26 AWG cable and offers about 92.5 percent penetration. However, the impulse noise conditions in one central office suggests a penetration of about 85 percent for a bit error rate (BER) of 10.0E-7.

In the proposed Italian system (11), (the 80 kbit/s, 64, Information + 8, signalling + 8, balancing, etc.) the BER objective of 10.0E-9 imposes a range of 3.15 km with the WAL II codes and 3.75 km with the CMI code. These distances correspond to about 95 percent and 98 percent of the Italian loop plant. Extensive investigation in light of the less severe bridged tap environment have still to be published. Laboratory tests indicate good eye openings for loops 2 and 3 km long without any equalization. The possibility of avoiding all equalization at the receiver has been proposed but not verified and documented. Two systems reported have burst rates of 256 kbit/s and 240 kbit/s with a burst frequency of 8 kHz. The burst length is 10 bits with eight information bits and one bit for synchronization and signalling each.

The British Post Office results (12) at 80 kbit/s (64 information + 8 or 16 information) information rate offer about 98 percent penetration with 37 dB loss limit at 100 kHz, even though the burst rate is 256 kbit/s, bursting at rate 4 kHz with 20 bits in each burst. The line code is AMI with decision feed back equalization and the transmit power level has been specified at 10 dBm.

In France, the PRANA network (13) offers 80 kbit/s and provides as one 64 (56 + 8) kbit/s telephone or data user channel (per CCITT Rec. 6.711) or telemetry service (per CCITT Rec. X.1) and a 16 kbit/s secondary channel typically used for subscriber-originated signalling and datagrams, etc. The system offers a 176 kbit/s with a burst frequency of 1 kHz and 176 bits/burst. A practical range of 5 km is envisaged penetrating about 88 percent of the French loop plant. Limited systems tests have yielded BER better than $5 \times 10.0E-6$ with loop having a loss of about 40 dB and a marked improvement for loops with reduced loss.

The Swedish tests (14) at 80 kbit/s use 88 bit bursts repeating every one ms at a burst rate of 200 kbit/s. A design objective of BER $< 10^{-7}$ is selected and 100 percent penetration is achieved with loops up to 4 km. (The adaptive hybrid system also meets this requirement and will be discussed in the next section.) The overall penetration is about 97 percent at BER $< 10^{-7}$ and about 99 percent at 10^{-5}. Here the system range appears to be impulse noise limited. An approximate range of 4 km is estimated but not substantiated in the reference. With this restriction the BER measured is of the order of 4×10^{-9} and increase to about 4×10^{-7} at about 7 km range reaching about 10^{-5} at 8.5 km. In the field test with 74 tests an overall penetration of 89 percent was achieved because the mean value of the field test population is 3 km whereas the national mean length is about 1.7 km.

The German PTT (15) (DBP) tests at 80 kbit/s has one B channel at 64 kbit/s, B data channel at 8 kbit/s and D signalling and synchronization channel at 8 kbit/s. Two ten-bit bursts are used at a burst rate of 256 kbit/s. Intrapolation of the presented data indicates that if a 33 dB equalizer gain is specified then the system offers a range of 3.8 km of .4 mm (finest gauge) cable. With 92 percent of the loops being less than 4 km, one would expect that the penetration would be at least 90 percent of subscriber loops. With no bridged taps and fewer discontinuities, the design should provide for a better penetration with longer bursts and lower burst rates.

The Norwegian efforts (16) are divided into two segments. An earlier (1979 publication reporting a project initiated in 1974) design reports a TCM system. Standard CMOS MSI designs are integrated with the adaptive equalizer pulse shaper, peak detector comparator and integrator to obtain a thick film circuit (23 × 35 mm) type of design. The penetration data of 93 percent with loop length of under 4 km with is also reported with the AMI code. These results are from the Norwegian Institute of Technology and the Regional Telecommunication Administration. The second effort (November 1981) from Electronics Research Laboratory in Trondheim and A/S Elektrisk Bureau, Nesbru in Norway report echo canceller system and scans the echo cancellation algorithms. A cost-effective implementation using a table loop up and a FIR filter for the generation of echo compensation signal is (also from A/S Elektrisk Bureau) reported in 1983.

6.5.2 Hybrid Echo Canceller (AEC) Systems

The line rate and the bit rate (i.e., the information plus overhead rate and the signal rate) are the same here. The echo canceller's ability to attenuate the reflected signal plays a dominate role in this system. A range of such an attenuation is possible with a quite easily achievable 30 dB lower limit. More recently the upper limit of 45–60 dB echo cancellation has become practical, making these hybrid system more and more attractive.

In the predivestiture Bell System the penetration of 64 kbit/s with no bridged taps in the loop plant is 91 percent with 30 dB cancellation. Reflection from the gauge discontinuities and the terminating impedance mismatch effects cause the reflections thus corrupting the received signal. With the bridged taps intact the penetration is reduced to about 68.8 percent. At 144 kbit/s the corresponding penetrations are 91.6 percent (no bridged taps) and 53.5 percent (with bridged taps). Simulation results with 35, 40, and 45 dB of cancellation are not published. Longer loops have a disproportionately large number of failures.

At the University of California (17), at Berkeley a new concept of echo cancellation has been developed to handle some of the possible sources of non linear distortion. The purpose of this design is to reduce the extent of memory requirement. The authors claim partial success and specify that some of the nonlinearities that can be handled by the new algorithm. An NMOS IC chip with 7 nm design rules has been constructed on a 3 mm × 3 mm chip. The experimental broadband modem using a 24 bit (2901 bit slice) microprocessor and an external 12 bit D/A converter having adequate linearity was tested in a 300 line local office. As specified by the authors the "relative ideal nature of the environment" is "not representative of the effects of crosstalk and impulse noise in a more severe environment." Such a severe environment portrays the loss limit of 40 dB at 72 kHz in XBAR office (impulse noise) and 45 dB at 72 kHz at 1 A ESS office (crosstalk). The maximum echo cancellation from these custom chips is about 40 dB whereas about 36–37 dB achieved at AT&T Bell Laboratories with discrete components under ideal conditions at 144 kbit/s. The University of California echo canceller chip has employed two eight tap programmable transversal filters. They operate in an interleaved fashion at twice the data rate. The system has been tested at 80 kbit/s and the circuit noise appears to be the limiting constraint (as it was at the AT&T Bell Laboratories) even though impulse noise may become the constraint in a realistic environment. Lines with loss up to 40 dB (at 40 kHz) have BER of 10^{-8} and tends to become 10^{-4} at 44 dB loss even in the ideal environment under which the tests were performed. These results are consistent with the simulation results from the AT&T Bell Telephone Laboratories for the TCM system with 40 and 45 dB loop loss at 72 kHz (corresponding to 144 kbit/s, burst rate). The convergence times are not available.

In Italy an experimental hybrid system at 88 kbit/s has been tried for a 4 km, .4mm cable using 6 bit D/A converters, 14 taps, and 16 bit representation. The details of the balancing network (matching impedance) are omitted and a theoretical limit of 66 dB of echo cancellation is claimed with convergence time of 40 ms. Since the only test was done on a uniform gauge wire without any bridged taps, very limited conclusion can be drawn from this experiment.

The British experiments (18) with the hybrid echo-canceller are limited to a few field tests. Only isolated examples of BER of about 5×10^{-8} with a 9 mV (corresponding to 40 dB loss) received signal have been recorded. Since no details of the impulse noise are available, it appears impossible to attach any significance to this isolated measurement. Additional tests were being carried out in the city of London and other rural, city, and industrial areas. Impulse noise appears to be the main restriction rather than crosstalk.

In France, the CIT Alcatel laboratory (19) is developing an echo canceller for the hybrid system at 80 kbit/s. The objective is to provide 10 second convergence time. An experimental discrete component system has been constructed. Test results presented constitute one single test result.

In Sweden, the hybrid echo cancellation system (20) is based on an adaptive transversal filter, which uses the table look up technique instead of computing the entire reflected wave shape. The length of the filter is 4 bits, the bit rate is 80 kbit/s and the code used is antipodal diphase. On the 74 selected loops, the design objective for the BER of less than 10^{-7} is met with a 100 percent penetration. The overall penetration is 98 percent with BER less than 10^{-7} and about 99 percent with BER less than 10^{-5}. The measured BER is 0 for loops up to 6.5 km. It starts to become quite erratic (greater than 10^{-4} at times) at about 7 km and falls out of synchronization at about 8 km (compare TCM results in Section 6.5.1). The overall penetration of the field trial suggests about 92 percent penetration, the major limitation being impulse noise in central offices. The average and maximum activation times are 1.35 to 2.3 seconds for start up and can increase to 8 seconds to 10 seconds for reconnect.

In Germany, an echo cancellation method (21) at 80 kbit/s uses the AMI code and offers a 33 dB, 4.9 km system range. By intrapolation it appears that this will cover about 98 percent of the loops. The effect of crosstalk and impulse noise was not found to be detrimental for loops with less than 45 dB of loss for the experimental measurement conditions. The system uses 16 coefficients for the filter with two scans per bit. Twelve-bit accuracy is used for the coefficients and the D/A conversion. The system is claimed to have achieved 50 dB echo suppression. However, the effect of impulse noise beyond the 100 μs duration (8 equivalent bits times 12.5 μs at 80 kbit/s) is not described, even though the lengths of some impulse noise event are expected to be between 20 to 200 μs. The convergence time for the echo canceller is not reported.

In Norway (22), a table look-up FIR filter (see Swedish echo canceller system) is used for the generation of the echo cancellation signal. The look-up is built around a memory compensation principle and can tolerate nonlinearities inherent in tapped delay line types of implementation. Easier integration is claimed to an advantage, thus reducing the cost of the device. A combination of received bits is used to access different parts of stored memory from where a prestored echo cancellation signal is retrieved and subtracted out. Such a straightforward procedure would employ memory where size is proportional to 2^N, and where N depends on the number of bits over which the echo cancellation needs to be extended. The transversal filling technique used in this paper needs memory which increases proportionally to N. Further, for 80–160 kbit/s the memory requirement is 1k–2k RAM. But the value of N is claimed to be "not too large" without assigning a number to it. Two different types of CMOS integration have been attempted. The CMOS Gate array integration is only partial at best. Components not integrated are D/A converter, RAM, line drives, equalizers, and the analogue hybrid. The chip area for the remaining components to 34 mm², with about 5000 transistors. The power consumption is 10–100 mw for the chip package 64 pin dual in-line package. The second CMOS standard-cell integration bit includes a 1 kbit RAM (7600 transistors) on the chip. The chip area is 36 mm² with 13500 transistors. The power consumption is 10 mw at a 5 volt supply level at 32 kbit/s for the chip packaged in 40 pin dual in-line package. The system has been tested on one ("the longest") 3.2 km, .6 mm cable at 160 kbit/s. The BER recorded was better than 10^{-8} in 41 days, even though the range is estimated at 6 km for .6 mm cable because of NEXT limitations and about 4 km because of noise in the local network. Additionally, three critical pieces of data are missing from the paper: (1) the value of N used in the device; (2) the convergence times; and (3) the performance of the system with composite, and preferably, loops with bridged taps.

6.5.3 Facts in Favor of the TCM System

The TCM systems were favored during the early days of CSDC development (23). These recommendations, based upon the extensive simulation studies and the experimented results both at AT&T-Bell Telephone Laboratories, have been instrumental in the TCM Systems now offered for CSDC (the forerunner to the ISDN) facility through the operating companies. For reasons not well published, the Japanese offer for the ISDN system also use the TCM systems. Both use burst lengths considerably longer than those experimented with in European countries. The shortest burst used is 10 bits long in Italy. Most of the countries have some TCM experience even though no country has denied that the AEC system does not offer the usual advantages. (See Section 6.5.4.)

However, there are five additional considerations. First, when the line losses are plotted against frequency, the higher losses at about 2.25 times (i.e., the TCM rate), the line rate (i.e., the AEC rate), are about 1.28 times the losses at the lower frequency, rather than 1.5 times the loss at the lower frequency. The $\sqrt{f}$ simplification suggests a multiplier of 1.5 ($1.5 = \sqrt{2.25}$), since the bursting line bit rate is 2.25 times the AEC line bit rate. The added losses do cause a 16 percent reduced penetration of the DSL in the loop plant, if a loss limit of 20 dB at 72 kHz, and at 32 kHz in the American loop plant is specified. The frequencies of 72 kHz and 32 kHz correspond to a bit rate of 144 kbit/s in the TCM mode with the bipolar code, and a bit rate of 64 kbit/s in a AEC mode with the bipolar code. If a higher line loss of 40 dB is specified in the same loop environment for the same line rates, then the DSL penetration is reduced by 9 percent. This difference arises because of the shape of the loss distribution of loops in any typical national loop survey loss.

One such typical distribution for loops in a typical loop survey at 32 kHz (for 64 kbit/s and at 72 kHz for 144 (2.25 × 64) kbit/s, is shown in Figure 6.37. However, both curves exhibit a rather steep rise in losses (between 90 to 100 percent penetration). The line losses for these lines is rather high. For such lines, impulse noise and crosstalk limitations dominate. This has a profound effect on the facts that favor the TCM system. Reiterating the prior discussion, it can be specified that if the crosstalk and impulse noise limitation were limiting the line losses to about 20 dB, about 16 percent additional penetration could be achieved by using the AEC system, whereas if the limitation is 40 dB (as it is most networks), then the additional penetration is only 8 or 9 percent. At 45 dB line loss limitation caused by crosstalk and impulse noise considerations, the margin tends to become even smaller (about 3 percent).

Second, sufficient experience for the equalizers has been generated to ascertain satisfactory performance between 0–60 dB of the line loss. Equalization has not been the threat for eye opening for the TCM Systems even though at such losses, crosstalk and impulse noise become disruptive.

Third, the impulse noise and crosstalk are central office and network dependent and arise because of imperfection (or realities) of the telephone network. This constraint cannot be easily altered and the burst synchronization technique partially eliminates the self NEXT even though FEXT and crosstalk into other systems is exaggerated by the higher bandwidth of the TCM system. Systems that suffer the most are the analogue carriers. These systems do not constitute a particular source of concern because in Australia and in the European countries the analogue carriers (such as those in the American environment) are less prevalent.

Fourth, even though the TCM mode demands a higher bandwidth, either terminal is in the receive mode for only about 44.4 (100/2.25) percent of the

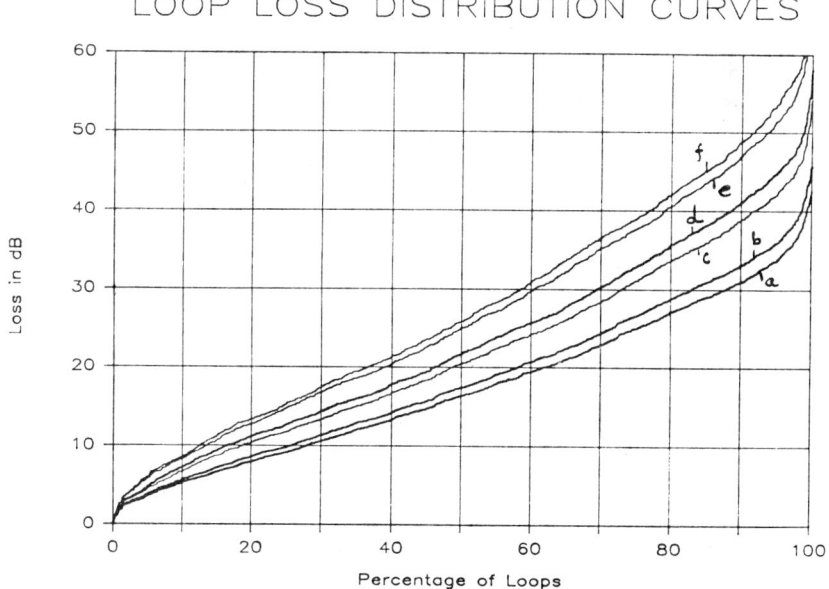

Figure 6.37 Loss distribution curves at (a) 32 kHz, (b) 40 kHz, (c) 72 kHz, (d) 90 kHz, (e) 144 kHz, and (f) 162 kHz for a typical loop survey.

time and even large impulse noise events arising during the transmit time do cause any errors in the system. Hence, even though the bandwidth is increased (by about 2.25 times), the duration of vulnerability is decreased (by about 55 percent). Therefore, the overall increase is far from profound. A strictly linear trade-off between the enhanced bandwidth and reduced duration is inaccurate. The spectral components of the typical central office noise events and the transfer functions of the particular class of equalizers influence the extent of the benefit accruing by adopting the AEC system over the TCM system.

Fifth, when the market projections are drawn for the possible ISDN users, there is a concentration of such users and small businesses located a short distance from the central office. The additional loss resulting from increased burst rate can be easily handled by improved designs for the equalizers. The added exposure to crosstalk and impulse noise because of the increased bandwidth is inconsequential at shorter distances. When there is indeed a cluster of ISDN users in a remote community, then the Carrier Serving Area (CSA) concept starts to gain an economic desirability, and high capacity digital carriers are used to serve these clusters. The subscriber loop length is still limited. Two consequences of this lower loss are: (a) lower loss in the TCM systems which enhances the signal-to-noise ratio, the noise being the crosstalk and central office impulse, and (b) lower loss in the AEC system which enhances the signal-to-noise ratio, the noise being the echos of the transmitted signal. The gain in either system cannot be predicted unless an exhaustive study of both the systems in CSA environments is undertaken.

To summarize, TCM employs a mature technology and devices with noncritical design requirements. The circuitry is simpler and less expensive. The start-up time is shorter. The burst synchronization has to be managed well if the advantage of the reduced near end crosstalk has to be realized. It does have a higher bandwidth requirement, implying that the noise (both the crosstalk and impulse noise) immunity is lower than in the AEC systems. Further, if the spectral power input into the loop plant is restricted by crosstalk into other systems, then the loss range for the subscribers may be reduced.

In spite of these attractive features of the TCM mode, the hybrid duplex system or the Adaptive Echo Canceller (AEC) systems have taken a strong foothold as the chosen mode for ISDN transmission. The basic reasoning behind this is that the procedures necessary and critical to the functioning of the AEC mode with the required dependability can be (and have been) achieved by an appropriate design of the VLSI components at the subscriber and the central office ends of the subscriber line.

6.5.4 Facts in Favor of the Hybrid Duplex (AEC) System

The line rate in the AEC system is equal to the information rate, plus the overhead. This fact alone reduces the bandwidth requirement considerably over the TCM bandwidth. The extent of the reduction depends upon the bursting block length, the guard space, and the maximum length of the loop for the TCM systems. The reduced bandwidth has profound ramifications on the noise (crosstalk and impulse) immunity, making the hybrid system more desirable.

The rate changing and buffering is also eliminated for hybrid systems. These circuits, though simple, are still necessary in the TCM systems. The additional delay involved because of the rate-change and buffering may sometimes be a concern in the echo return delay, if the DSS is being used for digitally encoded speech. For data applications the delay does not appear to cause any problems in most applications.

The progress in the development of echo canceller semiconductor devices has to be considered in complete detail. The major concerns over the satisfactory and prolonged operation of these echo cancellers is dwindling as the major vendors are publishing the results of the field trials. The caution necessary here is the actual use of the devices in the central office environment where impulse noise and crosstalk both pose a threat to the echo canceller function. The central office environments in Europe are considerably different from those in United States and only the #5ESS is quieter than other central offices (cross-bar and #1ESS). The rapid trend towards these completely digital offices is a definite point in favor of the DSS and ISDN, rather than a point in favor of the AEC systems. Only the actual field trial results of these systems will completely fulfill the 60 dB echo cancellers promised by the vendors.

The majority of customers who subscribe to the ISDN services are expected to be close to the central office or a data distribution center served by a high capacity data link. Shorter loops have a stronger received signal in comparison to the near end echo. If the proposed CSA data centers exhibit noise characteristics (over the enhanced bandwidth) that are far inferior to those of the #5ESS central offices now in use, then the AEC systems may well prove to be the most logical choice. As the CSA concept gains popularity in the next decade, the impact will be felt on both the TCM and the AEC systems. The hybrid system will become attractive when the echo canceller algorithms and the chip fabrication techniques become economically viable, the field trials prove to be trustworthy under all the environmental conditions, and the data distribution center's noise characteristics exhibit enhanced interference with the increased bandwidth. The present thinking appears to be very optimistic for the AEC systems, even though the noise studies in the data centers have not been initiated, except by a few operating companies in the U.S.

Network topology considerations play an important part here. Although the general consensus is towards designing the DSS for ISDN with the hybrid duplex or the AEC mode of transmission in the United States, the problem appears too complex to unconditionally favor the hybrid system for *all* the national and for all local loop environments.

6.6 SUMMARY

Digital Subscriber Systems are experiencing an explosive interest in the context of the ISDN applications. The main design conceptual breakthroughs, algorithms, implementations, VLSI circuits, and the channel units will be directly influenced by the considerations presented in Chapter 5 and 6.

In the United States, the TCM Systems are completely superseded by the AEC systems. This view is based upon the present status of the VLSI technology, the crosstalk considerations, and the well-established trend in the Western countries toward a digital telephone network. For other national and local networks the conditions may not be identical, thus leaving a suitable niche for the TCM systems to exist. Another consideration is the possible use of both the systems (TCM and AEC) to coexist, even in the mature telephone networks. The possibility of using the cheaper TCM system for about 70 percent of the loops and using the AEC systems for the loops with harsher conditions also deserves detailed consideration.

The DSS component and system optimization has been expedited with the introduction of the CSDC services at 56 kbit/s. Still, major issues remain. The newer AEC, DSS concepts and codes offer similar advantages for the TCM systems. The extent of the marginal gain by importing the concepts leading to the choice of the newer codes, the decision feedback equalizers, the digital signal processing, the newer digital filtering techniques, etc., still remains unexplored.

The major thrust of the recent ISDN research and development has become focused on the AEC techniques for the (2B&D) or 160 kbit/s rate with 2B1Q code. The echo cancellation algorithms and the decision feedback equalization techniques are being studied independently and interdependently. The digital filtering with the possible $(1-D)$ filter at the transmitter and a $(1+D)$ filter at the receiver is also under close scrutiny. The newer codes, especially the block codes, modified duobinary codes, and other classes of partial response codes with digital and analogue filtering to enhance the signal amplification in the selected bandwidth are being reviewed in the current and the future central office environments. The research topics appear numerous, challenging.

6.7 REFERENCES

1. M. T. Manford, G. A. Nelson, and C. H. Sharpless, "Loop Plant: Digital Loop Carrier Systems," *Bell System Technical Journal*, Vol. 57, pp. 1129–1156, 1978.
2. U.S. Patent, 4,316,061 "Minimal Delay Rate Change Circuits," February 16, 1982.
3. S. V. Ahamed, "A General Class of Zero or Minimal Delay Fractional Rate Change Circuits," *Bell System Technical Journal*, Vol. 61, pp. 327–346, 1982.
4. B. S. Bosik and S. V. Kartalopoulos, "Time Compression Multiplexing System for a Circuit Switch Digital Capability," *IEEE Trans. on Comm.*, Vol. COM. 30, pp. 2046–2052, 1982.
5. S. V. Ahamed, P. P. Bohn, N. L. Gottfried, "A Tutorial on Two-Wire Digital Transmission in the Loop Plant," *IEEE Trans. on Comm.*, Vol. COM. 29, pp. 1554–1564, 1981.
6. S. V. Ahamed, "Simulation and Design Studies of the Digital Subscriber Lines," *Bell System Technical Journal*, Vol. 61, pp. 1003–1077, 1982.
7. S. V. Ahamed and R. R. P. Singh, "Physical and Transmission Characteristics of Subscriber Loops for ISDN Services," 1986 International Conference on Communications, Toronto, June 21–24, 1986.
8. N. Inoue, R. Komiya, and Y. Inoue, "Time-Shared Two Wire Digital Subscriber Transmission System and Its Application to the Digital Telephone Set," *IEEE Trans. on Comm.*, Vol. COM-29, pp. 1565–1572, 1981.
9. H. Ogiwara and Y. Terada, "Design Philosophy and Hardware Implementation for Digital Subscriber Loops," *IEEE Trans. on Comm.*, Vol. COM-30, pp. 2057–2065, 1982.
10. E. Aron, E. A. Munter, S. C. Patel, P. A. Roddick, and P. W. Willcock, "Customer Access System Design," *IEEE Trans. on Comm.*, Vol. COM-30, pp. 2143–2149, 1982.
11. A. Borsio, V. De Julio, V. Lazzari, R. Ravaglia, and A. Tofanelli, "A Comparison of Digital Subscriber Line Transmission System Employing Different Line Codes," *IEEE Trans. on Comm.*, Vol. COM-29, pp. 1581–1588. Also see, R. Fossati, S. Galio, V. Lazzari, R. Ravagalia, "A Remote Powered Digital Telephone Set: Problems, Performance and Prospects," *ISSLS*, pp. 60–63, 1980.
12. E. C. Vogel and C. G. Taylor, "British Telecom's Experience of Digital Transmission in the Local Network," *IEEE*, Catalog No. 1686-5/82, pp. 126–130, 1982.
13. J-A Le Guillou, F. Marcel, and A. J. Schwartz, "PRANA at the Age of Four: Multiservice Loops Reach Out," *IEEE Trans. on Comm.*, Vol. COM-30, pp. 2185–2210, 1982.
14. J-O Anderson, B. Carlquist, and G. Nilsson, "A Field Trial with Three Methods of Digital Two-Wire Transmission," private communication.
15. D. Becker, L. Gasser, and F. Kaderali, "Digital Subscriber Loops; Concept, Realization and Field Experience of Digital Customer Access," *Proc. of ISS*, Montreal, 1981. Also see B. Aschrafi, P. Meschkat and K. Szechenyl, "Field Trial Results of a Comparison of Time Separation, Echo Compensation and Four-Wire Transmission on Digital Subscriber Loops," *IEEE*, Catalog No. CH1686-5/82, pp. 181–185, 1982.
16. J. Meyer, T. Roste, and R. Torbergsen, "A Digital Subscriber Set," *IEEE Trans. on Comm.*, Vol. COM-27, pp. 1096–1103, 1979.

17. O. Agazzi, D. A. Hodges, and D. G. Messerschmitt, "Large Scale Integration of Hybrid-Method Digital Subscriber Loops," *IEEE Trans. on Comm.*, Vol. COM-30, pp. 2095–2108, 1982. Also see O. Aggazi, D. G. Messerschmitt, and D. A. Hodges, "Nonlinear Echo Cancellation of Data Signals," *IEEE Trans. on Comm.*, Vol. COM-30, pp. 2421–2433, 1982.
18. E. C. Vogel and C. G. Taylor, "British Telecom's Experience of Digital Transmission in the Local Network," *IEEE*, Catalog No. 1686-5/82, pp. 126–130, 1982.
19. J-A Le Guillou, F. Marcel, and A. J. Schwartz, "PRANA at the Age of Four: Multiservice Loops Reach Out," *IEEE Trans. on Comm.*, Vol. COM-30, pp. 2185–2210, 1982.
20. J-O Anderson, B. Carlquist, G. Nilsson, private communication.
21. B. Aschrafi, P. Meschkat, and K. Szechenyl, "Field Trial Results of a Comparison of Time Separation, Echo Compensation and Four-Wire Transmission on Digital Subscriber Loops," *IEEE*, Catalog No. CH1686-5/82, pp. 181–185, 1982.
22. N. Holte and S. Stueflotten, "A New Digital Echo Canceler for Two-wire Subscriber Lines," *IEEE Trans. on Comm.*, Vol. COM-29, pp. 1573–1580, 1981. Also see S. Stueflotten, R. S. Olsen, and K. Ragnhildrod, "Cost-Effective Implementation and Application of Echo Cancellers in Duplex Two Wire Transmission," *Telecommunications '83*, Geneva.
23. B. S. Bosik, "The Case in Favor of Burst Mode Transmission for Digital Subscriber Loops," *Proc. of ISSLS*, pp. 26–30, 1980.

Volume II

INDEX

ABCD matrix, 241, 288
   loop matrix, 241
accuracy of clocks, 85
acknowledgement (ACK), 190
activity factor, 268
adaptation time, 282
adaptive decision feed back equalizer (DFE), 283
adaptive echo canceller (AEC) system, 221, 230, 271
   advantages, 334
   balancing network, 282
   echo cancellation circuit, 282
   hybrid, 281
   semiconductor chips, 271
adaptive differential pulse code modulation (ADPCM), 68
adaptive equalization, 52, 57
   IF amplitude compensation, 57
   time domain equalizer, 58
ADCCP, 130
ADPCM, 68
address, 117
   field, 131
addressing, 148
Advanced Communication System (ACS), 225
airline reservation, 150
algebraic decoding, 189
algorithms
   deterministic, 114
   flow control, 114
   least time delay, 136
   shortest path, 136
   stochastic, 114
aloha technique, 113
American
   loop plant, 240
   network, 252
   subscriber network loop statistics, 233

American Telephone and Telegraph (AT&T), 227
American wire gauge (AWG) #19, #22, #24, #26, 233
   primary constants, 254
Ameritech, 227
amplitude response in digital radio, 53
   frequency selective fade model, 53
analogue networks, 68
   architectures, 112
   carrier systems, 263
   exchanges, 69, 77
   FDM-FM radio systems, 1, 6, 11, 51, 59
   microwave radio, 1, 4
   radio link costs, 3
   radio links, 1, 41
   radio transmitter, 4
   telephone network, 216
antennas, 44
   gain, 44
   heights, 44
   size, 44
application layer, 119
architectural model, 118
ARPANET, 138
ARQ, see automatic-repeat-request
arrival rate, 82, 152
ASCII code, 123
atomic clocks, 86, 87
attenuation of loops, 250
Australia,
   telephone networks, 232
   loops, 251
   plant, 255
authorized spectrum, 6, 11,, 21
   mask, 21
automatic gain control (AGC), 283
automatic-repeat-request (ARQ) systems, 163, 190, 198
   go-back-N ARQ, 192

mixed-mode ARQ, 201
parity retransmission ARQ, 206
retransmission protocols, 191
stop and wait ARQ, 192
selective-repeat ARQ, 201
selective-repeat plus go-back-N
  (SR + GBN), 203
selective-repeat plus stutter
  (SR + ST), 203
SETRAN, 200
system reliability, 191
type I ARQ, 205
type II ARQ, 206
availability objective, 10, 44
  digital radio, 10, 44

balancing network for AEC system, 282
bandwidth
  digital radio, 7, 11, 19
  for an M-PSK signal, 19
  noise, 38
  Nyquist, 35
Barker codes, 108
BCH codes, 185
BDLC, 130
Bell Atlantic, 227
Bell Canada, 227
Bell Communications Research, 227
BellSouth, 227
binary synchronous communications
  (BSC) protocol, 125
bipolar (AMI) code, 249, 260, 275, 286
BISYNC protocol, 125
bit error probability, 35, 263, 270, 328
  objectives, 35, 263
bit error rate (BER), 35, 263, 270, 328
  Asia, 326
  Canada, 328
  digital radio, 35, 44
  digital subscriber loop systems, 263,
    270, 328
  Europe, 328
  Japan, 327
  USA, 325
bit-oriented protocols, 122, 130
bit stuffing, 131
bit synchronization, 130
block code, 167, 185
blocking probability, 81
Boltzmann's constant, 47

Bose-Chaudhuri-Hocquenghem (BCH)
  codes, 186
BPSK, 14
bridged tap, 232, 255, 280, 290, 299
  critical length, 300
  open-ended, 285
  quarter wave length, 249
  statistics, 234
British experiments, 330
  systems, 328
BSC protocol, 125, 192
buffers, 83
  overflow, 85
  underflow, 85
burst mode system, 280
burst transmission, 278
byte count protocols, 130
byte serial encoder, 180

cables, 231
  characteristic impedance, 254
  characterization, 277
  coaxial, 87
  data, 233
  discontinuities, 251
  pairs, 253, 281
  propagation constant, 254
call accepted packets, 141
call connected packets, 141
call request packets, 140
call time supervision, 73
calling rate, 82
Canada
  ISDN, 228
  loops, 239, 251, 255
  subscriber loop conditions, 328
canceller start up time, 283
carrier recovery systems, 23, 33
  synchronization, 33
carrier systems
  L4, 224
  T1, 223
carrier to noise ratio (C/N), 35
CCIR frequency plan, 6, 9
CCITT
  alphabet 5, 123
  objectives, 85
  recommendations, 92, 121, 140, 148,
    173

Index to Volume II 341

signalling system No. 7, 224
standards, 120
CCITT recommendations
   V.24, 121
   X.20, 121
   X.21, 121
   X.22, 121
   X.25, 140, 173
   X.75, 148
cellular networks, 61
central control, 76
centrally controlled network, 116
Cesium beam standard, 86
channel
   B, 220
   D, 220
   logical, 141
   units, 277
character-oriented protocols, 122
character synchronization, 130
check mode, 105, 106
circuit switched digital capability
     (CSDC), 223, 286, 325
   filtering, 321
   systems, 281
circuit switched networks, 223
circuit switching, 116, 117
class 4 partial response coding, 261
classification
   computer networks, 112
clear confirmation packets, 143
clear indication packets, 141
clear request packets, 141
clicks, 85
clock accuracy, 85
   crystal, 87
   distribution, 94
   extraction, 73
   frequency, 84
   frequency accuracy, 93
   maximum error, 93
   stability, 85
clock recovery, 276, 319
clock synchronisation, 322
clocks, 83
   atomic, 86
coaxial cable, 87
co-channel interference, 54
code
   ASCII, 123

BCH, 185
bipolar, 286
closure rule, 172
conversion, 115
cyclic, 175
D, 143
duobinary, 286
EBCDIC, 124
error correcting, 164
error detection, 173
4B3T, 275, 286
generator matrix, 169
Golay, 186
Hamming, 185
memory, 187
minimum distance, 174, 187
modified duobinary, 286
nonreturn to zero, 286
partial response, 286
Reed-Solomon, 186
selection, 272
systematic, 169, 177
translator, 89
tree, 187
trellis, 187
types, 185
weight distribution, 170
weight spectrum, 171
with memory, 187
word, 165, 167
word weight, 170
coding and modulation, 40
   combined with redundant signal sets,
     40, 187
   gain, 40, 187
coherent detector, 32, 39
combined modulation and coding, 40,
   187
common carrier interoffice switching
     (CCIS), 224
common control unit, 89
communication subnetwork, 116, 117,
   133
compatibility between systems, 260
computer aided design techniques, 285
computer communications networks,
   111
   classification, 112
   decentralized, 112
   structures, 115

computer networks, 111
concentrator, 68, 71, 112
   exchange unit, 71
   functions, 115
   subscribers' unit, 71
congestion control, 138
   bottleneck, 139
   deadlock, 139
   reassembly lockup, 139
connectors, 120
constant envelope signal, 13, 21, 28
contention, 125
contention control, 76, 125, 139
   central, 76
   end-to-end, 139
continuous phase frequency shift keying (CPFSK), 28
   minimum shift keying (MSK), 28
control
   centralized, 114
   characters, 127
   distributed, 114
   field, 131
   functions, 73
control units, 76
convergence process
   echo canceller, 283
convolutional codes, 40, 185
   decoding, 189
co-phase diversity combiner, 56
correction of erasures, 166
correlation detector, 32, 35, 55
cosine roll-off filtering, 19
cost, 3, 77
   analogue radio link, 3
   digital radio, 3
critical length bridged tap, 300
cross-polarization discrimination, 9, 51
crosspoints, 77
crosstalk, 253, 258, 321, 326
   interference, 258
   considerations, 286
   loss, 260
   loss advantage, 269
   noise power, 260
   power spectral density, 260
CSA data centers, 335
CSDC (Circuit Switched Digital Capability), 226
cyclic codes, 126, 175,
   error detection, 183

   polynomial representation, 175
   shortened, 183
cyclic redundancy checking (CRC), 126
crystal clock, 87

data above video (DAVID), 2, 60
data above voice (DAV), 2, 60
data bases, 115, 288
   accessing, 115
data-circuit-terminating equipment (DCE), 120
data communications, 111
data communication network, 111
   design, 150
   flow control, 149
data in voice (DIV), 2, 60
data link control (DLC), 130, 149
   protocols, 130
data link layer, 119
data networks, 115, 150, 163, 214
   design, 150
   error control, 163
data over voice (DOV), 2, 60
data packets, 143
data rates, 6, 9, 42, 220
   digital radio, 6, 9, 42
   ISDN, 220
data scrambler, 276
data storage systems, 164
data subscriber loop carrier, 225
data switching exchanges (DSE), 120
data terminal equipment (DTE), 120
data transmission, 85
data under voice (DUV), 2, 60
datagrams, 134, 140, 146
datagram service, 134, 147
   signal packet, 147
DATANET, 135
D channel, 220
dc restoration, 286
DDCMP protocol, 130
decentralized computer networks, 112
DECNET, 149
decoders, 164
   hard decision, 189
   soft decision, 187
decoding, 165
   algebraic, 189
   convolutional codes, 187, 189
   error detection, 180

Index

sequential, 190
soft decision, 187
table look-up, 189
techniques, 189
Viterbi, 190
delay, 138, 154, 198
modulation coding, 261
packet, 154
delta modulation (DM), 73
demodulator 16 QAM, 25, 26, 27
descrambler, 276
design
digital radio link system, 44
digital subscriber systems, 285
frame synchronization system, 105
rules loop selection, 299
despotic synchronization, 94
detection in digital radio systems, 30, 41, 55
detection of transmission errors, 168
detectors, 16, 32, 55
ASK, 32
coherent, 32, 55
correlation, 32, 55
decision regions, 16
FSK, 32
matched-filter, 32
M-PSK, 32
non-coherent, 41
optimum, 31
PSK, 32
QAM, 34
deterministic routing algorithm, 114, 137
fixed for session, 137
fixed indefinite, 137
detrimental loop, 285
dial tone receivers, 73
differential encoding, 23, 26
differential phase shift keying (DPSK), 23
digital exchanges, 69
group selectors, 71
network synchronization, 89
switching, 68
digital modulation, 14, 23
continuous phase frequency shift keying (CPFSK), 28
differential phase shift keying (DPSK), 23
frequency shift keying (FSK), 24

minimum shift keying (MSK), 28
phase-shift-keying (PSK), 14
quadrature amplitude modulation (QAM), 24
quadrature partial response signalling (QPRS), 27
digital processing unit, 25
digital radio, 1
advantages, 3
availability objectives, 10, 44
bandwidths, 7, 11, 19
channel spacing, 6, 9, 42
combining modulation and coding, 40, 187
concentrator systems, 2, 61
costs, 3
cross polarization, 9, 51
disadvantages, 3
equipment, 4, 25, 56
error rate performance, 35
fading, 1, 4, 10, 48, 51
frequency allocations, 7, 11, 19
frequency bands, 7, 11, 19
high capacity systems, 6
hybrid systems, 2, 60
immunity to noise, 4, 35
local networks, 2, 61
low capacity systems, 6
modulation, 14
performance objectives, 10
point-to-multipoint, 2, 61
polarization, 9, 51
short-haul, 2, 61
spectral efficiency, 11
supply power requirement, 4
terminals, 4, 25
transmission capacities, 7, 11, 19
transmission rates, 7, 11, 19
U.S. networks, 7
digital recording systems, 164
digital subscriber systems (DSS), 259, 275, 318
design, 285
filters, 318
digital subscriber line (DSL), 259, 277
digital telephone networks, 68, 217, 218
digital wire carriers T1 & T2, 223
directory routing table, 136
discontinuities in loops, 251
distributed systems, 112

diversity system
  space, 4, 56
DPSK, 23
drift rate, 85
DTE-DCE interface, 140
duobinary coding, 260
duplex connection, 75, 120

EBCDIC, 123
echo
  amplitude, 280
  cancellation, 282
  cancellation FIR filter, 331
  canceller, 221, 275
  canceller adaptation, 282
  canceller chip, 330
  canceller convergence process, 283
  canceller semiconductor devices, 335
echos, 278
effective receiving area, 45
effects of impulse noise, 295
elastic stores, 85
electrical characteristics
  subscriber networks, 240
electrical interference, 257
electronic industries association (EIA), 120
electronic mail, 135
electronic switching systems (ESS), 217, 224
encoders, 164, 177
  byte-serial, 180
  circuit, 179
  cyclic code, 177
  for CCITT Rec. X.25 code, 179
encryption, 149
end-to-end control, 139
energy per bit, 38
enhanced range equalizer, 311
equalizer, 4, 57, 275, 282
  analysis and design, 57, 301
  decision feedback, 57
  design, 57, 301
  enhanced range, 311
  four section, 304
  gains, 308
  optimization, 317
  performance, 302
  single pole, 307
  TCM mode, 309

  three section, 302
  transversal, 58, 301
erasures, 165, 189
  correction, 167
ERICSSON, 230
erlangs, 82
error checking, 122, 126
error control, 163
  in microprocessor systems, 165
  in random-access memory (RAM) systems, 165
  in storage systems, 164
error correcting codes, 164
error correction, 164
error detection, 163, 168, 181
  magnetic tapes, 164
  patterns undetectable, 171
  reliability, 171
error probability, 35, 263, 328
  2FSK, 36
  16QAM, 36
  DPSK, 36
  M-PSK, 36
  MSK, 30
  M-QAM, 36
  QPRS, 28
error protection, 91
error rates, 35, 263, 328
  objectives, 35, 263
error recovery process, 145
ESSEX model, 69
exchanges, 68, 83
  analogue, 69
  buffers, 83
  clock, 83
  congestion, 81
  digital, 69
  local, 68
  stored-program control, 68
  terminal unit, 73
  terminal units, 72
  trunk, 68
  unit, 62
  3 EAX, 224
  4 ESS, 224
  5 EAX, 224
exponential distribution, 152
eye diagrams, 286, 292, 314
  statistics, 292
eye statistics scatter plots, 292

Index

facsimile, 85
fade margins, 48
fading, 4, 48
fading channel
   digital radio, 4, 48
   flat, 48
   frequency selective, 51
   impulse response, 55
   rain, 52
   Rayleigh model, 50
far end crosstalk (FEXT), 259, 266
fast select, 140, 146
FDM-FM spectrum, 22
feedback shift registers, 179
FEXT signal-to-noise ratio, 267
fiber optic systems, 224
filters
   Nyquist, 19
   digital subscriber systems, 318
   FIR for echo cancellation, 331
first-in first-out (FIFO) buffer, 89
fixed for session method, 137
fixed indefinite method, 137
flag, 99, 130
   detector, 101
   field, 131
   sequences, 184
flat fading, 48
flooding, 136
flow control, 118
   algorithm, 114
forward error correction (FEC), 163, 185
   codes, 164
four section equalizer, 304
four-wire
   mode, 222
   switching, 75
   system, 278
frame, 131
   alignment signal, 100
   alignment signal choice, 107
   check sequence (FCS), 133
   checking sequence (FCS), 173
   information, 131
   structure, 88
   supervisory, 132
   synchronization, 98
   synchronization design criteria, 105

synchronization lock mode, 100
synchronization state diagram, 103
frames, 122
framing, 73
   sequence, 98
France
   loops, 239
   network, 328
   system, 330
free-space loss, 44
frequency
   accuracy, 85
   accuracy of clocks, 93
frequency allocations
   analogue radio, 6
   digital radio, 6, 42
frequency division multiplexing (FDM), 75, 222
frequency domain simulations, 288
frequency selective fading, 4, 51
   amplitude response, 53
   intersymbol interference, 54
   model, 51
   single echo model, 51
frequency shift keying (FSK), 24, 36
   spectra, 9, 19
FSK, 24, 36

gateways, 148
gauge conversion numbers, 306
gauge discontinuities loop plant, 232
general format identifier (GFI), 141
generator
   code, 169
   matrix, 169
   polynomial for cyclic codes, 176
German PTT system, 329
global status ISDN, 222
go-back-N ARQ, 192
Golay code, 186
grade of service, 82
graded switching stage, 77
grading, 78
group selector, 71, 75
GTE, 227
guard interval, 279, 299
   TMC mode, 292, 299

Hamming
  codes, 185
  distance, 174
  weight, 170
hard decision decoders, 189
HDLC, 130
header, 141
heterogeneous network, 112
hierarchical networks, 116
hierarchical synchronization system, 94
high capacity digital radio, 6
highways, 79
hitless switching combiners, 57
homogeneous network, 112
hospital information system, 150
host, 112
  computers, 116
hybrid, 221, 275
  balance method, 222
  echo canceller systems, 329
  impedance balance system, 278
  mode, 221
hybrid ARQ, 163, 191
  schemes, 205
  throughput analysis, 209
hybrid radio systems, 2, 60
hybrid subscriber systems, 278, 334
hybrid telephone network, 217
hypothesis testing, 31
hypothetical reference digital path
    (HRDP), 10

Illinois Bell, 227
impulse noise, 269, 286
  acceptance test, 270
  loop acceptance test, 270
  measure test set, 270
  signal to noise ratio, 269
  subscriber loops, 269
in-phase signals, 16
incoming call packets, 140
independent point-to-point
    synchronization, 88
information
  bits, 165
  frame, 131
integrated digital network (IDN), 227
integrated services digital network, 275
    (ISDN), 214

interarrival time, 152
interconnection networks, 148
interference crosstalk, 258
intermediate-frequency (IF), 4
international
  gateways, 98
  links, 94
International Radio Consultative
    Committee (CCIR), 6, 42
  frequency use recommendations, 6,
    42
International Standards Organization
    model for Open Systems
    Interconnection, 225
  (ISO-OSI) reference model, 225
International Standards Organization
    (ISO), 120, 225
intersymbol interference (ISI), 39, 51,
    249
  frequency selective fading, 39, 51
IMP (Interface Message Processor), 138
ISDN, 325
  data rates, 6, 220, 271
  early foundations, 215
  global status, 222
  hybrid mode, 221
  IC chips, 231
  in Britain, 229
  in Canada, 228
  in European countries, 228
  in France, 229
  in Japan, 229
  in Sweden, 230
  in United States and Canada, 224
  objectives, 223
  services, 229
  services offered, 218
  standards, 220
  subscriber loop, 231
  telephone networks, 216
  trends, 271
ISO architectural model, 114, 118
Italy
  network, 252
  systems, 328

Japan
  ISDN, 229
  loop loss, 251

Index

loop plant, 255
   systems, 326, 332
jitter, 83, 87, 322
junctors, 78
justification, 88
   control bits, 91
justifying bit, 91

layer
   in ISO-OSI model, 133, 225
   network, 133
   protocol, 120
layered architecture, 114
least time delay algorithm, 136
lightning surges, 253
limiter, 14
line circuits, 72
line codes, 272, 285
line equalizer, 282
line terminating network, 275
linear codes, 172
link set up, 135
Little's formula, 153
loading coils, 257
local area data networks, 71
local area data transport capability (LADT), 225
local exchange, 68
lock mode, 101
logical channel, 141
longitudinal redundancy checking (LRC), 126
long-term drift, 86
loops (subscriber lines), 71, 240, 251, 277
   ABCD matrix, 241, 288
   acceptance test impulse noise, 270
   attenuation, 242, 300, 308
   bridged taps removed, 241
   cable, 292
   capacitance, 240
   conductance, 240
   gauge, 290
   image impedance, 290
   image reactance, 241
   image resistance, 241
   inductance, 240
   length, 233
   loss, 226, 251, 306
   multi-gauge sections, 232
   nonloaded, 283
   plant, 231, 232, 275
   plant gauge discontinuities, 232
   plant penetration, 328
   plant physical design rules, 254
   reactance, 291
   reflections, 290
   resistance, 240, 291
   scatter plot, 290
   selection design rules, 299
   survey data, 277
   surveys, 233, 287, 290
   transfer function, 288
   water in, 256
loop loss, 251
loop plants
   Canada, 255
   Finland, 255
   France, 255
   Germany, 255
   Italy, 255
   Norway, 255
   Sweden, 255
   United Kingdom, 255
   United States, 255
losses
   loops, 251
   microwave feeders, branching filters, 44
   radiopath, 44
   subscriber lines, 251
low capacity digital radio systems, 6

magnetic
   disks, 173
   tape errors, 164
marker, 72
Markov models, 152
master clock, 92
master-slave, 92
   synchronization, 94
   systems, 111
maximum error in clocks, 93
maximum likelihood decoder, 190
median depression, 48
medium capacity systems, 6
MEGAROUTE service, 228
message arrival patterns, 151
message switching, 116, 117

metering, 73
metric, 190
metropolitan area networks, 226
microprocessor systems error control, 165
microwave radio systems, 1
  analogue, 1, 41
  digital, 1, 4, 19
  frequency bands, 7, 11, 41
minimum distance, 174
  code, 174, 187
minimum free distance, 187
minimum shift keying (MSK), 28
mixed-mode ARQ, 201
modulation
  continuous phase frequency shift keying (CPFSK), 28
  differential phase shift keying (DPSK), 23
  frequency shift keying (FSK), 24
  minimum shift keying (MSK), 28
  phase-shift-keying (PSK), 14
  quadrature amplitude modulation (QAM), 24
  quadrature partial response signalling (QPRS), 27
modulation type
  choice of, 14
modulator
  4PSK, 16
  8PSK, 18
  MSK, 29
modulo 2 addition, 167
Mountain Bell, 227
MSK, 28
  error performance, 30
  spectrum, 29
multidrop, 115
multi-path fading, 4, 51
  single-echo model, 51
multiple access, 113
multiplex
  higher order, 6
  primary, 6
multiplexing, 88, 113
multipoint, 115
  transmission, 129
mutual synchronization, 91, 96

near end crosstalk (NEXT), 259, 268
negative-acknowledgement (NAK), 190

network
  architectures, 112, 118
  centralized, 116
  centrally controlled, 116
  classification, 112
  data communication, 111
  delays, 118, 138
  design, 76, 150
  distributed, 116
  environment, 151
  heterogeneous network, 112
  hierarchical, 112
  hybrid, 113
  integrated, 113
  loop, 113
  mesh, 113
  nonhierarchical, 112, 113
  objectives, 150
  star, 113
  structure, 114
  synchronization, 83, 88, 92
  topologies, 116
  tree, 113
network addressable unit (NAU), 150
network layer, 111, 119, 133
network terminal equipment (NTE), 225
networks
  computer, 111
  control, 114
  interconnection, 148
  plesiochronous, 92
  problem oriented, 114
  synchronous, 83
  terminal oriented remote-access, 114
  United States, 325
  value-added, 114
New York Telephone, 227
NEXT (Near End Crosstalk), 286
Nippon Telegraph and Telephone Public Corporation, 231
node, 112
noise
  bandwidth, 38
  immunity, 286
  spectral density, 38
nonblocking network, 78
  switching stage, 77
non-coherent detectors, 41
nonlinear systems, 13
  AM/AM conversion, 13
  AM/PM conversion, 13

nonloaded loops, 283
nonreturn to zero (NRZ) codes, 261, 286
  coding, 261
Northern Telecom, 227
Norway, 252, 329
  network, 252
  systems, 329
notch frequency, 54
Nynex, 227
Nyquist
  bandwidth, 35
  filtering, 19

octets, 140
open-ended bridged taps, 285
open wires, 253
optimum detectors, 16, 32
  for channels with ISI, 39
outage, 10
  prediction techniques, 49
  probability, 44, 49
  Vigant's formula, 49

PABX, 68
Pacific Northwest Bell, 227
Pacific Telesis, 227
packet
  call accepted, 141, 143
  call connected, 141
  call request, 140
  call restart, 143
  clear confirmation, 143
  clear indication, 141
  clear request, 141
  data, 143
  datagram service signal, 147
  delay, 154
  header, 141
  incoming call, 140
  length, 229
  restart indication, 143
  restart request, 143
  size, 118
  types, 141
packet switched networks, 173, 223
packet-switched data networks, 99
packet switching, 112, 116, 117, 173
parabolic reflector, 44
parity check bits, 164

parity check equations, 167
parity check polynomial, 178
parity retransmission ARQ, 206
partial response
  codes, 286
  signalling, 27
path control, 149
path losses
  single hop radio system, 44
PCM, 68
  codec, 75
  convertors, 72
  hierarchical bit rates, 6
  multiplexer, 75
  switching, 79
  thirty-channel, 98
  twenty four-channel, 98
performance objectives, 35, 263, 302
  digital radio, 35
  ISDN, 263, 302
  long term, 35
  short term, 35
permanent virtual circuit, 134, 140
phase
  ambiguity, 33
  jitter, 322
  locked loop, 33, 319
  locked oscillator, 95
physical characteristics
  loop plant, 232
physical layer, 119
  service data unit, 122
plesiochronous, 91
  networks, 92
point-to-multipoint digital subscriber
  radio, 2, 61
point-to-point
  network, 115
  synchronization, 88
Poisson distribution, 152
polling, 115, 129
  message format, 129
Polloczek-Khinchine formula, 153
polynomial
  division, 176
  representation, 175
power spectral density for a PSK signal, 19
pre-emphasis, 6
PRELUDE project, 229

presentation layer, 119, 150
primary constants
  American Wire Gauge (AWG), 254
primary multiplex, 6
privacy, 276
probability
  of error, 35, 263, 320
  of undetected error, 184
problem oriented networks, 114
propagation, 44, 51
  experiments, 51
protocols
  binary synchronous communications, 125
  bit-oriented, 122, 130
  character-oriented, 122
pseudo-ray, 52
PSK (Phase-shift-keying) modulation, 14
  binary PSK, 14
  eight-phase PSK, 15
  four-phase PSK, 15
  M-ary PSK, 15
public switched digital capability (PSDC), 225
pulse amplitude modulation (PAM), 73
pulse code modulation, 68
pulse stuffing, 88
pulse widths, 290

QAM (quadrature-amplitude) modulation, 24
QPRS, 27
  error rate, 28
  spectral efficiency, 27
QPSK, 15
quadrature amplitude modulation (QAM), 24
quadrature signals, 15
quarter wave length bridged taps, 249
quartz crystal oscillators, 86
queue, 151
  lengths, 198
  size, 154
queueing
  model, 151
  time, 154
queueing system
  G/G/M, 152

M/G/1, 152
M/M/1, 152

radio
  digital, 1
  equipment system signature, 58
  link system design, 41
  relay systems, 1
radio networks
  packet-radio, 2
  point-to-multipoint, 2, 61
radio systems
  TD, 224
  TH, 224
  digital, 1
  spectral efficiency, 11
  terminal schematic, 4, 25
rain fading, 52
random-access-memory (RAM), 173
random routing, 137
Rayleigh fading model, 50
Rayleigh probability density function, 50
real-time applications, 118
receiver carrier phase reference, 33
receiver noise figure, 44
receiver phase ambiguity, 33
receiver sensitivity, 30
recovery of clock, 276
rectangular array, 78
Reed-Solomon codes, 186
reference model for open systems interconnection, 118
reflections
  bridged taps, 290
  loop, 290
regional control unit, 73, 76
reliability, 163
remote control units, 72
  switching stages, 68
repeater spacing, 4, 41
repeater unit, 61
reset procedure, 145
restart indication packets, 143
restart procedure, 145
restart request packets, 143
retransmission of data, 163
round-trip delays, 191
routing, 76, 118, 136
  deterministic, 136

random, 136
table, 136
rubidium vapour cell, 86

sampling, 74
satellite systems, 34, 191, 196
scrambling, 26, 262, 276, 322
scanner, 72, 322
scanning instant, 322
SDLC, 130
search mode, 100
selection procedures, 129
selective-repeat ARQ, 201
selective-repeat plus go-back-N (SR + GBN) ARQ, 203
selective-repeat plus stutter (SR + ST) ARQ, 203
self-organizing system, 94
self-protected system, 95
servers, 151
service channel, 26
service time, 82, 152
services X.25, 144
serving area multiplex sites (SAMS), 228
session layer, 119
sequence number, 132
sequential decoder, 190
SETRAN ARQ, 200
Shannon capacity, 157
shortened cyclic codes, 183
shortest path algorithm, 136
signal space vector diagrams, 19, 25
M-PSK, 19
16-QAM, 25
signal-to-noise ratio, 35, 38, 267
FEXT, 267
impulse noise, 269
radio systems, 35
satellite systems, 38
signalling, 71, 148, 224
processor, 72
signalling terminal equipment (STE), 148
signature, 58
significant instants, 92
simulation studies, 277, 301
digital subscriber systems (DSS), 277
equalizers, 301
frequency domain, 288

limit cycle condition, 288
time domain, 288
simulation systems, 287
software, 289
single pole equalizer, 307
single server model, 152
slave stations, 129
sliding-window ARQ, 192
slip, 83, 93
interval, 85, 93
rate, 85
slotted aloha, 113
SNA, 136, 149
system network architecture (SNA), 136, 149
soft decision decoding, 40, 187, 189
South-western, 227
space diversity systems, 4, 56
space-division switching, 77
space-shift switches, 80
space-time-space (S-T-S), 80
spectral density, 19, 22, 25
analogue radio, 22
digital radio, 19, 22, 25
M-PSK, 19, 22
MSK, 29
QPRS, 27
16-QAM, 25
spectral efficiency, 11, 19, 25
2PSK, 19
4PSK, 19
8PSK, 19
16 QAM, 25
spectrum, see spectral density
authorized, 21
FDM-FM, 22
speech circuits, 85
stability, 85
clock, 85
standards
CCIR, 6, 41
CCITT, 6, 85, 92, 120, 140, 173
computer networks, 114, 120, 140, 173,
digital radio, 6, 41
ISO, 114, 120, 225
V.10/X.26, 120
V.11/X.27, 120
V.24, 121
X.20, 121

X.21, 121
X.22, 121
X.25, 140, 173
X.75, 148
star configuration, 115
state diagrams, 103
 frame synchronization, 103
 transition probability, 104
stochastic
 algorithm, 114
 routing, 137
store-and-forward
 switching, 117
 techniques, 117
stored-program control (SPC), 68, 217, 224, 326
 crossbar offices, 326
 exchanges, 68
structure of networks, 114
 hybrid, 114
 loop, 114
 mesh, 114
 star, 114
 tree, 114
stuffing, 88
subjective synchronization, 94
subnetwork, communication, 117
subscriber drop out unit, 62
subscriber networks, 68, 223, 236, 269
 American, 238
 Australia, 236
 Canada, 236
 electrical characteristics, 240
 France, 238
 Germany, 239
 impulse noise, 269
 Italy, 236
 Japan, 236
 local loop, 68, 223, 231, 269
 loop, 68, 223, 231, 275
 loop carrier, 318
 Norway, 239
 Sweden, 238
 United Kingdom, 238
 United States, 238
subscriber radio concentrator systems, 2, 61
subscriber radio systems point-to-multipoint, 2, 61
subscriber systems, 275
subscriber unit, 62

subsets of networks, 112
supervisory
 frame, 132
 functions, 132
survey results, 288
 cable characteristics, 288
Sweden
 ISDN, 230
 systems, 328
switch network, 72
switched arrays, 78
switched star integrated services network, 230
switched virtual circuit, 134, 140
switching, 68,
 centers, 68
 circuit, 114, 117
 G-M-G, 81
 matrices, 78
 message, 114, 117
 M-G-M, 81
 packet, 114, 117
 principles, 77
 space-division, 77
 SSTSS, 81
 store-and-forward, 117
 S-T-S, 80
 system AXB 30, 230
 system AXE 10, 230
 system MD 100, 230
 three-stage, 78
 time-division, 77
 TSSST, 81
 T-S-T, 80
symbol timing, 321
 recovery, 27
synchronization, 26, 33, 83, 130
 bit, 130
 carrier, 26, 33
 frame, 98
 hierarchical, 95
 independent point-to-point, 88
 master-slave, 94
 mutual, 96
 pattern, 130
 slips, 83
 subjective, 94
synchronizer, 26
synchronous data link control (SDLC), 149
synchronous networks, 92

syndrome, 181
SYNTRAN, 228
system signature, 58
systematic code, 169, 177

T1 carrier systems, 256, 286
table look-up decoding, 189
tandem switch, 223
tapped delay lines, 276
TCM system, 226, 231, 325
   advantages, 332
   eye diagrams, 292
   guard space, 292, 299
   techniques, 272
   timing considerations, 280
telephone network, 68, 214
   analogue, 216
   digital, 217
teleprocessing systems, 112
terminal control units, 112
terminal oriented remote access
   networks, 114
terminals
   data communication, 111
   radio, 4, 25
terminating channel units, 277
testing, 76, 270
   impulse noise, 270
   telephone network, 76
text compression, 150
thirty-channel PCM, 6, 98
three section equalizer, 302
three-stage switch, 78
   switching, 78
threshold detector, 32, 276
throughput, 151, 164, 191
   analysis, 151
   ARQ system, 191
   efficiency, 164, 191
   Go-back-N system, 197
   hybrid ARQ schemes, 209
time compression muliplexing (TCM),
   221, 226, 278, 325
time division multiple access (TDMA),
   61
time-division switching, 77
time domain simulation, 288
time outs, 146
time shift switch, 79
time slot interchange (TSI), 80
time-space-time (T-S-T) switching, 80

timers, 146
timing recovery, 275, 321
traffic, 82, 138
   concentration, 76
   matrix table, 136
   rate, 152
training sequence, 282
transaction-oriented features X.25, 146
transfer function
   three-path fading model, 53
   two-ray fading model, 53
transmission
   burst, 278
   capacities of digital radio, 6, 42
   control, 149
   delays, 87
   errors, 163
transmit/receive (T/R) switch, 279
transmitter
   back off, 14
   digital radio, 4, 26
TRANSPAC, 135
transparency, 127
transparent mode, 129
transport layer, 119
tree
   diagrams, 187
   structure, 116
trellis
   code, 40, 187
   coding, 40
   diagrams, 187
tributary stations, 129
trunk exchanges, 68
T-S-T switching, 80
twenty-four channel PCM, 6
2B+D subscriber rate, 221
2BIQ code, 287
TYMNET, 137

UDLC, 130
undetectable error patterns, 172
United Kingdom
   subscriber networks, 232
   telephone networks, 232, 252
United States
   subscriber networks, 328
   telephone networks, 232
utilization factor, 153

value-added networks, 114
variable frequency oscillator (VFO), 96

vertical redundancy checking (VRC), 126
virtual call, 140
virtual circuits, 134, 140
   permanent, 134
   switched, 134
Viterbi algorithm, 39, 40, 189
Viterbi decoding, 189
   algorithm, 189
voltage controlled oscillator (VCO), 24

weight
   distribution of a code, 170
   of codeword, 170
   spectrum, 171
wire sizes, 233

X.25
   interface, 140
   packet characteristics, 140
   services, 144
   transaction-oriented features, 146
X.75 inter-network protocol, 148

zero crossings, 322

Volume I
INDEX

Active filter, 228
Adaptive delta modulation (ADM), 245, 253
Adaptive differential pulse code modulation (ADPCM), 9, 216, 245, 254
Adaptive equalizers, 140
Adaptive predictive coding (APC), 245, 257
Adaptive pulse code modulation (APCM), 245
Adaptive quantization, 254
Additive Gaussian noise, 152, 158, 196
ADPCM codec, 255
A-law companding, 239
  signal to noise ratio, 241
  segmented, 240
Algebraic representation of line signals, 37
Aliasing, 225
  distortion, 225
ALOHA network, 25
Alphabetic codes, 78
Alternate-mark-inversion (AMI) code, 40, 47
  auto correlation function, 69
  code properties, 47
  coded signal spectrum, 69
  coding, 175
  decoder, 47
  encoder, 47
Analogue repeater, 123
Analogue telecommunications, 4
  telephone networks, 12
Anti-aliasing filter, 225
Aperiodic signals, 54
Aperture effect, 226
Application level, 25
Architectural model for a computer network, 23

Area under
  Gaussian function, 161
  the probability density function, 160
ARPANET, 20
Accounting and statistics centre (ASC), 21
ASCII character set, 19
Asia, 19, 239
Asynchronous, 22
  character-mode, 22
  data, 28
Attenuation of lines, 133
Audio program, 29
AUSTPAC, 21
Australia, 19, 239
Auto-correlation function, 56, 58, 192
  from ensemble averaging, 56
  from time averaging, 58
  of AMI coded sequence, 69
  of random binary signal, 59
  of pseudorandom binary signal (PRBS), 192
  of pseudonoise (PN) signal, 192
Automatic equalizers, 140
Automatic-gain-control (AGC) amplifier, 129
Automatic-repeat-request (ARQ) systems, 19
Automatically synchronized error detector, 198
Availability, 203
  objectives, 211

Bandwidth
  efficiency, 100
  of cable system, 143
  of line codes, 62
Baseband, 11
  digital transmission, 33

Baseband (*Continued*)
  modems, 18
  signals, 34
Baseline wander, 88, 117
Basic pulse shape, 38
Baud rate, 16, 105
Binary signals, 38, 56, 143, 160
  pseudorandom, 174, 185
  random, 149
Bipolar code, 40
Bit-error rate (BER), 91, 143, 163, 204
  analysis, 155, 164
  analytical model, 163
  calculations, 155
  long-term mean, 197
  performance objectives, 204
  relation to probability of error (P), 173
Bit, 26, 29
  rate reduction techniques, 243
  rates, 26, 29
  timing, 100, 124
  timing recovery, 127
Block error rate (BLER), 173, 197
British 24-channel PCM system, 8

Cable
  attenuation, 133
  coaxial, 13, 132
  insertion loss, 132
  pairs, 13, 25, 36
  repeaters, 36
  systems, 13, 36
Canada, 19, 239
CCITT
  Plenary Assembly, 208
  standards, 208
  standards for quantization noise, 241
CCITT Recommendations, 208
  V21, V23, 16
  V26 bis, 16
  V29, 17
  V36, 17
  X.25, 21
Central limit theorem, 88, 158
Cepstrum modelling, 245
Channel, 8, 163
  coder, 33
  decoder, 34
  distortion, 89
  errors, 218, 236
  unit, 8
Character error rate (CER), 197
Characteristic polynomial, 190
Charge-coupled delay (CCD) line, 136
Circuit switching, 18
Class-4 partial response, 107, 118
  encoder, 118
  signalling, 107
Clock, 9
  recovery, 50, 89
  recovery procedures, 123
  signals, 9
Coaxial cable system, 13
Coaxial cables, 132
Code redundancy, 79
Codes
  AMI, 47
  bipolar, 47
  differential diphase (DD), 51
  error detecting and correcting, 33
  Gray, 16
  HDB3, 74
  Manchester, 50
  MBNT, 77
  quaternary-to-7-level, 113
  4B3T, 79
  5B4T, 79
  7B5T, 79
  8B6T, 79
  10B7T, 79
Coding of speech, 216,
  adaptive delta modulation, 245, 253
  adaptive predictive, 245, 257
  ADPCM, 9, 216, 245, 254
  delta modulation, 9, 216, 245
  PCM, 27, 237
  techniques for reduced bit rates, 243
Collisions, 26
Colour television, 30
Common equipment subassemblies, 9
Communication channel model, 169
Communication subnetwork, 24
Companding, 6, 7, 237
  A-law, 239
  compressor, 239
  expandor, 239
  -law, 239
Comparator, 47
Compression, 7

Compressor, 239
Computer, 19
  communications, 6
  networks, 22
Concentrators, 13
Conditional probability, 65
  density function, 164
Confidence limits, 201
Contention, 25
Control field, 24
Conversion A-law to -law, 243
Convolution, 59, 110, 224
Correlation
  between symbols, 117
  in the speech signals, 244
Correlative encoding, 106, 112
  class-4 partial response, 117
  duobinary, 107
  generalized, 117
  partial response, 106
  signalling, 106
Correlator circuit, 198
Coupling mechanisms for crosstalk, 184
Crosstalk, 72, 88, 132, 184
  in PCM systems, 158
  noise figure, 183
Cycle sets, 190
Cyclostationary signals, 71

Data, 3
  networks, 17, 20
  rates, 23
  services, 5
  terminals, 3
  transmission, 9, 14
Data circuit-terminating equipment (DCE), 21
Data compression, 243
Data link level, 24
Data terminal equipment (DTE), 19, 21
DATAPAC, 21
DATAROUTE, 19
DC wander, 40
DDX, 21
Decision circuits, 128
  threshold levels, 184, 207

Decision-directed equalizer, 141
Decoder, 47
  AMI code, 47
  HDB3 code, 76
  duobinary code, 115
Delta function, 9, 110
  sifting property, 63, 110
Delta modulation (DM), 9, 216, 245, 248
  hunting, 251
  modulation start-up, 250
  signal-to-quantization noise ratio, 251
  slope overload, 251
Design
  equalizer, 178
  feedback shift register, 191
  regenerator, 184
Destination address, 24, 25
Detection of errors, 49
Dibits, 16
Differential diphase (DD) code, 51
Differential pulse code modulation (DPCM), 245
Digital circuit reliability, 14
Digital communications systems, 5
Digital data service (DDS), 10, 19
Digital filter, 102, 107
Digital radio, 13, 25
Digital signal, 47, 74, 218
Digital speech encoding, 216
Digital switching, 2, 5
Digital telephone systems, 11
Digital transmission, 2, 34
  transmission rates, 29
Discrete Fourier transform (DFT), 61
  fast Fourier transform (FFT), 61, 258
Discrete random signal, 60, 160
  random variables, 160
Discrete-time random signal, 60
Distortion
  channel, 89
  criteria, 218
  errors, 218
  in PCM systems, 221
Duobinary, 107
  decoding, 113
  encoding, 111
  pulse shape, 111
  scheme decoding, 115
  signal spectrum, 116
  signalling, 107

Duty cycle, 39
Dynamic range, 237

EFS, 203
Electrical model of speech production, 244
Electronic mail, 6
Elliptic filter, 101
Encoding, 9, 47, 74, 243
　AMI, 47
　HDB3, 74
　of analogue signals, 9
　of speech, 216
　quarternary-to-7-level, 113
Encryption, 25, 197
Ensemble, 56, 159
　average, 56
Equalization, 129
Equalizer, 37, 89, 123, 129
　adaptive, 140
　adjustment algorithms, 141
　analogue filter type, 133
　automatic, 140
　convergence, 143
　decision-directed, 141
　design, 130, 178
　filter, 123
　frequency-domain approach, 133
　impulse response, 138
　increment step size, 142
　matrix form analysis, 139
　output waveform, 93
　peak distortion, 141
　preset, 140
　pulse response, 131
　simulation, 143
　structure, 135
　transfer function, 134
　zero-forcing, 138
Error, 19, 24, 47, 92
　bursts, 202
　counts, 201
　detection, 19, 24, 47, 198
　detector circuit, 199
　mean square, 220
　measure, 220
　occurrence statistics, 202
　patterns, 173
　performance objectives, 208, 210
　propagation, 47, 113
　rate estimates, 201
　rate measurements, 197
　rate test sets, 201
　rates, 92
　waveforms, 219
Error-free seconds, 173, 202
　performance objectives, 211
　relation to bit-error rate, 210
Error-free second runs (EFSR), 204
Error seconds, (ES), 203
Error second outage, (ESO), 203
Estimates of low error rates, 202
ETHERNET, 26, 51
EURONET, 21
Europe, 19, 239
European Economic Community (EEC), 21
Expandor, 7, 239
Eye-diagrams, 174
　measurement, 174
　opening, 177
　rate of closing, 177
　width, 177
Eye-pattern, 175

Facsimile (FAX), 3, 6, 221
Far-end crosstalk (FEXT), 183
Fast Fourier transform (FFT), 61, 258
FDM carrier systems, 12
Feed-forward adaption, 254
Feedback shift registers, 185
　feedback polynomials, 191
　feedback circuit design, 191
　pseudorandom sequence generators, 187
　structure, 190
FEP (Front-end processor), 21
FEXT noise figure, 183
Files, 25
Filter
　simulation, 143
　switched-capacitor, 119
Filtering allocation, 168
First-order multiplex, 27
Flat-top samples, 222
　spectrum of sampled signal, 224
　reconstruction filter, 226
Four-sigma loading, 234
Four-wire signals, 13
Fourier
　series, 52
　transform, 54

Fourier (*Continued*)
   transform pair, 55
   transform pairs—table of, 84
Frame, 7, 24
   check sequence (FCS), 24
   packet network, 24
   sequence number, 24
   structure, 7, 24
   synchronization, 7, 198
   24-channel PCM, 8
   30-channel PCM, 7
Frequency-division-multiplex (FDM), 4, 12
Frequency-shift-keying (FSK), 5, 15
Frequency-domain, 52, 245
   analysis, 52
   equalizer, 133
   speech coding, 245
Frequency spectra, 51
Full-wave rectifier, 127
Functions, 10
   autocorrelation, 56, 192
   delta, 41
   error function, 161
   Gamma probability density, 238
   Gaussian probability density, 152
   Laplace probability density, 159
   Nyquist raised-cosine, 101
   power spectral density, 53
   probability density, 156
   $Q(x)$, 161
   root-Nyquist, 169
   $\sin x / x$, 41
   sinc, 41
   unit impulse, 41
Functions of a telecommunications network, 10

Gamma density, 238
Gaussian
   noise model, 158
   noise signal, 196
   normal probability distribution, 158
   probability distribution, 152
   random number, 152
Generalized correlative encoding, 117
Global networks, 10
Granular noise, 233
Gray code, 16
Group band FDM, 16

Harmonic scaling, 245
HDB3 code, 74
   decoder, 76
Hierarchies, 26
   digital transmission, 26
High density bipolar (HDB) codes, 74
   HDB3 code, 74
High grade systems, 211
Hybrid transformer, 9
Hypothetical reference connection (HRX), 208
Hysteresis effects, 129

Idling noise, 233
Impairments
   sampling and reconstruction, 221
   quantization, 218
Impulse response
   equalizer, 138
   transversal filter, 103
Increment step size equalizer, 142
Information packet, 24
Input amplifier, 129
Insertion loss, 132, 181
Integral magnitude (im) error, 220
Integral square (is) error, 220
Integration of services, 14
Integrated services, 26
Integrated services digital network (ISDN), 2
Interface requirements, 23
Intermodulation measurements, 196
International performance standards, 204, 208
   hypothetical reference connection (HRX), 208
   reference circuit, 241
International Standards Organization (ISO), 22
Intersymbol interference (ISI), 45, 88, 179
   calculation, 40
   effect on error rates, 34
   measurement, 174
   simulation, 143
ISO model, 23

Japan, 8, 19, 21, 27
Jitter
   timing, 47, 91, 128, 155, 179, 181

Joint probabilities, 66
  matrix, 68
Junction cable, 132

Laplace probability density function, 159, 237
Laplacian random signal, 159
Leased-line, 19
Least mean square (LMS) algorithm, 142
Lender, 106
Linear digits, 241
Linear predictive coding (LPC), 216, 245
Linear timing extraction method, 126
Line, 19
  amplification, 124
  build-out network, 130
  code, 39
  code selection, 46
  coding—scrambling, 33
  encoder, 39
  leased, 19
  loss, 132, 181
  printer, 19
  signal spectrum, 51
  signals, 37
  signals algebraic representation, 37
  simulator, 184
  terminating units, 35
  transformer, 37
  transmission, 35
  waveforms, 40
Loading
  factor, 233
  value quantizer, 234
Local area networks, 25, 51
Local exchange, 11
Local grade systems, 210
Local networks, 12
Long-haul transmission, 13
Long-term mean bit-error rate, 197
Long-time predictor, 257

Manchester code, 50
  power spectrum, 74
Mark, 51, 75
  density, 51
M-ary signalling, 104

Matrix
  joint probabilities, 68
  method of equalizer analysis, 139
Maximal length
  pseudorandom binary sequence, 189
  sequence generator, 190
MBNT, 77
  alphabetic line code, 78
  codes, 77
Mean integral square (mis) error, 220
Mean square error, 220
Measurement
  bit error rates, 173
  error-free seconds, 197
  eye-diagrams, 174
  intermodulation, 196
  intersymbol interference, 174
  techniques, 173
  transfer function, 196
  triples fault location, 205
Medium grade systems, 210
Message switching, 18
Microwave digital radio, 13
Miniaturization, 14
Model
  bit-error rate analysis, 163
  channel, 155
  ISO model for open system interconnection, 23
  seven-layer, 23
  vocal system, 216
Modem
  data, 6
  pulse code modulation (PCM), 229
Modified duobinary, 107
  scheme, 118
  signalling, 107
Mu-law (μ-law), 239
Multilevel signalling, 104
Multiplexing, 6, 18
  frequency-division multiplexing (FDM), 4, 12
  time-division multiplexing (TDM), 5
MUX-NTU, 20

Near end crosstalk (NEXT), 132, 183
NETSTREAM, 20
Network, 18
  access protocol, 23
  analogue, 10
  architecture, 22

Index to Volume I

Network (*Continued*)
  computer, 22
  data, 17, 20
  design, 18
  digital, 5
  integrated services digital network (ISDN), 2
  level, 24
  local area data network, 25, 51
  local telephone network, 12
  multiplexer network terminating unit (MUX-NTU), 20
  packet-switched, 20
  telecommunications, 5, 10
  telephone, 5, 10
Network management centre (NMC), 21
Network terminating unit (NTU), 19
NEXT noise figure, 183
Node, 25
Noise, 88, 151
  additive Gaussian, 152
  crosstalk, 72, 88, 132, 158, 177, 184
  figure, 183
  Gaussian, 152, 158, 196
  Laplacian, 159, 237
  margin, 177
  margin degradation, 180
  regenerator, 152
  source, 183
  waveform, 156
Noncausal, 96
Nonswitched data networks, 19
Nonuniform quantization, 239
  quantizer, 239
Nyquist, 41
  criterion, 167
  filter, 95, 169
  pulse, 41
  pulse shaping, 95
  raised-cosine response, 101
  rate, 61, 96
  rate signalling theorem, 96
  root-Nyquist filter, 169
  sampling rate, 225
  transmission rate, 61, 96
  vestigial symmetry criterion, 97
  vestigial symmetry theorem, 97

Optical fibre systems, 13
Outage, 203
  performance standards, 208

Overload
  levels, 233
  noise, 233
  in quantizers, 233
Overvoltage protection, 125

Packet, 10, 18, 24
  assembler/dissassembler (PAD), 22
  exchanges, 22
  information packet, 24
  radio, 25
  structures, 24
Packet switched data networks, 20
Packet switched data service, 10
Packet switching, 18
PAD, 21
Partial-response signalling, 106
PCM
  encoder, 229
  hierarchies, 27
  repeater systems, 129
  system performance, 237
  24-channel, 6, 28
  30-channel, 6, 27, 229
Peak distortion
  algorithm, 141
  equalizer, 141
Performance
  ISI, 143
  measurement, 173
  monitoring, 18
  objectives, 204, 209
  objectives of the CCITT, 209
  of a data circuit, 203
  of ADPCM system, 256
  of DM and PCM systems, 253
  standards, 208
Phase-lock loop, 126
Phase-shift-keying (PSK), 5, 15
Phasor diagram, 16
Physical level, 23
Point-to-point, 20
  to-multipoint, 20
Poisson model, 202
Polar RZ, 37
Polarity digit, 241
Power spectral density, 53
  of AMI coded signal, 69
  of duobinary signal, 116
  of HDB3 coded signal, 77
  of line codes, 62

Power spectral density (*Continued*)
  of pseudonoise (PN) signal, 194
  of pseudorandom binary signal
    (PRBS), 194
  of random binary signal, 60
  of random signal, 55, 194
  of sampled signals, 224
PRBS
  frame synchronization, 198
  generator, 187
  sequence generator parameters, 191
  signal sequence length, 190
  signals, 192
  waveform spectral density, 194
Precoding, 114
  for M-ary inputs, 116
Predictor, 248, 255
  filter, 248
Presentation level, 25
Preset equalizers, 140
Primary multiplex, 27
  PCM, 132
Probability, 65
  conditional probability, 65, 164
  density function, 156, 164
  density function for a Gaussian
    process, 157
  density function for a Laplacian
    process, 159
  discrete random variables, 60, 160
  error, 91, 162, 197
  joint probability, 66
  model for speech, 238
  random processes, 56, 160
  random signals, 56, 160
  random variables, 38, 56, 160
Processes
  random, 56, 160
  stationary, 160
Propagation of errors, 47
Protocol, 22
Pseudo-error monitoring, 202
Pseudorandom binary sequence (PRBS),
    174, 185
  applications, 196
  auto correlation function, 192
  feedback shift register generator, 187
  frame synchronization, 198
  generator, 187
  maximal length, 189
  properties, 192
  sequence length, 190
  spectral density, 194
  test signals, 185
Pseudo-noise (PN), 174
Public data networks, 17
  ALOHA, 25
  ARPANET, 20
  AUSTPAC, 21
  DATAPAC, 21
  digital data service (DDS), 10, 19
  EURONET, 21
  packet-switched data networks, 20
  TELENET, 21
Pulse, 39, 88
  dispersion, 91, 130
  reconstruction, 124
  shaping, 39, 88, 167
  shaping circuits, 101, 128
  shaping filter, 70
  shaping using a transversal filter, 102
Pulse code modulation (PCM), 1, 27,
    216, 230
  encoder, 229
  fundamentals, 6
  hierarchies, 27
  repeater systems, 129
  system performance, 237
  24-channel, 6, 28
  30-channel, 6, 27, 229
Pure ALOHA, 26

QD units, 243
Quadrature amplitude modulation
    (QAM), 16
Quality, 203
  network performance objectives, 204
Quantization, 6, 218
  distortion, 218
  distortion limits, 241
  effects, 231
  noise, 7, 218, 231
  noise feedback, 257
  nonuniform, 239
  signal to noise ratio, 232
  uniform, 232
Quantizer, 6, 218
  loading value, 234
Quantizer distortion (QD) units, 242

Index to Volume I

Quaternary, 105
 signalling, 105
 to 7 level encoding, 112, 113

Radar systems, 196
Radio systems, 13, 25, 212
 digital, 13, 25
Raised cosine
 basic pulse, 176
 rolloff, 167
 spectrum, 99
Random binary signals 143, 149, 174, 186
 generator, 143
 simulation of, 143
 channel noise simulation, 143
Random channel noise, 88, 151
Random numbers, 149, 152
 binary, 38, 56, 143
 generators, 152
 sequence, 174, 186
Random variables, 38, 56, 160
 binary, 149
 discrete, 56, 160
 probability density function, 156
Rates
 bit rates, 26, 29
 error rates, 91, 143, 204
 rate of closing of eye, 177
Receive filter, 167
Reconstruction, 218
 distortion, 226
 filter, 226
 low-pass filter, 224, 236
Redundancy, 243
Regenerative repeaters, 37, 123
Regenerator, 34, 37, 88, 123, 163
 decision circuits, 124, 128
 design, 184
 equalizer, 152
 error, 157
 fault location tests, 205
 input SNR, 152
 input waveform, 157
 noise margin, 207
 simulation, 143, 153
 supervisory filter, 207
 test signal, 125
 waveforms, 125

Remote power feed unit, 36
 power feeding, 125
Remote switching stages, 13
Repeaters, 35, 123
 fault location tests, 205
 functions, 123
 regenerative, 34, 88, 123
 spacing, 36
Residual ISI, 138
Root-Nyquist filters, 169

Sampling, 6, 89, 124, 128, 217
 circuits, 128
 frequency spectra, 222
 function, 223
 instants, 45, 178
 Nyquist rate, 225
 process, 222
 rate, 226
Sampling and reconstruction, 218
 distortion, 221
 errors, 218
 filter, 119
 operations, 222
Satellites, 169, 212
 communication link, 212
Scrambler/descramblers, 198
SDM switch, 12
Segment digits, 241
Segmented A-law, 240
Self-synchronization, 169
Sequence, 174
Session level, 25
Short-time predictor, 257
Sifting property of delta function, 63, 110
Signals
 aperiodic, 55
 digital, 34, 218
 periodic, 52
 probability density function, 237
 pseudorandom, 174, 185
 speech, 244
 test signals, 125, 143
Signal-to-noise ratio (SNR), 151, 236
Signal distortion, 230
Signalling, 7, 9
Signal regeneration, 123
Signal to NEXT noise ratio, 185

Signal-to-quantization noise ratio, 237, 242
Simulated system, 150
Simulation, 143
  computer programs, 150
  of additive Gaussian noise, 143
  of a lowpass channel, 143
  of an equalized transmission system, 143, 150
  of crosstalk, 183
Sinc function, 41
Sinc pulse, 95
Slicing, 129
Slope overload, 251
Slotted ALOHA, 26
Sound program, 28
Source coding, 33
  decoder, 34
Source model coding, 216
Space-division-multiplex (SDM), 4, 12
  switching, 5
  switches, 13
Spacing between regenerators, 36, 183
  repeater, 36
Spectral density, 62
  of AMI coded signal, 69
  of aperiodic signal, 55
  of duobinary signal, 116
  of HDB3 coded signal, 77
  of line codes, 62
  of periodic signal, 52
  of pseudonoise (PN) signal, 194
  of pseudorandom binary signal (PRBS), 194
  of random binary signal, 60
  of random signal, 55, 194
  of sampled signals, 224
  of the line signal, 51, 64
Spectrum, 51, 62
  see Spectral density
Speech
  parameters, 245
  redundancy, 243
  signals, 244
  waveforms, 221
Speech coding, 216
  adaptive delta modulation, 245, 253
  adaptive predictive, 245, 257
  ADPCM, 9, 216, 245, 254
  delta modulation, 9, 216, 245

  hybrid or parametric methods, 245
  PCM, 27, 237
  techniques for reduced bit rates, 243
Speech compression, 245
Speech production
  electrical model, 244
Speed tolerance, 107
Split phase, 50
Spread spectrum radio, 197
Standards for performance, 208
  availability, 211
  bit-error rate, 204
  bit rate hierarchies, 26
  CCITT objectives, 209, 241
  CCITT recommendations, 16, 21, 208, 241
  data modems, 16, 21
  jitter, 107, 178
  quantization noise, 241
  timing, 107, 178
Stationary process, 56, 160
Statistics, 65, 159
  conditional probability, 65, 164
  discrete random variables, 60, 160
  error, 91, 162, 197
  joint probability, 66
  model for speech, 238
  probability density function, 156, 164
  probability density function for a Gaussian process, 157
  probability density function for a Laplacian process, 159
  random processes, 56, 160
  random signals, 56, 160
  random variables, 38, 56, 160
  sample values, 160
Stochastic process, 159
Stored program control (SPC), 12
Structural layers, 11
Subband coding (SBC), 245
Subscriber line termination, 3
Substitution mark, 75
Supervisory
  circuit, 125
  filter, 207
Switched-capacitor filters, 119, 227
Switching nodes, 22
Symbol
  clock frequency, 126
  error probability, 169

Index to Volume I

Symbol *(Continued)*
 packing rate, 97
 synchronization, 126
 timing extraction, 126
Synchronization, 7, 24, 198
 techniques, 127
Synchronizing error detector, 199
System identification, 196
System supervision, 125

Tap gains, 138
T1 carriers, 1, 28, 132
Telecommunications network, 10
 data networks, 17
 functions, 10
 models, 22
 telephone networks, 5, 11
 types, 12
TELENET, 21
Telephone
 analogue systems, 11
 digital systems, 11
 exchanges, 5
 networks, 5, 11
 signals, 244
 traffic, 5
 waveforms, 221
Teletex, 3
Teletypewriter, 19
Television, 29, 220
Telex, 28
Ternary line code, 40
Test instruments, 185
Test signals
 crosstalk, 183
 eye-diagram, 174
 noise, 183
 pseudorandom binary, 185
 triples test, 205
Testing procedures, 173, 205
Tests for randomness, 186
Text compression, 25
Thirty-channel PCM
 frame structure, 7
 hierarchy, 27
Threshold, 47
 decision circuit, 89
 detectors, 47
 levels, 47
 uncertainty effects, 129

Throughput, 26
Time-division multiplexing (TDM), 5
 pulse code modulation (TDM-PCM), 7
Time domain, 52
Time domain waveform coding, 245
Time out interval, 26
Time switching, 5
 switches, 13
Timing, 47
 information, 47
 jitter, 91, 128, 155, 179, 181
 offset, 179
 perturbation tolerance, 107
 recovery, 178
Traffic volume, 6
Transfer of a file, 25
Transmission
 delay, 130
 errors, 49
 line characteristics, 132
Transmit
 and receive filtering allocation, 166
 filter, 70, 88, 167
 filtering, 124
 pulse shapes, 131
TRANSPAC, 21
Transparency, 19
Transport level, 24
Transversal equalizer, 133, 135
Transversal filter, 102, 133, 246
 for encoding, 117
 for pulse shaping, 102
 impulse response, 103
Trio fault location system, 205
Triples
 fault location system, 205
 test, 205
 test signal generator, 205
Trunk exchanges, 12
Twenty four-channel PCM, 8
 frame, 8
 hierarchies, 28
Twinned binary, 50
Two-wire, 11

U.K., 19, 27, 239
Uniform quantizers, 232
Unipolar NRZ, 37
 RZ, 37

Unit impulse, 41
  sifting property, 63, 110
United States, 8, 19, 20, 27, 239
U.S. Bell T1 carrier system, 8, 28, 132
Unvoiced regions of speech, 244

Valid time, 204
Vestigial symmetry, 97
Video, 244
  transmission, 6
Violation, 75
VLSI technology, 14
  PCM codec, 229
Vocal
  filtering, 216
  tract, 216, 243
Vocoder, 216, 245
Voice-band, 14
  data modems, 14
  data transmission, 14
  services, 17

Voice excited vocoding, 245, 246
Voiced regions of speech, 244

Waveform coding, 216
  adaptive transform coding, 258
  error, 219
  time domain methods, 245
  tree encoding, 258
  vector quantizing, 258
White noise
  process, 160
  signal, 194
  test signal, 196
Word synchronization, 196

Zero crossings, 45
  detector, 126
Zero crossing distortion, 177
Zero-forcing equalizer, 138
Zero ISI requirements, 130
Zero memory systems, 107
Zero order multiplex, 28